Pipeline Engineering

Pipeline Engineering

Henry Liu

CRC Press
Taylor & Francis Group
Boca Raton London New York

CRC Press is an imprint of the
Taylor & Francis Group, an **informa** business

First published 2003 by Lewis Publishers
Taylor & Francis Group
6000 Broken Sound Parkway NW, Suite 300
Boca Raton, FL 33487-2742

Reissued 2018 by CRC Press

© 2003 by Taylor & Francis
CRC Press is an imprint of Taylor & Francis Group, an Informa business

No claim to original U.S. Government works

A Library of Congress record exists under LC control number: 2003047413

Publisher's Note
The publisher has gone to great lengths to ensure the quality of this reprint but points out that some imperfections in the original copies may be apparent.

Disclaimer
The publisher has made every effort to trace copyright holders and welcomes correspondence from those they have been unable to contact.

ISBN 13: 978-1-138-10571-3 (hbk)
ISBN 13: 978-1-138-56123-6 (pbk)
ISBN 13: 978-0-203-71101-9 (ebk)

Visit the Taylor & Francis Web site at http://www.taylorandfrancis.com and the
CRC Press Web site at http://www.crcpress.com

Foreword

Think about it. The U.S. has been one of the world leaders in developing increasingly more sophisticated pipeline systems for transportation of crude oil, natural gas, petroleum products, water, solids, and slurries. A vast network of pipelines literally blankets the U.S. as well as many other countries worldwide. Given these facts, it is amazing that there is no university curriculum, at least in the U.S., that recognizes pipeline engineering as a separate and distinct discipline. Equally amazing is the fact that there currently exists no comprehensive and recognized textbook that specifically addresses pipeline engineering. When I entered the pipeline industry in 1970 as an engineering trainee, my company supplied me with a copy of a textbook entitled *Hydraulics for Pipeliners* by C.B. Lester, which was published in 1958 and out of print at the time it was given to me. It is difficult to believe that more than 30 years have passed and no comprehensive text has been published that addresses the topics covered by Lester in his landmark book.

Dr. Henry Liu has filled this void with this book, which can be used by universities wishing to offer a course in pipeline engineering. It will be a valuable reference not only for students but also for practicing engineers who are confronted with pipeline engineering, construction, and/or operations issues in the real world. Even though Dr. Liu's book addresses a wide variety of topics in sufficient detail, it also provides an excellent yet concise list of references for those who wish and need to delve into particular areas in greater detail.

Dr. Liu has performed a valuable service by writing this book, which will be a tremendous asset to the pipeline industry.

James R. Beasley
President
Willbros Engineers, Inc.
Tulsa, OK

Preface

An extensive network of underground pipelines exists in every city, state, and nation to transport water, sewage, crude oil, petroleum products (such as gasoline, diesel, or jet fuel), natural gas, and many other liquids and gases. In-plant pipelines are also used extensively in most industrial or municipal plants for processing water, sewage, chemicals, food products, etc. Increasingly, pipelines are being used for transporting solids including minerals (such as coal, iron ore, phosphate, etc.); construction materials (sand, crushed rock, cement, and even wet concrete); refuse; municipal and industrial wastes; radioactive materials; grain; hospital supplies; and hundreds of other products. Pipelines are an indispensable and the preferred mode of freight transport in many situations.

Pipelines perform vital functions. They serve as arteries, bringing life-dependent supplies such as water, petroleum products, and natural gas to consumers through a dense underground network of transmission and distribution lines. They also serve as veins, transporting life-threatening waste (sewage) generated by households and industries to waste treatment plants for processing via a dense network of sewers. Because most pipelines are buried underground or underwater, they are out of sight and out of mind of the general public. The public pays little attention to pipelines unless and until a water main leaks, a sewer is clogged, or a natural gas pipeline causes an accident. However, as our highways and streets become increasingly congested with automobiles, and as the technology of freight pipelines (i.e., the pipelines that transport freight or solids) continues to improve, the public is beginning to realize the need to reduce the use of trucks and to shift more freight transport to underground pipelines. Underground freight transportation by pipelines not only reduces traffic on highways and streets, but also reduces noise and air pollution, accidents, and damage to highways and streets caused by trucks and other vehicles. It also minimizes the use of surface land. Surely, we can expect an increase in the use of pipelines in the 21st century.

Despite the long history and widespread application of pipelines, pipeline engineering has not emerged as a separate engineering discipline or field as have highway engineering and railroad engineering. This is due in part to the diverse industries and government organizations that use different kinds of pipelines, and in part to the lack of a single textbook or reference book that examines the general principles and different applications of pipelines. This situation has motivated the author to write this book.

The fragmentation of pipeline engineering can be seen from the number of different equations used to predict pressure drop along pipelines that carry different fluids such as water and oil. Yet, all these fluids are incompressible Newtonian fluids, which should be and can be treated by the same equations. As Professor Iraj Zandi of the University of Pennsylvania wrote in the editorial of the first issue of the *Journal of Pipelines*, the fragmentation of the pipeline field has impeded the diffusion of knowledge and transfer of manpower from one pipeline business to another, thereby creating an artificial barrier to technology transfer and job mobility (professional development). There is a strong need to unify the treatment of different types of pipelines by using a common approach, so that the next generation of engineers can be educated to understand a broad range of pipelines for a wide variety of applications.

In this book, pipeline is considered to be a common technology or a single transportation mode that has different applications. Pipeline engineering is defined here as the field that studies the various principles, technologies, and techniques that are used in the planning, design, analysis, construction, operation, and maintenance of pipelines for transportation of any cargo, be it liquid, gas, solid, or even packaged products.

This book will be useful not only to those employed by pipeline companies, but also to most mechanical, civil, chemical, mining, petroleum, nuclear, and agricultural engineers who must deal with piping or pipelines in their professional work. It provides the essentials of pipeline engineering — concepts, theories, calculations and facts — that all engineers working on pipelines should know. The book can be used as a reference book or as a college textbook. At the University of Missouri-Columbia, an early version of this manuscript was used as the basis for a 3-semester-hour course entitled "Pipeline Engineering." The prerequisite for this course was fluid mechanics. The course was taken by both graduate and undergraduate students from various engineering departments, especially civil and mechanical engineering. Student feedback was used to improve the original manuscript that evolved into this book.

This book consists of a total of 14 chapters, divided into two parts. Part I, Pipe Flows, consists of seven chapters that present the equations needed for the analysis of various types of pipe flows. It begins with a treatment of single-phase, incompressible Newtonian fluids, then follows with discussions on compressible flow, non-Newtonian fluids, flow of solid/liquid mixtures, flow of solid/air mixtures, and capsule flow, in that order. Part II, Engineering Considerations, consists of seven chapters that deal with nonfluid-mechanics-related engineering of pipelines required for the proper planning, design, construction, operation, and maintenance of pipelines. Topics include pipe materials, valves, pumps, blowers, compressors, pressure regulators, sensors, flowmeters, pigging, computer control of pipelines, protection against freezing, abrasion and corrosion, planning, design, construction, maintenance, rehabilitation, and integrity monitoring of pipelines.

As discussed above, this book examines the principles and important engineering aspects of all types of pipelines, and provides details on a wide range of subjects to broaden the reader's knowledge about the planning, design, construction, and operation of various types of modern pipeline systems. Practicing engineers will find the book useful for broadening their knowledge of pipelines, especially with respect to recent developments, such as freight pipelines and trenchless technologies. Professors may find this to be the most suitable textbook available for a new course in pipeline engineering. It is the author's belief that all engineering and mining colleges should offer pipeline engineering as a senior-level elective course offered by any engineering or mining department with an interest in the course and open to students from all engineering departments. This will greatly enhance the competence of future graduates involved in pipeline engineering work.

This book addresses only the most fundamental aspects of pipeline engineering. Consequently, it excludes discussion of various software systems that are used currently by design professionals. Nor does it include treatment of codes, standards, manual of practices, and current laws and regulations on pipelines, which not only differ from nation to nation, but also are in a state of constant change.

The book assumes that the reader has taken a college-level course in fluid mechanics. Even so, the book provides some review of fluid mechanics pertinent to pipe flow (see Chapter 2) to ensure a smooth transition to the more advanced subjects covered in this book. Today, most pipeline engineers in the U.S. still use the English (ft-lb) units in practice, although the SI units are being used increasingly. To be able to practice, to communicate with one another, and to effectively comprehend the literature on pipelines, the current generation of U.S. engineers and future generations must be familiar with both SI and ft-lb units. For this reason, some examples and homework problems in this book are given in SI units and others are given in ft-lb units. This will enable the reader to master both systems of units.

Henry Liu

Acknowledgments

First, I wish to express my gratitude to individuals who reviewed parts of this book and provided valuable input for improvement. They include John Miles, Professor Emeritus, University of Missouri-Columbia (UMC), on Chapter 3, Single-Phase Compressible Flow in Pipe; Thomas C. Aude, Principal, Pipeline Systems Incorporated, on Chapter 5, Flow of Solid–Liquid Mixture in Pipe (Slurry Pipelines); Sanai Kosugi, General Manager, Pipeline Department, Sumitomo Metal Industries, on Chapter 7, Capsule Pipelines; Charles W. Lenau, Professor Emeritus, UMC, on Chapter 9, Pumps and Turbines; Shankha Banerji, Professor Emeritus, UMC, on Chapter 11, Protection of Pipelines against Abrasion, Freezing, and Corrosion; Mohammad Najafi, Assistant Professor of Construction Management, Michigan State University, on Chapter 12, Planning and Construction of Pipelines; Russ Wolf and Tom Alexander, Willbros Engineers, Inc., on Chapter 13, Structural Design of Pipelines; and Robert M. O'Connell, Associate Professor of Electrical Engineering, UMC, on parts of Chapter 9 that deal with electric motors and electromagnetic pumps. I also wish to thank those individuals and organizations that provided photographs used in this book, or that allowed me to use their copyrighted materials; they are separately acknowledged in the figure captions. Three individuals helped type the manuscript: my wife, Susie Dou-Mei; my youngest son, Jeffrey H.; and the former Senior Secretary of Capsule Pipeline Research Center, Carla Roberts. Deep gratitude is due my wife Susie who, during the last few months of the manuscript preparation, freed me from most household chores so I could concentrate on the book project. All of the drawings in this book were done by Ying-Che (Joe) Hung, a freelance draftsman and industrial artist in Columbia, Missouri.

Finally, this book is dedicated to all those who share the belief that underground freight transport by pipelines is not a pipedream. It is realistic and innovative, and it should be promoted until it becomes the principal mode of freight transportation of the future, for the best interest of humankind.

The Author

Henry Liu, Ph.D., is Professor Emeritus of Civil Engineering, University of Missouri-Columbia (UMC). Dr. Liu has his B.S. from National Taiwan University, and his M.S. and Ph.D. from Colorado State University, Fort Collins. His main background and expertise are in fluid mechanics. Prior to retirement from UMC, he was a chaired Professor of Civil Engineering, and the founding Director of Capsule Pipeline Research Center (CPRC), a State/Industry University Cooperative Research Center (S/IUCRC) established by the National Science Foundation (NSF) in 1991.

At UMC, Dr. Liu taught many engineering courses including pipeline engineering, a course that he established at UMC.

Dr. Liu has served in leadership positions in professional organizations, such as Chairman, Pipeline Research Committee, American Society of Civil Engineers (ASCE); Chairman, Aerodynamics Committee, and Chairman, Executive Committee, Aerospace Division, ASCE; President, International Freight Pipeline Society (IFPS); and Steering Committee Chair, International Symposium on Underground Freight Transport (ISUFT). He is also a member of the American Society of Mechanical Engineers (ASME) and a member of the National Society of Professional Engineers. Dr. Liu has won prestigious national and international awards for his contributions to industrial aerodynamics and capsule pipelines, including the Bechtel Pipeline Engineering Award and the Aerospace Science and Technology Applications Award of ASCE, the Distinguished Lecture Award of IFPS, Missouri Energy Innovation Award, and three University of Missouri faculty awards for distinguished research.

Dr. Liu is the inventor or co-inventor listed on five U.S. patents dealing with various aspects of capsule pipelines. He has written more than 100 technical papers for professional journals and conference proceedings. He is the author of a book, *Wind Engineering: A Handbook for Structural Engineers*. Dr. Liu took early retirement from teaching to form the Freight Pipeline Company (FPC), headquartered in Columbia, Missouri, in order to bring capsule pipeline and other related new technologies to early commercial use in the U.S.

Dr. Liu has an extensive record of international involvement, including Fulbright Scholar (from Taiwan to the U.S.); Visiting Professor, National Taiwan University; Visiting Professor, National Chiao Tung University, Taiwan; Visiting Professor, Melbourne University, Australia; Visiting Fellow, National Institute for Resources and Environment, Japan; and Consultant, Taiwan Power Company. He has also conducted several lecture tours in China. He has served on the International Program Committee of, and given keynote speeches at, four international conferences organized by the Chinese Mechanical Engineering Society (CMES). He is the Chairman of the Steering Committee, International Symposium on Underground Freight Transport (ISUFT), which has held three symposia since 1999 in three countries — the U.S., the Netherlands, and Germany.

Contents

PART I Pipe Flows

PART II Engineering Considerations

Part I

Pipe Flows

1 Introduction

1.1 DEFINITION AND SCOPE

The term **pipe** is defined herein as a closed conduit, usually of circular cross section. It can be made of any appropriate material such as steel or plastic. The term **pipeline** refers to a long line of connected segments of pipe, with pumps, valves, control devices, and other equipment/facilities needed for operating the system. It is intended for transporting a fluid (liquid or gas), mixture of fluids, solids, fluid–solid mixture, or capsules (freight-laden vessels or vehicles moved by fluids through a pipe). The term pipeline also implies a relatively large pipe spanning a long distance. Unless otherwise specified, the pipelines discussed in this book generally have a minimum diameter of 4 inches (102 mm) and a minimum length of 1 mi (1.6 km). The largest and longest pipelines discussed may have a diameter of over 10 ft (3.05 m) and a length of over 1000 mi (1609 km). This book treats all important aspects, parts, and types of pipelines.

1.2 BRIEF HISTORY OF PIPELINES

The use of pipelines has a long history. For instance, more than 1,000 years ago, the Romans used lead pipes in their aqueduct system to supply water to Rome. As early as 400 B.C., the Chinese used bamboo pipes wrapped with waxed cloth to transport natural gas to their capital Beijing for lighting. Clay pipes were used as early as 4000 B.C. for drainage purposes in Egypt and certain other countries.

An important improvement of pipeline technology occurred in the 18th century when cast-iron pipes were manufactured for use as water lines, sewers, and gas pipelines. A subsequent major event was the introduction of steel pipe in the 19th century, which greatly increased the strength of pipes of all sizes. In 1879, following the discovery of oil in Pennsylvania, the first long-distance oil pipeline was built in this state. It was a 6-inch-diameter, 109-mi-long steel pipeline. Nine years later, an 87-mi-long, 8-inch-diameter pipeline was built to transport natural gas from Kane, Pennsylvania to Buffalo, New York. The development of high-strength steel pipes made it possible to transport fluids such as natural gas, crude oil, and petroleum products over long distances. Initially, all steel pipes had to be threaded together, which was difficult to do for large pipes, and they often leaked under high pressure. The development of electric arc welding to join pipes in the late 1920s made it possible to construct leakproof, high-pressure, large-diameter pipelines. Today, virtually all high-pressure piping consists of steel pipe with welded joints. Large seamless steel pipe was another major milestone achieved in the 1920s.

Major innovations in pipeline technology made since 1950 include:

- Introduction of new pipeline materials such as ductile iron and large-diameter concrete pressure pipes for water, and PVC (polyvinyl chloride) pipe for sewers
- Use of *pigs* to clean the interior of pipelines and to perform other functions
- *Batching* of different petroleum products in a common pipeline
- Application of cathodic protection to reduce corrosion and extend pipeline life
- Use of large side booms to lay pipes, machines to drill or bore under rivers and roads for crossing, machines to bend large pipes in the field, x-rays to detect welding flaws, and so forth.

Since 1970, major strides have been made in new pipeline technologies including trenchless construction (e.g., directional drillings, which allow pipelines to be laid easily under rivers, lakes, and other obstacles, without having to dig long trenches), pipeline integrity monitoring (e.g., sending intelligent *pigs* through pipes to detect pipe wall corrosion, cracks, and other pipe flaws), computers to control and operate pipelines, microwave stations and satellites to communicate between headquarters and remote stations, and new pipeline technologies to transport solids over long distances (e.g., slurry pipelines for transporting coal and other minerals, and capsule pipelines for bulk materials transport).

More about the history of pipelines can be found in the references listed at the end of this chapter [1–3].

1.3 EXISTING MAJOR PIPELINES

Most of the major oil and gas pipelines that exist today around the world were constructed either during or after World War II. In most cases, they were built to meet compelling national or international needs. For instance, the U.S. built the **Big Inch** and the **Little Big Inch** pipelines during World War II to counter the threat of German submarine attacks on coastal tankers. In the 1960s the Colonial Pipeline Company built a large product pipeline from Houston, Texas to New York City to counter a strike of the maritime union. The Arab oil embargo in 1973 prompted the construction of the Trans-Alaska pipeline to bring crude oil from the oil-rich fields of Prudhoe Bay located on the Artic Ocean north of Alaska to an ice-free port at Valdez on the south shore of Alaska.

The Big Inch was a 24-inch (61-cm) line designed to transport 300,000 bpd (barrels per day) of crude oil, and the Little Big Inch was a 20-inch (51-cm) product pipeline designed to deliver 235,000 bpd. Both lines extend from Texas to the East Coast. They were built between 1942 and 1943 (during World War II) by the U.S. government, but were sold after the war (in 1947) to the Texas Eastern Transmission Corporation (TETCO), and converted to transport natural gas. Later, TETCO expanded both lines, and converted the Little Big Inch back to a petroleum products

FIGURE 1.1 The Big Inch pipeline under construction in 1942; historic photograph showing the pipeline crossing a cotton field near the Arkansas–Missouri border. (Courtesy of Duke Energy.)

line operated by a different company — the Texas Eastern Products Pipeline Company (TEPPCO). Now it carries about 70 types of gasoline and 4 types of fuel oil, in addition to kerosene, jet fuel, butane, propane, and alkylate. At present (2003), both TETCO and TEPPCO are under Duke Energy, which has published an interesting booklet on the history of the Big Inch and Little Big Inch pipelines [4]. Figure 1.1 is a historic photograph of the Big Inch during construction.

The Colonial Pipeline is telescoping from 36 inches (91 cm) to 30 inches (76 cm). It transports approximately 1.2 million barrels of petroleum products per day. The pipeline was constructed between 1962 and 1964 by the Colonial Pipeline Company, which had been incorporated by a consortium of nine oil companies. The name Colonial was chosen because this pipeline from Texas to New York crosses most of the original 13 colonies. From 1967 to 1987, the company has greatly expanded the Colonial Pipeline System — adding main pipelines along the existing mainline, and adding lateral lines and pump stations. Currently, it is the largest pipeline system for transporting petroleum products — approximately 1.8 million bpd, which is equivalent to about 10% of the petroleum used daily in the U.S. Figure 1.2 is a photograph of the Colonial Pipeline.

The Trans-Alaska pipeline is a crude oil pipeline completed in 1977. It is 48 inches (1.22 m) in diameter and 798 mi (1284 km) long, transporting approximately 1.7 million barrels of oil a day, which is equivalent to about 9% of the oil consumed in the entire U.S. Due to the extreme arctic climate, rugged mountain terrain, earthquake regions (geological faults), stringent standards to preserve the Arctic environment, and lengthy delays during construction caused by law suits filed by opponents of the pipeline, the construction cost of the Trans-Alaska pipeline approached $9 billion, making it by far the most costly pipeline project in the world.

FIGURE 1.2 Clint A. Brist is an operator for Colonial Pipeline in Houston. The facility is a starting point for delivery of petroleum products along 5500 mi of the pipeline. (Courtesy of Colonial Pipeline Company; photograph by Robin Hood.)

Despite the cost, the pipeline has been profitable, and it serves vital national interests. Figure 1.3 is a photograph of the Trans-Alaska Pipeline.

The U.S. has far more oil and natural gas pipelines than any other nation in the world: approximately 1.3 million mi (2.1 million km) of gas pipeline and 0.25 million mi (0.4 million km) of oil pipeline. The amount of oil transported by pipeline in the U.S. in 2000 was approximately 500 billion tons, which constitutes about half of the oil transported in the nation. The largest natural gas producing state in the U.S. is Texas. A network of pipelines totaling 4300 mi (6920 km) transports natural gas from Texas to the Central and Eastern U.S., and California.

The former Soviet Union (FSU) is second only to the U.S. in the total length of oil and gas pipelines constructed. It has approximately 30,000 mi (50,000 km) of oil pipelines, most of which were built after 1940. The Comecon Pipeline transports oil from the Urals to Eastern Europe over a distance of 3800 mi (6115 km) — the longest pipeline in the world. FSU also has some of the longest and largest natural gas pipelines in the world, such as the 3400-mi (5500-km)-long Middle Asia-Central Zone line. It transports approximately 880 billion ft^3 (25 billion m^3) of gas per year.

In Canada, the Interprovincial Pipeline, with a diameter up to 34 inches (86 cm) and a length of 2000 mi (3220 km), carries oil from Manitoba and Saskatchewan to Ontario. Another major crude oil pipeline in Canada is the Trans-Mountain pipeline that links Edmonton, Alberta to refineries in British Columbia and the State of Washington in the U.S. over a distance of 825 mi (1328-km). Canada also has major gas pipelines, such as the 36-inch (91-cm) diameter Trans-Canada Pipeline, which is 2300 mi (3700 km) long, from Alberta to Montreal.

Major oil and gas pipelines exist in many other parts of the world. For instance, the oil-rich nations in the Middle East rely exclusively on pipelines to bring their oil from inland oil fields to seaports for export. Countries such as Iraq, Iran, and

FIGURE 1.3 An aboveground portion of the Trans-Alaska Pipeline about 5 mi north of the Yukon River. The pipeline was designed to meet the most stringent environmental standards for protecting the arctic environment. (Courtesy of Alyeska Pipeline Service Company.)

Saudi Arabia have thousands of miles of major oil pipelines. A notable example is the 1000-mi (1610-km) Trans-Arabian pipeline, built in 1950, which connects the oil fields of Saudi Arabia to the Mediterranean port of Sidon. Newer lines include a 1055-mi (1700-km) pipeline from southern Iran to the Turkish port of Iskenderun, and an Iraqi pipeline to the Turkish port of Yurmurtalik.

Offshore (submarine) pipelines are more expensive and more difficult to build than overland pipelines (except for the Trans-Alaska Pipeline). The most notable offshore oil pipeline is the one linking the British North Sea oil fields to the Shetland Islands. The line is 36 inches (91 cm) in diameter and 94 mi (151 km) in length, transporting 1 million barrels of crude oil per day, about half of Britain's consumption. It is the world's deepest large-diameter submarine pipeline — lying under 525 ft (160 m) of water. The world's largest underwater pipeline project for natural gas, located off the Louisiana coast, was completed in 1976. It is a submarine network of pipes and rigs extending 125 mi (200 km) offshore. A total of 269 mi (433 km) of pipe is used, with a maximum diameter of 36 inches (91 cm). The system can deliver up to 1 billion ft^3 (28 million m^3) of natural gas a day.

Although statistics on water pipes and sewers are lacking because they are too numerous, there is little doubt that the U.S. has the world's most extensive network of water pipelines and sewers. According to estimates of the American Water Work

Association (AWWA), the distribution network for large water supply systems in the U.S. comprises about 600,000 mi of pipe. Adding water pipelines for irrigation and various industrial uses, and adding pipelines of countless small distribution networks, it is reasonable to expect that the U.S. has well over 1 million mi of water pipelines, and about 1 million mi of sewers. Together they exceed the mileage of oil and natural gas pipelines combined.

The largest diameter pipelines in the world are water pipelines. For instance, it is not uncommon to see penstocks (i.e., water pipe for hydropower generation) of diameter greater than 10 ft or 3 m. Aqueducts for long-distance transport of water may also use larger pipes either for the entire length of the aqueduct or portions of it. Note that aqueducts are mostly open channels (canals), which also use pipes (circular conduits) as inverted siphons to cross existing watercourses from beneath. These inverted siphons vary in length from 0.4 to 3.2 km. For example, a part of the aqueduct that brings water from the Colorado River to central Arizona uses prestressed concrete pipes of 6.4 m (21 ft) inner diameter (see Figure 1.4). The "Great Man-Made River," an underground pipeline built in Libya, uses prestressed concrete cylinder pipes of diameters varying from 1.6 to 4.0 m (5.2 to 13 ft). The total length of the pipeline is 1900 km (1180 mi). It was designed to transport water collected from aquifers in the southern part of Libya to the cities and agricultural area in the north. It is the world's most ambitious and costly pipeline project for water supply and irrigation. The cost of the first half of the project exceeded $10 billion.

FIGURE 1.4 One of the world's largest diameter concrete pipes (21 ft I.D.) used by the U.S. Bureau of Reclamation to convey water as a part of an aqueduct in central Arizona. (Courtesy of Bureau of Reclamation.)

1.4 IMPORTANCE OF PIPELINES

Pipelines are the least understood and least appreciated mode of transport. Pipelines are poorly understood by the general public because they are most often underground and invisible — out of sight, out of mind! Despite the low degree of recognition by the public, pipelines are vitally important to the economic well-being and security of most nations. All modern nations rely almost exclusively on pipelines to transport the following commodities:

- Water from treatment plants to individual homes and other buildings
- Sewage from homes to treatment plants
- Natural gas all the way from wells to the consumers who may be located more than a thousand miles away — be it a home, a factory, a school, or a power plant
- Crude oil from oil fields to refineries
- Refined petroleum products (gasoline, diesel, jet fuel, heating oil, etc.) from refineries to various cities over hundreds of miles

In addition, hundreds of other liquid, gas, and solid commodities (freight) are transported via pipeline over long and short distances.

In the U.S., pipelines of various types transport a total of about 2.5 trillion ton-miles of cargo in liquid, gas, and solid form, more than the total quantity of freight transported by trucks and trains combined. As discussed in Section 1.3, major pipelines in the U.S., including the Big Inch, the Little Big Inch, the Colonial, and the Trans-Alaskan pipelines, were all built to solve problems caused by national crises. The U.S. has a dense network of underground pipelines in every state and under every city, mirroring the network of roads, highways, and streets above ground. Maps of such pipelines can be obtained from various state, federal, and local agencies. It can be said that pipelines are the lifelines of modern nations.

1.5 FREIGHT (SOLIDS) TRANSPORT BY PIPELINES

During the past century, pipelines have been used extensively to transport solids of various kinds, including coal and other minerals, gravels and sand, grain, cement, mail and parcels, etc. They are transported over short as well as long distances. Pipelines that transport solids are usually referred to as **freight pipelines.** The three general types of freight pipelines are **slurry pipeline, pneumatic pipeline** (also called **pneumo conveying**), and **capsule pipeline**.

The slurry pipeline is used to transport fine particles of solids mixed with a liquid, usually water, to form a paste (slurry) that can be pumped through the pipeline. It is used commonly in mining for transporting both minerals and mine wastes (tailings). An example is the Black Mesa Coal Slurry Pipeline, which transports 5 million tons of coal each year from Arizona to Nevada, over a distance of 273 mi (438 km), using 18-inch-diameter steel pipe. The slurry in this pipeline is a mixture of fine coal particles (of less than 1 mm size) and water, at the ratio of

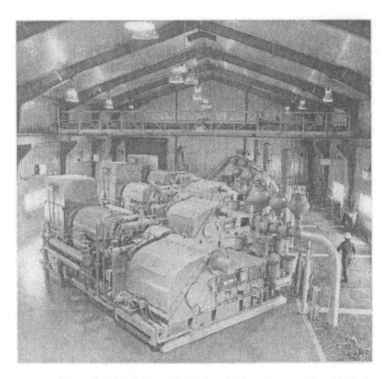

FIGURE 1.5 A pump house of the 273-mi-long Black Mesa Coal Slurry Pipeline that transports 5 million tons of coal per year from Arizona to Nevada. (Courtesy of Black Mesa Pipeline Company.)

approximately 1 to 1 by weight. Figure 1.5 is a photograph of a pump station of the Black Mesa Pipeline.

Pneumatic pipelines are used by various industries to transport hundreds of solids such as grain, cement, plastics, and even fish for short distances — usually less than 1 mi or 1 km. They are used extensively at train stations to load and unload trains, at ports to load and unload ships, and in factories to receive and convey bulk materials.

Capsule pipelines use either water or air to move freight-laden capsules (containers or vessels) through pipelines. Small diameter pneumatic capsule pipelines (PCPs) have been used for transporting mail, money, documents, and many other lightweight products. Large (1 m) diameter PCPs have been used for transporting limestone (e.g., from mine to a cement plant) and, in one instance, for transporting excavation and construction materials during the construction of a long tunnel in Japan. Figure 1.6 is a photograph of the loading station of a Japanese PCP used for transporting limestone, the raw material needed for manufacturing cement, to a cement plant. More about this and other capsule pipeline projects will be discussed in Chapter 7. The current status and anticipated future use of freight pipelines are discussed in an American Society of Civil Engineers (ASCE) Task Committee report [5]. According to this report, the advancement in new pipeline technologies such as capsule pipelines will cause more kinds and greater quantities of solids to

FIGURE 1.6 The cargo loading station of a pneumatic capsule pipeline used successfully in Japan by Sumitomo Metal Industries, Ltd. Each capsule in this 1-m-diameter, 3.2-km (one-way length) pipeline carries 1.6 tonnes of limestone from a mine to a cement plant. (Courtesy of Sumitomo Metal Industries.)

be transported by pipelines in the future, further enhancing the importance of pipelines as a means of freight transport in the 21st century. The advantages of pipelines to transport freight over truck and rail shipment will be discussed in Section 1.8.

1.6 TYPES OF PIPELINES

Pipelines can be categorized in many different ways. Depending on the commodity transported, there are water pipelines, sewer, natural-gas pipelines, oil pipelines (for crude oil), product pipelines (for refined petroleum products such as gasoline, diesel, or jet fuel), solid pipelines (freight pipelines) for various solids, etc.

According to fluid mechanics or the types of flow encountered, pipelines can be classified as single-phase incompressible flow (such as water pipelines, oil pipelines, and sewers*), single-phase compressible flow (natural gas pipelines, air pipelines, etc.), two-phase flow of solid–liquid mixture (hydrotransport), two-phase flow of solid–gas mixture (pneumotransport), two-phase flow of liquid–gas mixture (oil–gas pipelines), non-Newtonian fluids, and finally, the flow of capsules. This type of classification is the best one from a scientific (analytical) standpoint since different pipelines of the same flow type are covered by the same fluid mechanic equations. The fluid mechanics part of this book is based on the flow types; different types are treated separately in different chapters.

---

* Sewage is treated in hydraulics as a single-phase Newtonian fluid because it usually contains more than 98% water and only less than 2% solids.

Other methods of classifying pipelines also exist. For instance, depending on the environment or where pipelines are used, there are offshore pipelines, inland pipelines, in-plant pipelines, cross-mountain pipelines, etc. Depending on the type of burial or support, pipelines may also be classified as underground, aboveground, elevated, and underwater (submarine) types. Depending on pipe material, there are steel, cast iron, plastic, concrete, and other types. Table 1.1 lists the classification of pipelines in various ways. References 1 through 3 and 5 contain information on the various types of pipelines.

1.7 COMPONENTS OF PIPELINES

A pipeline is a complex transportation system. It includes components such as pipe, fittings (valves, couplings, etc.), inlet and outlet structures, pumps (for liquid) or compressors (for gas), and auxiliary equipment (flowmeters, pigs, transducers, cathodic protection systems, and automatic control systems including computers and programmable logic controllers).

1.8 ADVANTAGES OF PIPELINES

For the transport of large quantities of fluid (liquid or gas), a pipeline is undisputedly the most favored mode of transportation. Even for solids, there are many instances that favor the pipeline over other modes of transportation. The advantages of pipelines are:

1. **Economical in many circumstances.** Factors that favor pipelines include large throughput, rugged terrain and hostile environment (such as transportation through swamps). Under ordinary conditions, pipelines can transport fluids (liquids or gases) at a fraction of the cost of transportation by truck or train (see Figure 1.7). Solid transport by pipeline is far more complex and costly than fluid transport. Still, in many cases, pipelines are used to transport solids because the cost is lower than for other modes of transportation, such as trucks.

2. **Low energy consumption.** The energy intensiveness of large pipelines is much lower than that of trucks, and is even lower than that of rail. The energy intensiveness is defined as the energy consumed in transporting unit weight of cargo over unit distance, in units such as **Btu per ton-mile**. Table 1.2 compares the energy intensiveness of pipelines to those for other modes of transport.

3. **Friendly to environment.** This is due mainly to the fact that most pipelines are underground. They do not pose most of the environmental problems associated with trucks and trains, such as air pollution, noise, traffic jams on highways and at rail crossings, and killing animals that strayed on highways and railroads. Oil pipelines may pollute land and rivers when a leak or rupture develops. However, far more spills would occur if trucks and trains transported the same oil.

TABLE 1.1
Taxonomy of Pipelines

According to Commodity Transported

1. Water pipelines
2. Sewage pipelines (sewers)
3. Natural gas pipelines
4. Oil pipelines (for crude oil)
5. Product pipelines (for petroleum products such as gasoline, diesel, jet fuel, etc.)
6. Solid pipelines (coal, other minerals, sand, solid wastes, wood pulp, mail, parcels, consumer goods, etc.)
7. Others (air, chemicals, hazardous waste, etc.)

According to Fluid Mechanics

1. Single-phase incompressible flow
2. Single-phase compressible flow
3. Two-phase flow solid-liquid mixture (hydrotransport)
4. Two-phase flow of solid–gas mixture (pneumotransport)
5. Two-phase flow of liquid–gas mixture
6. Two-phase flow of capsules
7. Non-Newtonian fluids

According to Environment

1. Offshore pipelines
2. Inland pipelines
3. In-plant pipelines
4. Mountain (or cross-mountain) pipelines
5. Space pipelines (pipelines to be built in outer space, such as for space exploration on another planet)

According to the Type of Burial or Support

1. Underground pipelines
2. Aboveground pipelines
3. Elevated pipelines
4. Underwater (submarine) pipelines

According to Pipe Material

1. Steel pipelines
2. Cast-iron pipelines
3. Plastic pipelines
4. Concrete pipelines
5. Others (clay, glass, woodstave, etc.)

4. **Safe for humans.** This is especially true for liquid pipelines and liquid–solid pipelines. The safety of natural gas pipelines is always of strong concern. Gas pipelines under high pressure can explode; however, if trucks and trains transported the same natural gas, it would be much more dangerous to the public. So, in general, it can be said that pipelines are much safer than all other land-based modes of freight transport. For

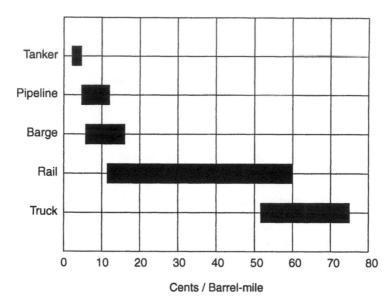

Cents / Barrel-mile

FIGURE 1.7 Comparison of petroleum transportation cost by pipeline with those by other modes. (From Kennedy, J.L., *Oil and Gas Pipeline Fundamentals,* PennWell Publishing, Tulsa, OK, 1984, p. 2. With permission from PennWell Publishing.)

TABLE 1.2
Energy Intensiveness (EI) of Various Modes of Freight Transport

Mode	EI[a] (Btu/TM)	Circumstances	Ref.[b]
Airplane	37,000	Average	6
Truck	2300	Average	6
Railroad	680	Average	6
Waterway (Barge)	540	Average	6
Pipeline (Oil)	450	Average	6
Coal Slurry Pipeline	1500	3-ft diameter, 300 mi	7
Hydraulic Capsule Pipeline (HCP)	700	3-ft diameter, 300 mi	7
Pneumatic Capsule Pipeline (PCP)	1800	20-inch diameter, Tubexpress System	8

[a] The unit of EI used is Btu/TM, which stands for British thermal units of energy consumed per ton-mile of cargo transported. Depending on circumstances, the value of EI within each mode may deviate considerably from the average values listed. For all pipelines, the value of EI decreases with increasing pipeline diameter, and increases with increasing flow velocity. For coal slurry pipelines, EI increases when pipeline length is decreased because of the need for much energy in the slurrification and dewatering processes.

[b] Numbers refer to numbered references at the end of this chapter.

instance, based on statistics published by the U.S. Department of Transportation, during the 12-year period between 1988 and 1999, the average number of people killed (injured) by pipelines per year was 23 (107), which includes 21 (92) for natural gas pipelines, and 2 (15) for hazardous liquid pipelines. In contrast, the number of people killed (injured) by large trucks per year during the same period was 5,162 (133,167). This shows that there were 200 times more people killed and 1000 times more people injured by trucks than by pipelines. It can be concluded that pipelines are enormously safer than trucks and trains.

5. **Unaffected by weather.** Weather does not affect pipelines because most of them are buried underground below the frostline.

6. **High degree of automation.** This makes pipelines the least labor-intensive of all transportation modes. Note that labor-intensive societies generally have low living standards. The high living standard in the U.S. would not be possible without automation.

7. **High reliability.** Because pipeline operation is continuous, automatic, and unaffected by weather, pipelines are highly reliable. Furthermore, they are least affected by labor strikes, holidays, delivery schedules, etc. The system operates continuously around the clock without stop.

8. **Less sensitive to inflation.** Due to high capital cost and low operational cost, pipeline tariffs are less sensitive to inflation than tariffs for trucks and trains. However, high capital cost is great when the interest rate is high.

9. **Convenience.** Water and gas pipelines transport commodities directly to homes, a great convenience to the public. Oil pipelines bring crude oil to refineries and bring refined petroleum products, such as gasoline and diesel fuel, to the market without the products leaving the pipelines. Even when one puts gasoline in a car at a filling station, the gasoline moves through a short pipe (hose) fitted with a nozzle.

10. **Less susceptible to theft.** Because pipelines are mostly underground and enclosed, the commodities transported by pipelines are less susceptible to theft than those transported by truck and train.

11. **Efficient land use.** Underground pipelines allow surface land to be used for other purposes. This results in more efficient land use.

12. **High degree of security.** Because pipelines are underground and fixed to the ground, terrorists cannot hijack a pipeline, as they can trucks and aircraft, and use it as a lethal weapon to destroy a major building or other important target. Also, it is far more difficult for terrorists to attack an underground pipeline and inflict catastrophic damage to it than to an aboveground structure such as a bridge or a power plant. Moreover, underground pipelines are inaccessible to people except at the inlet and outlet. Thus, they can be more easily guarded against attack or sabotage. Even though any unguarded long pipeline right-of-way may be vulnerable to sabotage, the damage that can be achieved is rather limited. Pipeline companies have the ability to repair a damaged underground pipe and return it to service within hours. Such sabotage activities can also be detected easily by spy satellites and other means of remote sensing. For

these reasons, pipelines must be low on the priority lists of targets of terrorists. This is not to say that security should not be of concern to pipeline companies. Two types of pipelines that require the greatest protection in terms of security are pipelines that supply drinking water, and natural gas pipelines that pass through densely populated areas.

REFERENCES

1. Kennedy, J.L., *Oil and Gas Pipeline Fundamentals*, PennWell Publishing, Tulsa, OK, 1984, chap. 1.
2. Liu, H., Freight pipelines, in *Encyclopedia Britannica*, R. McHenry, Ed., Chicago, IL, 1993, pp. 861–864.
3. Pipeline, in *McGraw-Hill Yearbook of Science and Technology*, McGraw-Hill, New York, 1994, pp. 305–307.
4. *The Big Inch and Little Big Inch Pipelines: The Most Amazing Government–Industry Cooperation Ever Achieved,* Duke Energy Gas Transmission, Houston, TX, 2000.
5. ASCE, Freight pipelines current status and anticipated future use, report of the Task Committee on Freight Pipelines, American Society of Civil Engineers, *Journal of Transportation Engineering*, 124(4), 300–310, 1998.
6. Hirst, E., Energy intensiveness of transportation, *Transportation Engineering Journal*, *ASCE*, 99(1), 111–122, 1973.
7. Liu, H. and Assadollahbaik, M., Feasibility of using hydraulic capsule pipeline to transport coal, *Journal of Pipelines*, 1(4), 295–306, 1981.
8. Carstens, M.R. and Freeze, B.E., *Pneumatic Capsule Pipeline*, Tubexpress Systems, Inc., Houston, TX.

2 Single-Phase Incompressible Flow of Newtonian Fluid

2.1 INTRODUCTION

A **multiphase flow** contains at least two separate phases, such as a liquid and a solid, a gas and a solid, a liquid and a gas, or two immiscible liquids. A **single-phase flow**, on the other hand, contains either a single liquid or gas without solids in it, or without any other immiscible liquid or gas. The flows of water, oil, natural gas, air, etc. are all examples of single-phase flow. Water laden with sediment particles or air bubbles is a two-phase flow. If the flow of water contains both air bubbles and sediment, it is a three-phase flow and so forth. A liquid with dissolved gas or another dissolved liquid, or with homogeneous suspension of very fine particles of solids, can be considered and treated as a single-phase flow, although in reality two phases are involved.

A flow is said to be **incompressible** if the density of any particle in the flow, be it a fluid or a solid particle, remains constant as the particle travels with the flow. A flow is said to be **homogeneous** if the density is constant throughout the flow. A single-phase incompressible flow is a homogeneous flow, whereas a multiphase incompressible flow is not homogeneous. For instance, for a pipe flow of water carrying gravel, the density of the flow is not the same everywhere at a given time, depending on whether water or gravel exists at the location at a given time. Normally, both liquid and gas are treated as incompressible flow. However, when the speed of a gas approaches, equals to, or exceeds the velocity of sound, large density changes occur in the flow within short distances and the flow can no longer be treated as incompressible. Also, when any gas is flowing through a long pipeline, there can be substantial change of the density of the gas over a long distance due to pressure change along the pipe even when the speed involved is low. Therefore, not all gas pipelines can be treated as incompressible, even when the velocity is low.

From elementary fluid mechanics, the shear stress τ in a two-dimensional laminar flow in the x direction as shown in Figure 2.1 is

$$\tau = \mu \frac{du}{dy} \tag{2.1}$$

17

where τ is the shear stress; u is the velocity at a distance y from the wall; du/dy is the derivative of u with y; and μ is the dynamic viscosity. Equation 2.1 is often referred to as Newton's law of viscosity.

When τ is plotted against du/dy, if the result turns out to be a straight line passing through the origin of the graph, the fluid is a **Newtonian fluid**. Otherwise, it is **non-Newtonian**. Figure 2.2 illustrates this concept. Various types of non-Newtonian fluids are discussed in Chapter 4; the present chapter deals with Newtonian fluids only. Note that the slope of the straight line representing a Newtonian fluid in Figure 2.2 is the dynamic viscosity μ of the fluid. Each graph given in Figure 2.2 is called a **rheogram**; it depicts the rheological properties of a Newtonian or non-Newtonian fluid.

Most single-phase fluids encountered in engineering practice, such as air, water, and oil, are Newtonian fluids. Some others, such as a paint, glue, and mud, are non-Newtonian. In general, Newtonian fluids are pure liquids or pure gases, whereas non-Newtonian fluids are fluids that contain a large concentration of fine particles of solids or another immiscible fluid. Because the particles in a non-Newtonian fluid are very small and uniformly distributed in the fluid, the mixture is considered to be single-

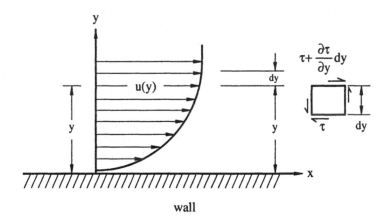

FIGURE 2.1 Velocity variation from wall and shear stress for a parallel flow.

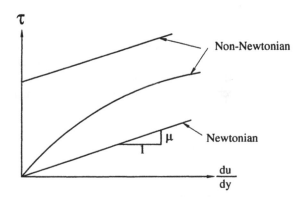

FIGURE 2.2 Rheograms of Newtonian and non-Newtonian fluids.

phase. When the particles are large, they can settle out easily and the flow is considered a two-phase or multiphase flow rather than a non-Newtonian fluid. Therefore, whether a fluid is considered as non-Newtonian or multiphase depends on the size of the particles in suspension, and how uniformly the particles are mixed with the fluid.

The most common type of flow encountered in engineering practice is single-phase incompressible flow of Newtonian fluid — a subject treated in ordinary fluid mechanics. Since the readers are assumed to have taken a course in fluid mechanics, only a brief review of the subject is provided herein. The review focuses on pipe flow only.

2.2 FLOW REGIMES

The flow in a pipe may be either laminar or turbulent, depending on the Reynolds number, $\Re$, and the amount of perturbation. The Reynolds number is defined as

$$\Re = \frac{\rho VD}{\mu} = \frac{VD}{\nu} \tag{2.2}$$

where ρ is the density of the fluid; μ is the dynamic viscosity; ν is the kinematic viscosity; V is the cross-sectional mean velocity of the flow in the pipe; and D is the inner diameter of the pipe.

The flow in a pipe is laminar when the Reynolds number is low. The flow becomes turbulent when the Reynolds number exceeds a critical value called the **critical Reynolds number, $\Re_c$**. The value of $\Re_c$ depends on the perturbation in the pipe. With large perturbation, $\Re_c$ is approximately 2100. Much higher values of $\Re_c$ can be obtained when the perturbation is small. Therefore, the value $\Re_c = 2100$ should be considered as a minimum. It happens in high perturbation situations such as in a pipe with an abrupt (sharp-cornered) entrance and rough pipe interior. When a pipe is smooth and the entrance is streamlined, $\Re_c$ is greater than 2100. The highest $\Re_c$ recorded is on the order of 40,000. Laminar flow can exist at such a high Reynolds number only in carefully prepared laboratory experiments, using smooth, straight pipe with a well-streamlined (bell-shaped) entrance, and with little disturbance of the flow. Any pipe flow at a high Reynolds number (exceeding 10^4) is highly unstable. A small disturbance, such as that produced by shaking or hitting the pipe or residual turbulence in the intake tank can cause the flow to change to turbulent. Generally, the higher the Reynolds number, the more unstable a laminar flow becomes, and the smaller the perturbation needed to change it to turbulent.

When the flow in a pipe is at or near the critical Reynolds number, the flow may be intermittent — oscillating back and forth between laminar and turbulence. In this case, the flow is said to be in the **transition region**.

2.3 LOCAL MEAN VELOCITY AND ITS DISTRIBUTION (VELOCITY PROFILE)

The velocity, u, of the fluid at any given location or point in a turbulent pipe flow is highly fluctuating. This velocity is referred to in fluid mechanics as the **local**

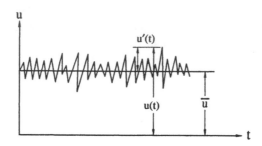

FIGURE 2.3 Mean and fluctuating components of velocity of turbulent flow.

velocity or **point velocity**. It has a time-averaged mean component, $\bar{u}$, and a fluctuating component, u', as shown in Figure 2.3.

What is normally measured with a Pitot tube in pipe is the velocity $\bar{u}$. The fluctuating component u' can be measured with special instruments such as a hot-wire or hot-film anemometer, or a laser-Doppler anemometer. The mean component $\bar{u}$ is the average over time of the instantaneous velocity $u(t)$ at a given location in the pipe. It is called the **temporal mean velocity** or the **local mean velocity**. Mathematically,

$$\bar{u} = \frac{1}{t_o} \int_0^{t_o} u(t)dt \tag{2.3}$$

where t_o is a long time. Note that while both u and u' are functions of time, $\bar{u}$ does not change with time in a steady flow.

Integrating the local mean velocity $\bar{u}$ across any pipe or conduit yields the discharge of the flow, Q, as follows:

$$Q = \int_A \bar{u}\,ds \tag{2.4}$$

where A is the cross-sectional area of the pipe (conduit), and ds is an infinitesimal area on A.

For pipes of circular cross section and when the velocity $\bar{u}$ is axisymmetric (i.e., when $\bar{u}$ is constant along any radius r), the integral in Equation 2.4 can be carried out by assuming that $\bar{u}$ is constant over an annulus of radius r and infinitesimal width dr as shown in Figure 2.4. The result is

$$Q = 2\pi \int_0^a r\bar{u}(r)dr \tag{2.5}$$

where a is the pipe radius. The cross-sectional mean velocity of the flow is then

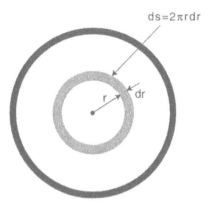

FIGURE 2.4 Integration of axisymmetric flow in a pipe.

$$V = \frac{Q}{A} = \frac{2\int_o^a r\bar{u}(r)dr}{a^2} \tag{2.6}$$

In most applications in pipeline engineering, it is the local mean velocity $\bar{u}$ rather than the instantaneous velocity u or the fluctuating component u' that is of interest. For this reason, in this book we will refrain from further discussing u and u', and will concentrate on $\bar{u}$. Also, in laminar flow, $\bar{u}$ reduces to u since u' is zero. To simplify notation, hereafter we will use u to denote $\bar{u}$, the local mean velocity, whether the flow is laminar or turbulent.

2.3.1 VARIATION OF VELOCITY ALONG PIPE

The variation of the local mean velocity u in a pipe depends on whether the flow is laminar or turbulent, and on the distance to the pipe entrance. At the pipeline entrance, the velocity across the pipe is usually rather uniform except in the region of flow separation near an abrupt (sharp-cornered) entrance (see Figure 2.5). Note that in regions of flow separation, flow reversal occurs, and eddies and turbulence are generated. On the other hand, at a distance downstream from the pipe entrance, the boundary layer* along the pipe has reached the centerline of the pipe, and the velocity profile has become fully developed (see Figure 2.6). Depending on whether the flow is laminar or turbulent, the boundary layer is laminar or turbulent.

The entrance region distance L' in Figure 2.6 (i.e., the distance required for the pipe flow to become fully developed) differs whether the flow is laminar or turbulent. In a turbulent flow, the boundary layer grows rapidly, and the flow becomes fully developed in a short distance from the pipe entrance, of the order of $L' = 10D$. On the other hand, for laminar flow of large Reynolds number (say, $\Re > 10^3$), the boundary layer grows more slowly, and hence a much longer distance, of the order of $L' = 100D$, may be required for the flow to become fully developed. More

---

* By definition, the boundary layer is the layer of fluid along the pipe where the velocity of the fluid has reached 99% of the free stream value outside the boundary layer.

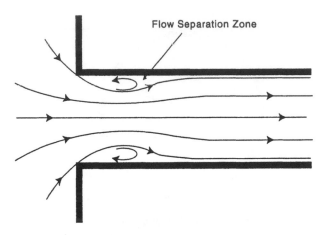

FIGURE 2.5 Flow pattern near a sharp-cornered entrance.

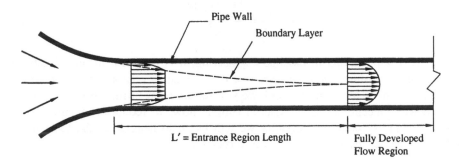

FIGURE 2.6 Concept of fully developed flow in pipe.

precisely, the distance L' for laminar flow can be calculated from a theoretical formula derived by Langhaar as follows:

$$L' = 0.058D\Re \tag{2.7}$$

Example 2.1 A man wants to conduct an experiment on laminar flow in a pipe in the fully developed region. He selects a 1-inch-diameter pipe and uses a glycerin/water mixture as the fluid, which has a kinematic viscosity equal to 10^{-4} ft^2/s. He also selects a mean flow velocity of 2 fps for the test. If he needs a test section of 10 ft of fully developed flow, what should be the minimum length of the pipe used for the experiment?

[Solution] With $D = 1$ inch $= 0.08333$ ft, $V = 2$ fps and $\nu = 10^{-4}$ ft^2/s, the Reynolds number of the flow is $\Re = VD/\nu = 1667$, which shows that the flow is laminar. From Equation 2.7, the length of the entrance region to establish fully developed flow for this case is $L' = 0.058D\Re = 8.06$ ft. Therefore, the total length of the pipe needed for the experiment is $8.06 + 10 = 18$ ft approximately.

2.3.2 VELOCITY PROFILE OF FULLY DEVELOPED FLOW

The shape of the velocity profile of a fully developed pipe flow depends on whether the flow is laminar or turbulent.

For laminar pipe flow, the velocity profile in the fully developed region is a parabola governed by the following equation:

$$u = V_c \left(1 - \frac{r^2}{a^2} \right) \tag{2.8}$$

in which V_c is the centerline velocity in the pipe; a is the pipe radius; and r is the radial distance. Equation 2.5 is derived in most fluid mechanics textbooks such as Reference 1 and hence is not repeated here.

Example 2.2 Determine the relationship between the mean velocity, V, in a pipe and the maximum velocity, V_c, at the centerline of the pipe, for any fully developed laminar flow.

[Solution] The velocity profile for a fully developed laminar flow is given by Equation 2.8. From Equations 2.5 and 2.8,

$$Q = 2\pi \int_o^a r u\,dr = \frac{\pi a^2}{2} V_c = \frac{AV_c}{2} \tag{a}$$

From Equation 2.6,

$$Q = VA \tag{b}$$

Comparing Equation a with Equation b yields $V_c = 2V$. This shows that for fully developed laminar flow in pipe, the maximum velocity (centerline velocity, V_c) is twice the main velocity V.

For turbulent flow, the velocity profile in the fully developed region in the pipe is logarithmic as follows:

$$\frac{u}{u_*} = \frac{1}{\kappa} \ell n \frac{y u_*}{v} + 5.5 \tag{2.9}$$

where u_* is the shear velocity defined as $u_* = \sqrt{\tau_o/\rho}$; τ_o is the wall shear stress; κ is the von Karman constant equal to 0.40, approximately; and y is the radial distance from the wall, namely, $y = a - r$.

Equation 2.9 is often referred to in the literature as the **Prandtl's universal velocity distribution,** which is valid not only for pipe flow but also for other types of flow such as the flow along a flat plate. Substituting $\kappa = 0.40$ into Equation 2.9 and changing the natural logarithm, $\ln$, to common logarithm, log yields

$$\frac{u}{u_*} = 5.76\log\frac{yu_*}{\nu} + 5.5 \tag{2.10}$$

It should be realized that while Equation 2.8 is valid for laminar flow at any distance r or y in the pipe, Equations 2.9 and 2.10 are valid for turbulent flow only in places not too close to the wall. For instance, if one uses Equation 2.9 or 2.10 at $y = 0$, the velocity u becomes minus infinity, which is ridiculous. From Equation 2.9, when $u = 0$, $y = 0.11\,\nu/u_*$. This means Equations 2.9 and 2.10 are valid only at a distance from the wall much greater than $y = 0.11\,\nu/u_*$, or when $y^+ = yu_*/\nu \gg 0.11$. In practice, Equations 2.9 and 2.10 are used only for the region where y^+ is greater than 70. In the region where y^+ is between 5 and 70, the following equation by Spalding can be used:

$$y^+ = u^+ + e^{-\kappa b}\left[e^{\kappa u^+} - 1 - \kappa u^+ - \frac{\left(\kappa u^+\right)^2}{2} - \frac{\left(\kappa u^+\right)^3}{6}\right] \tag{2.11}$$

where $\kappa = 0.40$ and $b = 5.5$.

For the region $y^+ < 5$, the velocity profile is almost linear according to the following law:

$$u = \left(\frac{\rho u_*^2}{\mu}\right)y \tag{2.12}$$

Equation 2.12 can be written in the following alternative forms:

$$u = u_* y^+ \tag{2.13}$$

or

$$u^+ = y^+ \tag{2.14}$$

where $y^+ = \rho u_* y/\mu = u_* y/\nu$, and $u^+ = u/u_*$.

The region $y^+ < 5$, where Equations 2.12 to 2.14 are applicable, is called the **viscous sublayer** of the flow. It is also often referred to in the literature as the **laminar sublayer.** However, the term laminar sublayer implies that the region $y^+ < 5$ is free of turbulence, which is not the case. Many researchers have demonstrated that the flow is still turbulent in this region. Therefore, the term laminar sublayer should be regarded as a misnomer and not be used in the future.

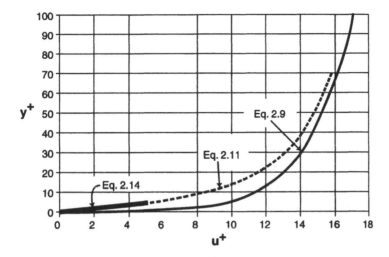

FIGURE 2.7 Comparison of various formulas used to predict velocity profile in turbulent pipe flow. Equation 2.9 is plotted only in the region $y^+ > 0.11$.

Equations 2.9, 2.11, and 2.14 are plotted in Figure 2.7. It can be seen from the figure that Spalding's equation provides a smooth transition for the region $5 < y^+ < 70$.

As will be proved later, the shear stress at the pipe wall, τ_0, can be determined from $\tau_0 = f\rho V^2/8$, where f is the Darcy-Weisbach friction factor, which can be found from the Moody diagram once the Reynolds number $\Re$ and the relative roughness e/D are known. Therefore,

$$u_* = \sqrt{\frac{\tau_0}{\rho}} = V\sqrt{\frac{f}{8}} \tag{2.15}$$

Equation 2.15 can be used to calculate the shear velocity, u_*, in Equations 2.9 through 2.14.

Example 2.3 Water at 80°F flows through a 3-ft-diameter pipe at the mean velocity of 5 fps. The velocity profile is fully developed. Find the local velocities of the flow at $y = 0, 0.001, 0.01, 0.1, 1, 10,$ and 18 inches.

[Solution] From the no-slip condition in fluid mechanics, the local velocity u at $y = 0$ (at wall) must be zero. At the other distances indicated in this example, the velocity should increase with y in a manner as calculated below.

The Reynolds number of the flow in the pipe is

$$\Re = \frac{VD}{\nu} = \frac{5 \times 1.5}{0.93 \times 10^{-5}} = 8.06 \times 10^5$$

The relative roughness is

$$\frac{e}{D} = \frac{0.00015}{3} = 0.00005$$

From the Moody diagram, $f = 0.0130$. Therefore, Equation 2.15 yields

$$u_* = V\sqrt{\frac{f}{8}} = 0.202\, fps$$

Using the above value of u_*, for $y = 0.001, 0.01, 0.1, 1, 10$, and 18 inches, the dimensionless distance $y^+ = yu_*/\nu$ is 1.81, 18.1, 181, 1810, 18,100, and 32,600, respectively.

For $y^+ = 1.81$, Equation 2.14 is applicable, which yields $u^+ = 1.81$ and $u = 1.81u_* = 0.366$ fps. For $y^+ = 18.1$, Equation 2.11 is applicable, which yields $u^+ = 11.3$ and $u = 2.28$ fps. For $y^+ = 181$, Equation 2.10 is applicable, which yields $u^+ = 5.76 \log y^+ + 5.5 = 18.5$, and $u = 18.5\, u_* = 3.74$ fps. For the last three cases where $y^+ = 1810, 18,100$, and 32,600, Equation 2.10 is applicable. The respective solutions for u^+ are 24.3, 30.0, and 31.5, and the respective solutions for u are 4.90, 6.06, and 6.36 fps. A summary of the results is listed below:

y (inch)	0	0.001	0.01	0.01	1.0	10	18
y^+	0	1.81	18.1	181	1810	18,100	32,600
u^+	0	1.81	11.3	18.5	24.3	30.0	31.5
u (fps)	0	0.366	2.28	3.74	4.90	6.06	6.36

From the results, it can be seen that the relation between the centerline velocity and the mean velocity of the flow is $V_c/V = 6.36/5 = 1.27$. This means that the centerline velocity V_c is greater than the mean velocity V by about $(6.36 - 5)/5 = 27\%$.

2.4 FLOW EQUATIONS FOR ONE-DIMENSIONAL ANALYSIS

2.4.1 CONTINUITY EQUATION

The continuity equation for incompressible flow in a pipe is

$$V_1 A_1 = V_2 A_2 = Q = \text{constant} \tag{2.16}$$

where A_1 and A_2 are the cross-sectional areas of the pipe at sections 1 and 2; V_1 and V_2 are the cross-sectional average velocities (mean velocities) at sections 1 and 2; and Q is the discharge (i.e., volumetric flow rate) at either section. Note that Equation 2.16 is applicable to not only steady flow but also unsteady flow through pipes. The pipe can have a cross section of circular or any other shape. Even when the fluid is non-Newtonian or multiphase, the equation still holds as long as the velocity V refers to the average velocity of the different phases at any pipe cross section, and when the flow is incompressible. The equation does not hold if the fluid leaves or enters the pipe between sections 1 and 2, as for instance when a branch exists between the two sections.

2.4.2 ENERGY EQUATION

The energy equation for incompressible flow along a pipeline can be written as

$$\alpha_1 \frac{V_1^2}{2g} + \frac{p_1}{\gamma} + z_1 = \alpha_2 \frac{V_2^2}{2g} + \frac{p_2}{\gamma} + z_2 + h_L + h_t - h_p \tag{2.17}$$

where subscripts 1 and 2 refer to an upstream section and a downstream section, respectively; p is the average pressure at a cross section; z is the average elevation of a cross section; g is the gravitational acceleration; γ is the specific weight of the fluid; α is the energy correction factor to be defined later; h_L is the headloss along the pipe between sections 1 and 2; h_t is the turbine head, i.e., the energy per unit weight of the fluid extracted by turbines installed in the pipeline between sections 1 and 2, if such turbines exist in the line; and h_p is the pump head, i.e., the energy per unit weight of the fluid imparted by pumps that may exist in the pipeline between sections 1 and 2.

The energy correction factor is

$$\alpha = \frac{\int u^3 ds}{V^3 A} \tag{2.18}$$

The integral in Equation 2.18 is to be performed over the entire cross-sectional area. Thus, α depends on the cross-sectional distribution of the local velocity u. For fully developed laminar flow in pipe, which has a velocity profile given by Equation 2.8, $\alpha = 2.0$; for fully developed turbulent flow, α is in the neighborhood of 1.05.

Normally, the term h_t in Equation 2.17 is zero because no turbine exists in the pipe between sections 1 and 2. Also, α_1 and α_2 are normally taken to be unity for turbulent flow. Under such conditions, Equation 2.17 becomes

$$\frac{V_1^2}{2g} + \frac{p_1}{\gamma} + z_1 = \frac{V_2^2}{2g} + \frac{p_2}{\gamma} + z_2 + h_L - h_p \tag{2.19}$$

Note that the headloss, h_L, gives the energy dissipated in the flow by a unit weight of the fluid moving from section 1 to section 2. The power lost or dissipated is

$$P_L = \gamma Q h_L \tag{2.20}$$

If γ, Q, and h_L are in **lb/ft³**, **cfs**, and **ft**, respectively, then P_L is in **ft-lb/s**. In this case, dividing $\gamma Q h_L$ by 550 gives the power dissipated in **horsepower**. On the other hand, if γ, Q, and h_L are in **N/m³**, **m³/s**, and **m**, respectively, the P_L obtained from Equation 2.20 will be in **watts (W)**.

The power dissipated in a flow causes the temperature of the flow to rise, and causes heat loss through the pipe wall. For a well-insulated pipe (i.e., adiabatic wall), the rise of temperature along the pipe in the flow direction is

$$\Delta T = T_2 - T_1 = \frac{g h_L}{c_v} \tag{2.21}$$

or

$$g h_L = c_v \left(T_2 - T_1 \right) = i_2 - i_1 \tag{2.22}$$

where c_v is the **specific heat capacity at constant volume**, and i_1 and i_2 are **specific internal energy**, which is the internal energy per unit mass of the fluid, at sections 1 and 2, respectively. For air under standard atmospheric condition, c_v and c_p (**specific heat capacity at constant pressure**) are 0.171 and 0.240 Btu/lbm °R, and $k = c_p/c_v = 1.40$, respectively. For liquids, c_v and c_p are the same, and the quantity c_v in Equations 2.21 and 2.22 can be written simply as c — the **specific heat capacity**. For water, $c = 1$ cal/gm °C = 4180 joules/kg °C = 1 Btu/lbm °F = 32.2 Btu/slug °F. Note that in using Equations 2.21 and 2.22, the units of g, h_L, and, c_v must be consistent. For instance, if g is in ft/s² and h_L is in ft, c_v must be in ft-lb/slug °F. For water, $c = 25,050$ ft-lb/slug °F.

If the pipeline is not insulated, the heat generated by friction is readily lost to the environment (normally the atmosphere or ground), resulting in a constant temperature along the pipe — the same temperature as that of the environment. In this case, the rate of heat loss through the pipe wall per unit length of the pipe is

$$q = \frac{P_L}{L} = \frac{\gamma Q h_L}{L} \tag{2.23}$$

or

$$g h_L = \frac{q L}{\rho Q} = \frac{Q_c}{\rho Q} \tag{2.24}$$

where Q_c is the rate of heat loss throughout the pipe between sections 1 and 2.

In using Equations 2.23 and 2.24, the units of q must be consistent with those of γ, Q, h_L and L. If γ, Q, h_L, and L are in ft-lb units, q must be in ft-lb/s/ft. To get q in BTU/s/ft, the result must be divided by 778 because 1 Btu = 778 ft-lb. When there is both heat transfer through the pipe and temperature rise caused by frictional loss, the combination of Equations 2.22 and 2.24 yields

$$h_L = \frac{c_v \Delta T}{g} + \frac{Q_c}{\rho g Q} \tag{2.25}$$

Example 2.4 Water is drained by gravity from tank 1 to tank 2 through a steel pipe. The discharge through the pipe is 2 cfs. The difference in the water levels in the two tanks is 100 ft. (a) What is the power dissipated due to headloss? (b) Assuming that the pipe is well insulated, what is the increase in the temperature of the water flowing through the pipe? (c) Assuming that the pipe is not insulated and is ready to conduct heat, what is the rate of the heat loss through the pipe?

[Solution] (a) Using Equation 2.19 between two points 1 and 2 at the free surfaces of the two tanks yields

$$\frac{V_1^2}{2g} + \frac{b_1}{\gamma} + z_1 = \frac{V_2^2}{2g} + \frac{p_2}{\gamma} + z_2 + h_L \tag{a}$$

Because $V_1 = 0$, $p_1, = 0$, $V_2 = 0$, and $p_2 = 0$, the above equation reduces to

$$z_1 - z_2 = 100 \text{ ft} = h_L \tag{b}$$

This shows that regardless of the pipe size and length, the headloss in this case is always the same as the water level difference in the two tanks, which is 100 ft. Using Equation 2.20, the power dissipated by the flow is $P_L = \gamma Q h_L$ = 62.4 × 2 × 100 = 12,480 ft-lb/s = 22.7 Hp.

(b) For the case of a well-insulated pipe, the temperature rise calculated from Equation 2.21 is $\Delta T = 32.2 \times 100/25{,}050 = 0.13°F$. This case shows that a rather large headloss results only in a slight increase in the temperature of the water moving through the pipe.

(c) From Equation 2.23, the rate of heat loss through the entire length of the pipe is the same as P_L, which is 22.7 Hp. The rate of heat loss per unit length of the pipe is thus $q = 22.7/L$ (in horsepower/ft) where L is the pipe length.

2.4.3 MOMENTUM EQUATION

The momentum equation for any steady incompressible flow is

$$\vec{F} = \rho \int \vec{V} \vec{V} \cdot d\vec{s} \tag{2.26}$$

where $\vec{F}$ is the resultant force on the control volume; ρ is the fluid density; and $d\vec{s}$ is an infinitesimal surface of the control volume. Equation 2.26 is written in vector form because it has three components in three mutually perpendicular directions, such as x, y, and z when rectangular coordinates are used. Each quantity with an arrow above, such as $\vec{F}$, is a vector quantity. The direction of $d\vec{s}$ is perpendicular to the control surface, and is outward from the control volume. The integral is carried out over the surface of the control volume. More about the control volume approach can be found in modern textbooks on fluid mechanics, such as Reference 1.

Application of Equation 2.26 to an incompressible flow between two sections, 1 and 2, of a straight pipe yields

$$F = \rho\left(\beta_2 V_2^2 A_2 - \beta_1 V_1^2 A_1\right) = \rho Q\left(\beta_2 V_2 - \beta_1 V_1\right) \tag{2.27}$$

where F is the force on the fluid in the flow direction; A is the cross-sectional area; V is the cross-sectional mean velocity; Q is the discharge; β is the momentum correction factor; and subscripts 1 and 2 refer to an upstream and a downstream cross section, respectively. The **momentum correction factor** is

$$\beta = \frac{\int u^2 ds}{V^2 A} \tag{2.28}$$

For fully developed laminar and turbulent flows in pipes, the values of β are respectively 4/3 and 1.02, approximately. In ordinary practice, especially in ordinary turbulent pipe flow, β is taken to be 1.0, and so β_1 and β_2 in Equation 2.27 disappear from the equation.

Note that Equation 2.27 holds for all steady incompressible flows in a straight pipe. The force F includes all the forces acting on the fluid in the region between sections 1 and 2, such as gravity (if the pipe is not horizontal), pressure, and wall shear. The pressure and shear forces are those created by pipe wall and the neighboring fluid at or between sections 1 and 2, and by any other object in contact with the flow between sections 1 and 2, such as a valve or pump.

When the pipe between sections 1 and 2 changes direction as shown in Figure 2.8, Equation 2.26 becomes

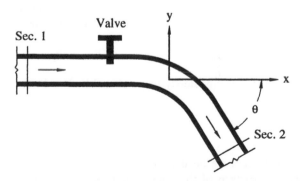

FIGURE 2.8 Analysis of fluid forces acting on a pipe bend.

$$F_x = \rho\left(\beta_2 V_2^2 A_2 \cos\theta - \beta_1 V_1^2 A_1\right)$$
$$= \rho Q\left(\beta_2 V_2 \cos\theta - \beta_1 V_1\right) \tag{2.29}$$

$$F_y = -\rho\beta_2 V_2^2 A_2 \sin\theta$$
$$= -\rho\beta_2 Q V_2 \sin\theta \tag{2.30}$$

Note that the minus sign in Equation 2.30 is due to the direction of the pipe bend shown in Figure 2.8 — in the opposite direction of the y coordinate. Had the bend been in the positive direction of y, the minus sign in Equation 2.30 would disappear.

Example 2.5 A 1-ft-diameter water pipe has a horizontal 90° bend followed by a valve at the outlet. The situation is similar to that shown in Figure 2.8 except that the angle θ is 90° and the valve is located near Section 2. When the valve is open, the discharge of water through the pipe is 5 cfs, and the pressure of the water outside the bend (at Sections 1 and 2 in Figure 2.8) is approximately atmospheric. When the valve is closed, the discharge goes to zero but the pressure in the pipe rises to 100 psig. Determine the thrust by the fluid on the bend before and after the valve closure.

[Solution] Equations 2.29 and 2.30 are applicable to this case with $\beta_1 = \beta_2 = 1.0$, and $\theta = 90°$ (see Figure 2.8).

When the valve is open, $Q = 5$ cfs, $A = 0.785$ ft², $V_1 = V_2 = Q/A = 6.37$ fps, and the two equations reduce to

$$F_x = 1.94 \times 5\,(0 - 6.37) = -61.8 \text{ lb}$$
$$F_y = -1.94 \times 5 \times 6.37 \times 1 = -61.8 \text{ lb}$$

The above forces F_x and F_y are those generated on the control volume of the fluid between sections 1 and 2. In this case, because $p_1 = p_2 = 0$ (atmospheric) and the shear force along the wall is negligible, the F_x and F_y found are the forces that the bend exerts on the control volume of fluid. Since action is equal to reaction, the force on the bend by the fluid is the same as F_x and F_y except for the opposite direction or sign. Consequently, when the valve is open, the flow exerts 61.8 lb of force in both the x and y directions, acting on the bend.

When the valve is closed, $Q = 0$, $V_1 = V_2 = 0$, but $p_1 = p_2 = 100$ psig $= 14,400$ psfg. From Equations 2.29 and 2.30, $F_x = 0$ and $F_y = 0$. However, in this case $F_x = p_1 A_1 + R_x = 0$, and $F_y = p_2 A_2 + R_y = 0$, where R_x and R_y are the x and y components, respectively, of the force acting by the bend on the fluid.

From the above, $R_x = -p_1A_1 = -100 \times 144 \times 0.785 = -11,300$ lb, and $R_y = -p_2A_2 = -11,300$ lb. Again, because the action is equal to reaction except for opposite direction or sign, the x and y components of the force by the fluid on the bend are both 11,300 lb in the x and y directions.

This example shows that whether the valve is open or closed, there is a force by the fluid acting on the pipe bend. When the valve is open, the force is caused by the momentum change. In contrast, when the valve is closed, the force is caused by the high pressure in the bend. In this case, higher force is generated on the bend when the valve is closed than when it is open, due to the high pressure in the pipe when the valve is closed. In general, fluids flowing around a bend may generate a strong force on the bend caused by both high pressure and momentum change. This will be seen in a homework problem at the end of this chapter.

2.4.4 HEADLOSS FORMULAS

The headloss term, h_L, in Equations 2.17 and 2.19 includes both the loss distributed uniformly along a pipe — the so-called **major loss** — and the loss due to localized disturbances — the so-called **minor losses**. The terms major loss and minor losses are misnomers because for a short pipe containing fittings that disturb the flow, such as valves or bends, the minor losses can be much greater than the major loss. This is especially true when a valve is partially closed, or a fitting greatly restricts and disturbs the flow. For this reason, hereafter major loss will be referred to as **pipe loss**, and the minor loss as **local loss** or **fitting loss**.

2.4.4.1 Fitting Loss

The local (fitting) loss can be determined simply from

$$h_L = K \frac{V^2}{2g} \tag{2.31}$$

where K is the **local loss coefficient,** which is approximately constant for a given fitting as long as the flow is turbulent. Tables 2.1 and 2.2, and Figures 2.9 and 2.10 contain K values for various fittings.

The values of K given in Tables 2.1 and 2.2 and Figures 2.9 and 2.10 are for turbulent flow. They depend not only on the types of fittings but also on pipe size (diameter), pipe schedule (wall thickness), type of connections to pipe (whether threaded, flanged, or welded), and design details. Generally, larger fittings have lower values of K, due to smaller relative roughness. For instance, the gate valve for a pipe of 12.5 mm may have a K value equal to 0.5 when the valve is fully open. However, when the pipe size is increased to 200 mm, the value of K decreases to 0.08. Also, for steel pipes of diameter greater than 12 inches, larger schedule number (i.e., thicker pipe wall) causes larger mismatch between the inside diameter of the fitting and that of the pipe. Consequently, the value of K increases somewhat with increased schedule number for large pipes. Finally, different manufacturers design fittings somewhat

TABLE 2.1
Typical Values of Head Loss Coefficient for Some Fittings

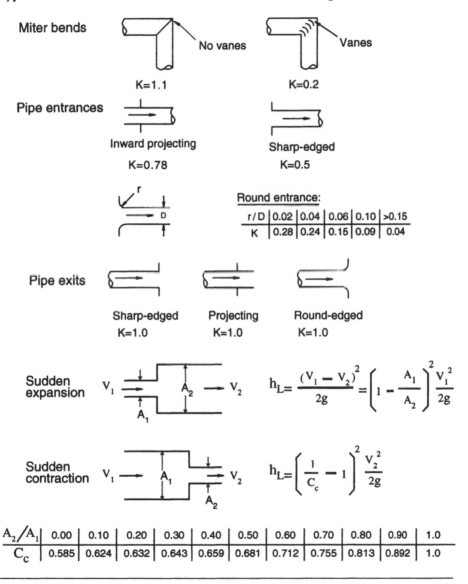

Miter bends

No vanes $K=1.1$

Vanes $K=0.2$

Pipe entrances

Inward projecting $K=0.78$

Sharp-edged $K=0.5$

Round entrance:

r/D	0.02	0.04	0.06	0.10	>0.15
K	0.28	0.24	0.15	0.09	0.04

Pipe exits

Sharp-edged $K=1.0$

Projecting $K=1.0$

Round-edged $K=1.0$

Sudden expansion

$$h_L = \frac{(V_1 - V_2)^2}{2g} = \left(1 - \frac{A_1}{A_2}\right)^2 \frac{V_1^2}{2g}$$

Sudden contraction

$$h_L = \left(\frac{1}{C_c} - 1\right)^2 \frac{V_2^2}{2g}$$

A_2/A_1	0.00	0.10	0.20	0.30	0.40	0.50	0.60	0.70	0.80	0.90	1.0
C_c	0.585	0.624	0.632	0.643	0.659	0.681	0.712	0.755	0.813	0.892	1.0

TABLE 2.2
Typical Values of Headloss Coefficient K for Selected Fittings

Fitting	K Value	Fitting	K Value
Check valves		Standard T	
Ball type	70	Side outlet	1.8
Disc type	10	Straight-through flow	0.4
Swing type	2		
Other valves		Elbows (90°)	
Foot valve	10	Regular	1.0
Globe valve	8	Long radius	0.4
Angle valve	3	Elbows (45°)	
Diaphragm valve	2	Regular	0.3
Gate valve	1.5	Long radius	0.2
Butterfly valve	0.2		
Full-bore ball valve	Negligible (<0.1)	Return bend	2.2

Note: All the K values given for valves are for fully open valves. For a given type of fitting, the K value may differ considerably for products of different manufacturers. It also depends on other factors such as whether the fittings are flanged, threaded, or welded to the pipe. Values given in this table should be considered approximate typical values. In practice, one should use test values supplied by manufacturers.

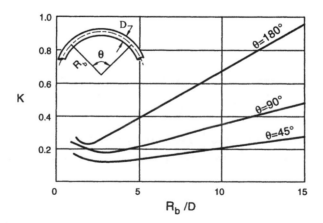

FIGURE 2.9 Headloss coefficient K for a smooth pipe bend ($\Re = 2 \times 10^5$).

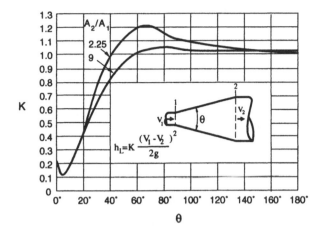

FIGURE 2.10 Headloss coefficient K for a conical enlargement (diffuser).

differently, causing the K value to differ for the same kind of valve made by different manufacturers. Due to the foregoing complexities, the values of K given herein are all approximate values. They should be used only for exercise purposes and in practices where the result of the computation is insensitive to the K values, such as for relatively long and straight pipes containing only a few fittings. In cases where the result of the calculation depends on accurate values of K, such as for short pipes containing fittings of large K values, accurate values of K should be determined from information supplied by the fitting manufacturers.

Another common way to determine fitting loss is through the use of the following equation:

$$h_L = f \frac{L_e}{D} \frac{V^2}{2g} \tag{2.32}$$

in which L_e is the **equivalent pipe length,** and f is the Darcy-Weisbach friction factor (resistance coefficient) to be discussed in the next paragraph. Many books and fitting manufacturers list the values of L_e/D instead of K for various fittings. From Equations 2.31 and 2.32, $L_e/D = K/f$, where f is the Darcy-Weisbach friction factor. Therefore, if $f = 0.02$, then $L_e/D = 50K$. This shows that the value of L_e/D depends on the friction factor f, which in turn depends on the Reynolds number $\Re$ and relative roughness of the pipe. It is better to use K rather than L_e/D, because K is less dependent on these factors. Treating L_e/D as a constant for a given fitting irrespective of the Reynolds number yields larger errors than treating K as a constant.

Example 2.6 The headloss caused by the sudden expansion of a pipe can be expressed as

$$h_L = K \frac{V_1^2}{2g} \tag{a}$$

where V_1 is the velocity of the approach upstream. Prove that the value of K is

$$K = \left(1 - \frac{A_1}{A_2}\right)^2 \tag{b}$$

where A_1 and A_2 are the cross-sectional areas of the pipe upstream and downstream of the expansion, respectively.

Also prove that

$$h_L = \frac{\left(V_1 - V_2\right)^2}{2g} \tag{c}$$

[Solution] Referring to the sketch of sudden expansion shown in Table 2.1, section 3 is selected at the location of the sudden expansion. Application of the momentum equation to the fluid between sections 3 and 2 yields:

$$\Sigma F_x = p_3 A_2 - p_2 A_2 = \rho V_2^2 A_2 - \rho V_1^2 A_1 \tag{d}$$

From the continuity equation,

$$V_2 = V_1 \frac{A_1}{A_2} \tag{e}$$

Substituting Equation e into Equation d, and dividing by A_2 yields

$$p_3 - p_2 = \rho V_1^2 \frac{A_1}{A_2}\left(\frac{A_1}{A_2} - 1\right) \tag{f}$$

Since there is little headloss between sections 1 and 3, $p_1 \approx p_3$, and Equation f can be rewritten as

$$p_1 - p_2 = \rho V_1^2 \frac{A_1}{A_2}\left(\frac{A_1}{A_2} - 1\right) \tag{g}$$

Next, from the energy equation,

$$\frac{V_1^2 - V_2^2}{2g} + \frac{p_1 - p_2}{\rho g} = h_L \tag{h}$$

Substituting Equations a and e into Equation h and rearranging terms yields

$$p_1 - p_2 = \left(K - 1 + \frac{A_1^2}{A_2^2} \right) \frac{\rho V_1^2}{2} \tag{i}$$

Combining Equations g and i yields Equation b.

Finally, substituting Equation e into Equation c yields

$$h_L = \frac{\left(V_1^2 - V_2 \right)^2}{2g} = \left(1 - \frac{A_1}{A_2} \right)^2 \frac{V_1^2}{2g} \tag{j}$$

Comparing Equation j with Equation a yields Equation b. This proves that Equation c is an alternative form of Equation a for the headloss generated by a sudden expansion. This result is listed in Table 2.1.

2.4.4.2 Pipe Loss

The most rational formula to calculate the pipe loss for single-phase incompressible flow of Newtonian fluids is the Darcy-Weisbach formula as follows:

$$h_L = f \frac{L}{D} \frac{V^2}{2g} \tag{2.33}$$

where L is the length and D is the inner diameter of the pipe. The quantity f is commonly known as the **friction factor** or **resistance factor**.

For laminar flow, the friction factor f does *not* depend on the roughness of the pipe. It is only a function of the Reynolds number $\Re$ as follows:

$$f = \frac{64}{\Re} \tag{2.34}$$

On the other hand, for turbulent flow, f is a function of $\Re$, and the relative roughness, e/D, where e is the roughness of the pipe interior surface. As readers know from elementary fluid mechanics, the value of f can be determined either from the Moody diagram (Figure 2.11), or the Colebrook formula as follows:

$$\frac{1}{\sqrt{f}} = -0.86 \ell n \left(\frac{e/D}{3.7} + \frac{2.51}{\Re \sqrt{f}} \right) \tag{2.35}$$

where ℓn is the natural logarithm. Note that the minus sign on the right of Equation 2.35 is correct because the sum of the two terms in the brackets of the equation is less than 1, and the logarithm of any quantity less than 1 is negative. Thus, the two minus signs cancel out, and the right side of Equation 2.35 is positive.

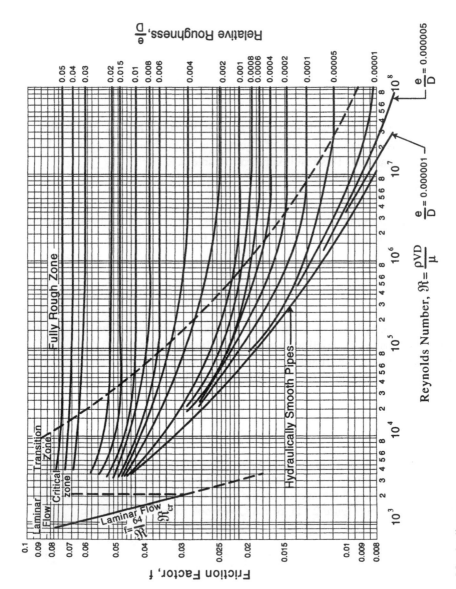

FIGURE 2.11 The Moody diagram.

TABLE 2.3
Absolute Roughness, e, of Various Types of Pipes When They Are New and Clean

Pipeline Material	Absolute Roughness, e	
	ft	mm
Glass and various plastics (e.g., PVC and PE pipes)	0 (hydraulically smooth)	0 (hydraulically smooth)
Drawn tubings (e.g., copper or aluminum pipes or tubings)	5×10^{-6}	1.5×10^{-3}
Commercial steel or wrought iron	1.5×10^{-4}	4.6×10^{-2}
Cast iron with asphalt lining	4×10^{-4}	0.12
Galvanized iron	5×10^{-4}	0.15
Cast iron	8.5×10^{-4}	0.25
Wood stave	$6 \times 10^{-4} – 3 \times 10^{-3}$	0.18–0.9
Concrete	$1 \times 10^{-3} – 1 \times 10^{-2}$	0.3–3.0
Riveted steel	$3 \times 10^{-3} – 3 \times 10^{-2}$	0.9–9.0

In terms of the common logarithm, **log**, the Colebrook formula can be rewritten as

$$\frac{1}{\sqrt{f}} = 1.14 - 2.0 \log \left[\frac{e}{D} + \frac{9.35}{\Re \sqrt{f}} \right] \qquad (2.36)$$

Values of e to be used in the two foregoing equations for various types of pipes are listed in Table 2.3.

Note that in the Colebrook formula, the factor f appears on both sides of the equation. Such equations are called **implicit equations** since they cannot be solved for f directly. An iteration (successive approximation) procedure is required to solve for the value of f in terms of $\Re$ and e/D, which is inconvenient. A simpler **explicit equation** is given by Haaland as follows:

$$\frac{1}{\sqrt{f}} = -0.782 \ell n \left[\frac{6.9}{\Re} + \left(\frac{e}{3.7D} \right)^{1.11} \right] \qquad (2.37)$$

For turbulent flow of Reynolds numbers up to 10^5, the following simple equation by Blasius holds for smooth pipes:

$$f = \frac{0.316}{\Re^{1/4}} \qquad (2.38)$$

In the completely rough zone of turbulent flow, which happens at high Reynolds number, f does not depend on $\Re$, but depends only on e/D. In this region, Equation 2.35 reduces to

$$f = \frac{1}{\left[1.14 - 2.0\log(e / D)\right]^2} \qquad (2.39)$$

A different form of the Darcy-Weisbach formula is the Fanning's equation as follows:

$$h_L = f' \frac{L}{R_H} \frac{V^2}{2g} \qquad (2.40)$$

where f' is the Fanning's resistance factor, and R_H is the hydraulic radius, which is the area A divided by the wetted perimeter P. For a circular cross-sectional pipe flowing full, $R_H = D/4$. Therefore, $f = 4f'$. Henceforth, whenever *friction factor* is mentioned in this book, it will mean f rather than f'. Equation 2.40 will not be used in this book unless otherwise indicated. Note that in the literature, the Fanning's friction factor f' is often used, written without the prime, and referred to simply as the *friction factor*, which is very confusing to the uninitiated. The reader should be aware of that! This book is consistent with modern textbooks in fluid mechanics, which use only the Darcy-Weisbach friction factor, f.

In the design of water pipes in the U.S., engineers often use the **Hazen-Williams** formula as follows:

$$V = 0.550\, C_H D^{0.63} S_e^{0.54} \qquad (2.41)$$

where C_H is the **Hazen-Williams coefficient**, and S_e is the energy slope, which is the headloss per unit length of pipe, h_L/L. Ordinarily, engineers consider C_H to be a function of pipe materials only, having a constant value for any given pipe material. For smooth pipe interior, the value of C_H used is 140; C_H decreases as the pipe gets rougher. With severely tuberculated old water pipes, values of C_H less than 80 may be used. Actually, C_H depends on not only the surface roughness of pipe interior but also the size of the pipe and the flow — velocity or Reynolds number. Therefore, using constant values of C_H without regard to pipe size and velocity can result in significant error. It would be more accurate to use the Darcy-Weisbach formula for the design of any incompressible flow in pipe including water flow. Using a personal computer, designing pipes based on f is as easy as and more accurate than using C_H. For these reasons, readers of this book are advised not to use the Hazen-Williams formula except for making rough estimates.

The same engineers who design water pipes, when designing sewers, use the Manning formula as follows:

$$V = \frac{1.486}{n} R_H^{2/3} S_e^{1/2} \qquad (2.42)$$

where n is the **Manning's roughness coefficient**, which is approximately equal to 0.014 for an unpolished concrete surface. The Manning formula was originally developed for designing open channels, not pipe flow. It is not as accurate for sewer pipes flowing full as is the Darcy-Weisbach formula.

Note that both the Hazen-Williams formula and the Manning formula are dimensionally incorrect. They can be used only in the ft-lb-s units. To use them in SI units without changing the values of C_H and n, the factors 0.550 (in Hazen-Williams) and 1.486 (in Manning) must be changed. Their corresponding values in SI units are 0.354 for the Hazen-Williams formula and 1.0 for the Manning formula. Many other equations or formulas exist for determining pipe headloss for incompressible flow, such as the Scobey formula used in irrigation. They will not be introduced here.

In summary, it should be said that the Darcy-Weisbach formula is both the most scientific and the most accurate equation for calculating the headloss in incompressible flow of any Newtonian fluid, whether the fluid is water, sewage, crude oil, or other liquid products. Even for low-speed gas flow in pipe over short distances without significant change in the gas density, the Darcy-Weisbach formula is still applicable.

2.4.4.3 Total Loss

From Equations 2.31 and 2.33, the total headloss along a pipeline, including both the local loss and the pipe loss, is

$$h_L = \sum K \frac{V^2}{2g} + f \frac{L}{D} \frac{V^2}{2g} = \left(\sum K + f \frac{L}{D} \right) \frac{V^2}{2g} \tag{2.43}$$

The summation sign, $\sum$, used in the above equation indicates that usually there are more than one fitting in a pipe that cause significant headloss. The K value of each fitting should be separately determined and summed together to determine the total fitting losses, $\sum K$, in Equation (2.43).

Example 2.7 An 8-inch-diameter commercial steel pipe of 1000 ft length is connected between two reservoirs having water surface elevations of 800 ft and 720 ft, respectively. The pipe consists of a re-entrant (inward-projecting) inlet, a sharp-edged exit, a globe valve, and three 90° regular elbows. When the valve is fully open, what is the discharge Q through the pipe?

[Solution] Using the one-dimensional energy equation between the two water surfaces in the reservoirs yields

$$z_1 - z_2 = 800 - 720 = 80 = h_L = \left(f \frac{L}{D} + \sum K \right) \frac{V^2}{2g}$$

or

$$V = \sqrt{\frac{2g \times 80}{f(L/D) + \sum K}} \tag{a}$$

Knowing that $g = 32.2$ ft/s, $L/D = 1000 \times 12/8 = 1500$, and $\Sigma K = 0.78 + 1.0 + 5 + 3 \times 0.3 = 7.68$, Equation a becomes

$$V = \sqrt{\frac{5152}{1500f + 7.68}} \tag{b}$$

Because V is not known *a priori,* the value of f must be found from successive approximation (iteration) as follows.

The relative roughness of the pipe is $e/D = 0.00015 \times 12/8 = 0.000225$. From the Moody diagram, the friction factor f for this pipe can be anywhere from 0.014 at a very high Reynolds number $\Re$, to 0.03 when $\Re$ is as low as 10^4. In the first approximation, it is assumed that $f = 0.02$. Using this f value in Equation b yields a value of $V = 11.7$ fps. Using this value of V and a kinematic viscosity of 1.059×10^{-5} ft²/s, which is for water at 70°F, the Reynolds number is $\Re = 11.7 \times (8/12)/(1.059 \times 10^{-5}) = 7.37 \times 10^5$. From the Moody diagram, this $\Re$ value yields $f = 0.0152$.

Therefore, in the second iteration, f is assumed to be 0.0152. From Equation b, $V = 1.0$ fps. This yields $\Re = 8.18 \times 10^5$, and $f = 0.0151$, which is very close to that determined in the previous iteration. Thus 0.0151 is used for f, and the velocity V calculated from equation b is still $V = 13.0$ fps.

The foregoing calculation shows that by assuming a reasonable value of f guided by the Moody diagram to start iteration, Equation b can be solved accurately (within 3 significant figures) in only two or three iterations. Once the value of V is found, the discharge of the flow is $Q = VA = 13.0 \times 0.349 = 4.54$ cfs.

This problem can also be solved by using the Hazen-Williams formula. By using $C_H = 130$, $D = 0.667$ ft and $S_e = 80/1000 = 0.08$, Equation 2.41 yields $V = 14.2$ fps, which is approximately 9% higher than the value given by the Darcy-Weisbach formula. Even though it is easier here, without using a computer, to use the Hazen-Williams formula than the Darcy-Weisbach formula because the former requires no iteration, the result obtained from the Hazen-Williams formula is less accurate (9% higher) than that obtained from the Darcy-Weisbach formula.

2.4.5 SHEAR ON PIPE WALL

The shear stress on the pipe wall can be determined by applying the energy equation and the momentum equation to a horizontal straight segment of pipe of constant cross-section. The segment lies between two cross sections, 1 and 2, with 1 being upstream of 2, as shown in Figure 2.12.

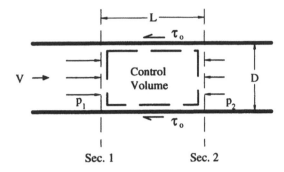

FIGURE 2.12 Force balance analysis to determine the shear stress on pipe wall.

From the energy equation (namely, Equation 2.19),

$$\frac{p_1 - p_2}{\gamma} = h_L = f \frac{L}{D} \frac{V^2}{2g} \qquad (2.44)$$

From the momentum equation (Equation 2.27) and a consideration of the pressure force and the shear force that contribute to the total horizontal force F on the fluid between two sections 1 and 2,

$$F = (p_1 - p_2)A - \tau_0 \pi DL = 0 \qquad (2.45)$$

which yields

$$\tau_0 = \frac{(p_1 - p_2)A}{\pi DL} = \frac{(p_1 - p_2)D}{4L} \qquad (2.46)$$

Substituting Equation 2.44 into Equation 2.46 yields

$$\tau_0 = \frac{f\rho V^2}{8} \qquad (2.47)$$

Note that Equation 2.47 holds for both turbulent and laminar flows. Although for simplicity it was derived herein for horizontal pipe, the same holds for pipes of any slope. Furthermore, though the equation is strictly correct only for steady incompressible flow, even for unsteady flow and compressible flow the same equation is often used — as will be seen later. Therefore, the equation is rather general and important.

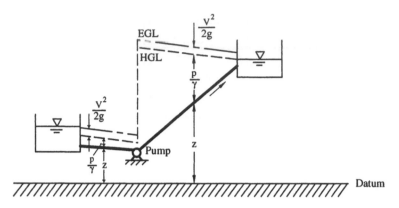

FIGURE 2.13 Energy grade line (EGL) and hydraulic grade line (HGL) along a pipeline.

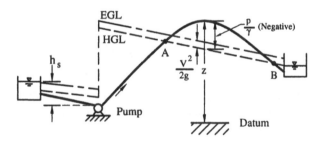

FIGURE 2.14 Energy grade line (EGL) and hydraulic grade line (HGL) along an elevated pipe.

2.5 HYDRAULIC AND ENERGY GRADE LINES

Referring to Equation 2.19, the term $V^2/2g$ is the **velocity head**; p/γ is the **pressure head**; z is the **elevation head**; the sum of the three terms is the **total head**; and the sum of the elevation head and the pressure head is the **piezometric head**.

A line drawn along a pipe to represent the total head variation is called the **energy grade line (EGL)**, and a line drawn to represent the piezometric head variation is called the **hydraulic grade line (HGL)**. An example of these lines are shown in Figure 2.13. It is important to know that *an energy grade line can never rise in the direction of the flow, unless outside energy (such as from a pump) is introduced.* While a continued decline in an EGL represents the continued dissipation of energy (i.e., the pipe loss), a sudden drop in the line represents a local loss. When an HGL is above a pipeline, the pressure in the pipe is positive. On the other hand, when an HGL is below a pipeline, the pressure in the pipe is negative (i.e., below atmospheric). The grade lines are illustrated in Figure 2.14.

2.6 CAVITATION IN PIPELINE SYSTEMS

Whenever the pressure of the liquid flowing in a pipe drops to vapor pressure, p_v, the liquid evaporates or boils. The phenomenon is called **cavitation** because it not only interrupts the flow but also cavitates — causing damage (cavities) to the interior of the pipe and its fittings, and to pump and turbine blades or vanes. Places where cavitation may take place in a pipeline are regions of low pressure (or high suction). These places include the summits of pipelines that cross mountains or hills, partially closed valves, and pumps or turbines. In the design of pipelines that carry any liquid, care must be taken to avoid cavitation.

When pumping liquids, the pump must be placed relatively close to the intake tank or reservoir and not too high above the tank or reservoir. Otherwise, cavitation may occur. To ensure that cavitation will not take place in a pump, the designer must calculate the **available NPSH (net positive suction head)** and compare it with the **required NPSH** for the pump. The available NPSH must be greater than the required NPSH in order to prevent cavitation. The required NPSH of any pump is usually given by the pump manufacturer in pump catalogs. The available NPSH can be calculated as follows:

$$\text{NPSH (available)} = h_a + h_s - h_L - h_v \qquad (2.48)$$

where h_a is the absolute pressure head of the atmosphere (approximately 34 ft or 10.4 m of water under standard atmospheric conditions); h_s is the static head, which is the height of the liquid in the intake reservoir above the pump elevation (see Figure 2.14), considered negative if the pump is above the liquid surface in the reservoir; h_L is the headloss in the pipe from the pipe entrance to the suction side of the pump, including both pipe loss and local losses; and h_v is the vapor pressure head of the liquid, p_v/γ, which is a function of temperature. Table C.3 in Appendix C gives the vapor pressure and other fluid properties of water as a function of temperature. For instance, for water at 90°F, $p_v = 0.7$ psia and $h_v = 1.6$ ft. The available NPSH is the available pressure head above the vapor pressure head in a pump.

Example 2.8 Referring to the sketch below, water is pumped from a lake to supply a town through a steel pipeline 10.5 km long. The I.D. of the pipe is 254 mm (10 inches), and the discharge is 0.1754 m³/s. The water tower of the town is 30 m high, and the ground elevation at the tower is 20 m above the lake water surface. There are two gate valves ($K = 0.2$ each, when fully open) and five 90° bends ($K = 1.0$ each) along the pipeline. Entrance and exit loss coefficients for this pipeline are respectively 0.5 and 1.0. (a) Calculate the pump power and motor power, assuming that the efficiencies of the pump and the motor are respectively 80% and 90%. (b) How far can the pump be from the pipeline intake without causing cavitation? Assume that the pump is placed 2 m above the lake water surface, and that the required NPSH is 3 m. (c) If the pipeline were well insulated, how much would the temperature of the water rise in traveling through the pipe from the lake to the water tower?

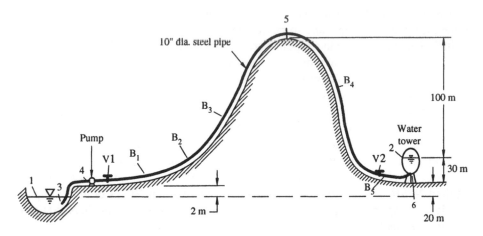

Note: 1 - water surface in upstream reservoir

2 - water surface in downstream tank

3 - pipe inlet

4 - pump suction side

5 - hill top

6 - pipe outlet

B_1, B_2, B_3, B_4 and B_5 — 90° bends (K=1.0 for each)

$L_{35} = 9.3$ km, $L_{56} = 1.2$ km, $L_{36} = 10.5$ km, and L_{34} is to be determined.

V1 and V2 — gate valves (K=0.2 for each)

[**Solution**] Consider the entire pipeline system from point 3 to point 6,

$L = 10,500$ m, $\gamma = 9810$ N/m³

$D = 0.254$ m, $A = \pi D^2/4 = 0.0507$ m², $Q = 0.1754$ m³/s

$V = Q/A = 3.46$ m/s, $g = 9.81$ m/s², $V^2/2g = 0.610$ m

Local loss coefficients $= \Sigma K = 2 \times 0.2 + 5 \times 1 + 1 + 0.5 = 6.9$

Assume $T = 20°C$, then $\nu = 1.0 \times 10^{-6}$ m²/s

$\Re = VD/\nu = 8.8 \times 10^5$, $e = 0.00015$ ft $= 0.0000457$m, $e/D = 0.00018$

From the Moody diagram, $f = 0.0147$.

(a) From the energy equation,

$$h_p = (z_2 - z_1) + h_L$$

$$z_2 - z_1 = 30 + 20 = 50 \text{ m} \qquad\qquad\qquad (a)$$

$$h_L = f \frac{L}{D} \frac{V^2}{2g} + \Sigma K \frac{V^2}{2g} = 371 + 4.2 = 375 \text{ m}$$

Therefore, (a) yields $h = 50 + 375 = 425$ m; power output of pump $= \gamma Q h_p$ $= 731,000$ W $= 731$ kW; power input of pump $= 731/0.80 = 914$ kW; power lost in pump $= 914 - 731 = 183$ kW $= 183,000$ W; efficiency of motor/pump combination $= 0.8 \times 0.9 = 0.72$; power input of motor $= 731/0.72 = 1015$ kW.

(b) NPSH (available) = $h_a + h_s - h_L - h_v$

$h_a = 10.37$ m, $h_s = -2$ m,

$$h_L = \left(0.5 + 0.014 \frac{L_{34}}{D}\right)\frac{V^2}{2g} = \left(0.5 + 0.0147 \frac{L_{34}}{D}\right) \times 0.610$$

h_v (at 20°C) = 0.25 m

NPSH (available) = $7.83 = 0.0090 \dfrac{L_{34}}{D}$

Setting NPSH (available) = NPSH (required) yields

$\dfrac{L_{34}}{D} = 537$. Therefore, $L_{34} = 136$ m.

(c) Both the headloss in the flow and that in the pump are turned into heat, resulting in a temperature rise of the fluid in a well-insulated pipe. Headloss along the pipeline = $h_L = 375$ m; headloss in pump = $183,000/(\gamma Q) = 106$ m; total headloss = $375 + 106 = 481$ m. Assuming that all the heat generated in the pump goes into the water, then from Equation 2.25, $\Delta T = gh_L/c_v$. For water, $c_v = 1$ cal/gm/°C = 4180 joules/kg/°C, therefore $\Delta T = 9.81 \times 481/4180 = 1.1$°C. The foregoing calculation shows that even in a perfectly well-insulated pipeline, the rise of temperature of the fluid along the pipe caused by dissipation is usually rather small. For this reason, and because long pipelines are normally not insulated, incompressible flow in pipe is usually regarded as constant temperature.

In this example problem, the effect of pipe crossing a hill (as shown in the sketch), which may cause cavitation, is not analyzed. It will be analyzed in a homework problem (see Problem 2.12).

2.7 PIPES IN SERIES AND PARALLEL

2.7.1 PIPES IN SERIES

A pipeline in series refers to the connection of pipes of different diameters in series as shown in Figure 2.15. The same flow rate (discharge Q) goes through the pipes connected in series. Pipes in series are often treated as a single pipe in using the energy equation, except that the headloss along each pipe must be evaluated separately and then summed to get the total headloss along the entire pipeline. For the three pipes in Figure 2.15, the total headloss is

$$\left(h_L\right)_{AB} = f_1 \frac{L_1}{D_1}\frac{V_1^2}{2g} + f_2 \frac{L_2}{D_2}\frac{V_2^2}{2g} + f_3 \frac{L_3}{D_3}\frac{V_3^2}{2g} + \frac{\left(V_1 - V_2\right)^2}{2g} + \frac{\left(V_2 - V_3\right)^2}{2g} \quad (2.49)$$

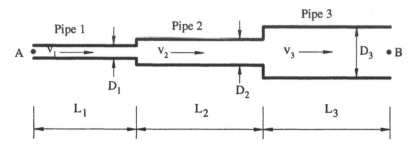

FIGURE 2.15 Pipes in series.

The subscripts 1, 2, and 3 refer to pipes 1, 2, and 3, respectively. The last two terms in the above equation are the expansion headlosses between pipes of different diameters (see Table 2.1). It is assumed in this case that the only fitting (local) headloss is that due to pipe expansion. Otherwise, additional local losses must be added to Equation 2.49.

The continuity equation for incompressible flow is

$$Q = V_1 A_1 = V_2 A_2 = V_3 A_3 \tag{2.50}$$

If the discharge Q in the pipe is known, Equation 2.50 can be used to calculate the velocities V_1, V_2, and V_3, which can be substituted into Equation 2.49 to determine the headloss $(h_L)_{AB}$. The values of f_1, f_2, and f_3 can be obtained from the Moody diagram using calculated Reynolds numbers $\Re_1$, $\Re_2$, and $\Re_3$, and roughness e_1, e_2, and e_3.

If, on the other hand, the pressure and the elevation at the two ends of the pipe (A and B) are known, the two above equations can be used with the energy equation to solve for the velocities V_1, V_2, and V_3. The solution involves iterations, which first assume reasonable values of f_1, f_2, and f_3, such as all being 0.01. The values of V_1, V_2, and V_3 can then be solved from the equations. Then, the Reynolds numbers for the three pipes are calculated to determine f_1, f_2, and f_3 from the Moody diagram. These calculated values of f_1, f_2, and f_3 are used in a second iteration to determine a second set of more accurate values of V_1, V_2, and V_3. Continuing the process for two to five iterations should yield accurate values of V and f for each pipe. The more accurate are the values of f_1, f_2, and f_3 assumed in the first iteration, the fewer iterations are needed to obtain accurate final results of V_1, V_2, and V_3.

2.7.2 PARALLEL PIPES

For parallel pipes as in Figure 2.16, the discharge through each pipe is different, whereas the pressure drops across the parallel pipes are the same. Continuity equation yields

$$Q_1 + Q_2 + Q_3 = Q \tag{2.51}$$

where Q is the total discharge in the main pipe upstream and downstream of the branches. The velocity in each pipe is

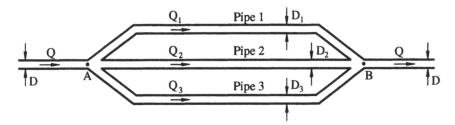

FIGURE 2.16 Pipes in parallel.

$$V_1 = \frac{Q_1}{A_1}, \quad V_2 = \frac{Q_2}{A_2}, \quad V_3 = \frac{Q_3}{A_3} \tag{2.52}$$

The energy equation yields the following:

$$\left.\begin{aligned}
h_{L1} &= f_1 \frac{L_1}{D_1} \frac{V_1^2}{2g} + \sum K_1 \frac{V_1^2}{2g} \\
h_{L2} &= f_2 \frac{L_2}{D_2} \frac{V_2^2}{2g} + \sum K_2 \frac{V_2^2}{2g} \\
h_{L3} &= f_3 \frac{L_3}{D_3} \frac{V_3^2}{2g} + \sum K_3 \frac{V_3^2}{2g} \\
h_{L1} &= h_{L2} = h_{L3} = \frac{p_A - p_B}{\rho g}
\end{aligned}\right\} \tag{2.53}$$

where ΣK_1 is the sum of all the local (fitting) losses of pipe 1 and so forth.

Again, an iteration procedure is required to solve the problem when V_1, V_2, and V_3 are unknown because they affect the values of f_1, f_2, and f_3. One can solve the problem by first assuming the value of f for each individual pipe (say, equal to 0.01). After V_1, V_2, and V_3 have been found from the equations, the f values for each pipe can be determined from the Moody diagram, and used in the second iteration and so forth. This in a nutshell explains how to handle pipes in parallel and series.

Example 2.9 For the three parallel pipes in Figure 2.16, assume that water flows through the pipes at 80°F, the discharge Q is 2 cfs, the pipe diameters D_1, D_2, and D_3 are 4, 6, and 8 inches, respectively, the pipe lengths L_1, L_2, and L_3 are each 1000 ft, the pipe material is steel, and local headlosses are negligible. Find the discharges Q_1, Q_2, and Q_3, and the headloss across the parallel pipes.

[Solution] Equation 2.51 yields

$$Q_1 + Q_2 + Q_3 = 2 \qquad \text{(a)}$$

Because $A_1 = 0.0873$ ft^2, $A_2 = 0.1963$ ft^2, and $A_3 = 0.349$ ft^2, Equation 2.52 yields

$$V_1 = \frac{Q_1}{0.0873}, \quad V_2 = \frac{Q_2}{0.1963}, \quad V_3 = \frac{Q_3}{0.349} \qquad \text{(b)}$$

Because $L_1/D_1 = 3000$, $L_2/D_2 = 2000$, and $L_3/D_3 = 1500$, Equation 2.53 reduces to

$$6110\, f_1 Q_1^2 = 806\, f_2 Q_2^2 = 191\, f_3 Q_3^2 \qquad \text{(c)}$$

Note that Equation c is a set of two independent equations. They can be solved algebraically along with Equation a to yield Q_1, Q_2, and Q_3, provided that f_1, f_2, and f_3 are known. They can be solved as follows:

From Equation c,

$$Q_2 = 2.75\sqrt{\frac{f_1}{f_2}}\, Q_1, \quad Q_3 = 5.66\sqrt{\frac{f_1}{f_3}}\, Q_1 \qquad \text{(d)}$$

Substituting Equation d into Equation a yields

$$Q_1 = \frac{2}{1 + 2.75\sqrt{\dfrac{f_1}{f_2}} + 5.66\sqrt{\dfrac{f_1}{f_3}}} \qquad \text{(e)}$$

For the first iteration, assume that $f_1 = f_2 = f_3 = 0.01$. Equation e yields $Q_1 = 0.213$ cfs. From Equation d, $Q_2 = 0.586$ cfs and $Q_3 = 1.21$ cfs. Thus, $Q_1 + Q_2 + Q_3 = 2.01$ cfs, which shows that Equation a is approximately satisfied. Next, from Equation b, $V_1 = 2.44$ fps, $V_2 = 2.99$ fps and $V_3 = 3.47$ fps. The kinematic viscosity of water at 80°F is $v = 0.930 \times 10^{-5}$ ft^2/s. Therefore, the Reynolds numbers of the three pipes are $\Re_1 = 8.75 \times 10^4$, $\Re_2 = 1.61 \times 10^5$, and $\Re_3 = 2.49 \times 10^5$. The relative roughness of the three pipes are $e_1/D_1 = 0.00045$, $e_2/D_2 = 0.00030$, and $e_3/D_3 = 0.000225$. From the Moody diagram, $f_1 = 0.0208$, $f_2 = 0.0185$, and $f_3 = 0.0170$.

For the second iteration, use the values of f_1, f_2, and f_3 obtained at the end of the first iteration. From Equation e, $Q_1 = 0.1965$ cfs; from Equation d, $Q_2 = 0.573$ cfs and $Q_3 = 1.23$ cfs. Thus, $Q_1 + Q_2 + Q_3 = 2.0$ cfs, which shows that Equation a is satisfied. This yields $V_1 = 2.25$ fps, $V_2 = 2.92$ fps, $V_3 = 3.52$ fps, $\Re_1 = 8.06 \times 10^4$, $\Re_2 = 1.57 \times 10^5$, and $\Re_3 = 2.52 \times 10^5$. From the Moody diagram, $f_1 = 0.0210$, $f_2 = 0.0182$, and $f_3 = 0.0170$.

Since the values of f_1, f_2, and f_3 obtained at the end of the second iteration are within 2% of those assumed in the beginning of the second iteration, they are accepted as the final correct values. Based on these final values, $Q_1 = 0.1953$ cfs, $Q_2 = 0.577$ cfs, and $Q_3 = 1.23$ cfs, and $Q_1 + Q_2 + Q_3 = 2.0$ cfs.

Because the headloss is the same across each of the three parallel pipes, it can be calculated from

$$h_L = f_1 \frac{L_1}{D_1} \frac{V_1^2}{2g} = 0.0210 \times 3000 \times \frac{V_1^2}{2g} = 63 \frac{V_1^2}{2g} \tag{f}$$

But, $V_1 = Q_1/A_1 = 0.1953/0.0873 = 2.24$ fps. Therefore, from Equation f, $h_L = 4.90$ ft.

2.8 INTERCONNECTED RESERVOIRS

Sometimes the need arises to transport liquid from one reservoir to others through interconnecting pipes. An example is shown in Figure 2.17, in which water is pumped from reservoir 1 to reservoirs 2 and 3, which are at different elevations. The steady flow through the pipes of this system can be solved using the following steps:

1. Assume a discharge Q_1 through the pump and through Pipe 1.
2. Use the energy equation (Equation 2.19) between reservoir 1 and the junction of the pipes. Find the piezometric head at the junction, h_j. Note that the pump head in the energy equation can be determined from the

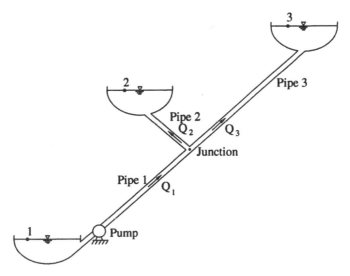

FIGURE 2.17 Flow through pipes connecting multiple reservoirs.

pump characteristic curve (see Section 9.2), which gives the pump head as a function of discharge.

3. Compare the piezometric head at the junction with the water surface elevation of reservoirs 2 and 3. If the piezometric head at the junction is higher than the height of water in any reservoir, the flow goes from the junction to the reservoir. Otherwise, the flow will be in the reverse direction.

4. Using energy equation between the pipe junction and each reservoir, the velocity and the discharge in pipes 2 and 3 can be calculated.

5. Check if continuity equation ($Q_1 = Q_2 + Q_3$) is satisfied at the junction. If not, a new discharge must be assumed in pipe 1 and the same six steps followed until $Q_1 = Q_2 + Q_3$.

6. Check the required and available NPSH to make sure that the pump does not cavitate.

Example 2.10 Refer to Figure 2.17 and assume that the elevations of reservoirs 1, 2, and 3 are 30, 40, and 50 ft, respectively, the pipe lengths for pipes 1, 2, and 3 are 1000, 500, and 1500 ft, respectively, and all three pipes are made of steel and have a diameter of 1 ft. Furthermore, assume that the water temperature is 80°F, the headloss coefficient for the entrance of each pipe is 0.5 and for the exit is 1.0, there is a pump in pipe 1, and a globe valve of headloss coefficient of 5 on the discharge side of the pump in pipe 1, the pipes are straight and there are no other fitting losses, and the pump head, h_p, is a function of the discharge through the pump Q_1, given as

$$h_p = 120 - 10 \, Q_1^2 \tag{a}$$

(a) Determine the directions and magnitudes of the discharges Q_1, Q_2, and Q_3 through the three pipes.

(b) If the pump is located 100 ft from the reservoir at an elevation of 10 ft, determine whether the pump will cavitate. The required NPSH of the pump is 10 ft.

[Solution] (a) The problem can be solved by following the steps for solving interconnected reservoirs:

1. From Equation a, when $h_p = 0$, $Q_1 = 3.464$ cfs. This means that the discharge through the pump must be smaller than 3.464 cfs in order for the pump to work. Assume that the discharge through the pump is $Q_1 = 1$ cfs.

2. Using Equation 2.19 (the energy equation) between the surface of reservoir 1 and the pipe junction yields

$$h_j = h_p + z_1 - \frac{V_1^2}{2g} - h_L \tag{b}$$

From Equation a,

$$h_p = 120 - 10V_1^2 A_1^2 = 120 - 6.169V_1^2 \tag{c}$$

From Equation 2.43,

$$h_L = \left(\sum K_1 + f_1 \frac{L_1}{D_1} \right) \frac{V_1^2}{2g} = \left(0.5 + 5 + f_1 \frac{1000}{1} \right) \frac{V_1^2}{2g} = \tag{d}$$

$$(5.5 + 1000 f_1) \frac{V_1^2}{64.4} = (0.0854 + 15.53 f) V_1^2$$

Substituting Equations c and d into Equation b, and knowing that $z_1 = 30$ ft, Equation b yields

$$h_j = 150 - (6.270 + 15.53 f_1) V_1^2 \tag{e}$$

$V_1 = Q_1/A_1 = 1.0/0.7854 = 1.273$ fps, $\Re_1 = V_1 D_1/v = 1.273 \times 1/(0.93 \times 10^{-5}) = 1.37 \times 10^5$, $e/D = 0.00015$, and from the Moody diagram $f_1 = 0.0178$. Therefore, Equation e yields $h_j = 139.4$ ft.

3. Since h_j is larger than the water surface elevation of both reservoirs 2 and 3, the flow must be from the junction to these two reservoirs.

4. Using the energy equation between the pipe junction and the surface of reservoir 2 yields

$$h_j = z_2 + \left(\sum K_2 + f_2 \frac{L_2}{D_2} \right) \frac{V_2^2}{2g}$$

$$139.4 = 40 + \left(1 + f_2 \frac{500}{1} \right) \frac{V_2^2}{2g}$$

and

$$V_2 = \sqrt{\frac{6401}{1 + 500 f_2}} \tag{f}$$

Using iterations and the Moody diagram, Equation f yields $f_2 = 0.0134$, $V_2 = 28.8$ fps, and $Q_2 = 22.6$ cfs. Likewise, using the energy equation between the pipe junction and the surface of reservoir 3 yields

$$V_3 = \sqrt{\frac{5757}{1 + 1500 f_3}} \tag{g}$$

From Equation g, $f_3 = 0.0136$, $V_3 = 16.40$ fps, and $Q_3 = 12.9$ cfs.

5. Because the sum $Q_2 + Q_3 = 35.5$ cfs is much greater than Q_1, which is 1 cfs, the value of Q_1 assumed is too small. Therefore, in the second trial (iteration), a much larger discharge is assumed, $Q_1 = 2$ cfs.

6. Repeating steps 1 through 5 by using $Q_1 = 2$ cfs yields $V_1 = 2.55$ fps, $\Re_1 = 2.74 \times 10^5$, $f_1 = 0.0161$, and from Equation e, $h_j = 107.6$ ft. Equation f becomes

$$V_2 = \sqrt{\frac{4353}{1 + 500 f_2}} \tag{h}$$

Solving Equation h yields $f_2 = 0.0134$, $V_2 = 23.8$ fps, and $Q_2 = 18.7$ cfs. Equation g becomes

$$V_3 = \sqrt{\frac{3709}{1 + 1500 f_3}} \tag{i}$$

which yields $f_3 = 0.0137$, $V_3 = 13.1$ fps, and $Q_3 = 10.3$ cfs.

Since $Q_2 + Q_3 = 29$ cfs is still much larger than Q_1, which is 2 cfs, the value of Q_1 must be much greater than 2 cfs but less than 3.464 cfs.

7. Next, assume $Q_1 = 3$ cfs. Repeating steps 1 through 5 yields: $V_1 = 3.82$ fps, $\Re_1 = 4.1 \times 10^5$, $f_1 = 0.0154$, $h_j = 55.0$ ft, $f_2 = 0.014$, $V_2 = 11.0$ fps, $Q_2 = 8.64$ cfs, $f_3 = 0.0153$, $V_3 = 3.67$ fps, and $Q_3 = 2.88$ cfs. Since $Q_2 + Q_3 = 11.52$ cfs is still larger than Q_1, the value of Q_1 must be further increased.

8. Next, assume $Q_1 = 3.2$ cfs. Repeating steps 1 through 5 yields $V_1 = 4.07$ fps, $\Re_1 = 4.4 \times 10^5$, $f_1 = 0.0151$, and $h_j = 42.0$ ft. Since h_j is higher than the water elevation in reservoir 2 but lower than the water in reservoir 3, flow enters reservoir 2 but leaves reservoir 3. In this case, $f_2 = 0.0153$, $V_2 = 3.86$ fps and $Q_2 = 3.03$ cfs.

For the flow leaving reservoir 3, the energy equation yields

$$z_3 = h_j + \left(\sum K_3 + f_3 \frac{L_3}{D_3} \right) \frac{V_3^2}{2g} \tag{j}$$

which yields $V_3 = \sqrt{8g / (1.5 + 1500 f_3)}$, $f_3 = 0.0149$, $V_3 = 4.65$ fps, and $Q_3 = 3.65$ cfs.

Because Q_3 is an outflow from reservoir, in this case $Q_1 + Q_3 = 6.85$ cfs should equal to Q_2. However, Q_2 is only 3.03 cfs. This means the assumed $Q_1 = 3.2$ cfs is too large. It should be between 3.0 and 3.2 cfs.

9. Continuing the above iterative approach will yield the final values of $Q_1 = 3.13$ cfs, $Q_2 = 5.58$ cfs, and $Q_3 = 2.39$ cfs, which approximately satisfies the condition $Q_1 + Q_3 = Q_2$. This completes the analysis.

In a homework problem at the end of this chapter, the reader will be asked to program the foregoing problem and solve it by computer with greater precision than presented herein.

(b) Now that the final discharge through pipe 1 is found to be $Q_1 = 3.13$ cfs, we have $V_1 = 3.99$ fps, and $V_1^2/(2g) = 0.247$ ft.

To determine whether the pump will cavitate, the following equation is applicable:

$$\text{NPSH (available)} = h_a + h_s - h_L - h_v \qquad (k)$$

where $h_a = 34$ ft, $h_s = 30 - 10 = 20$ ft, $h_v = p_v/\gamma = 1.17$ ft, and

$$h_L = \left(0.5 + f_1 \frac{100}{1}\right) \frac{V_1^2}{2g} = (0.5 + 0.0152 \times 100) \times 0.247 = 0.50 \text{ ft}$$

Therefore, Equation k yields NPSH (available) = $34 + 20 - 0.5 - 1.17 = 52.3$ ft. Because the NPSH (required) is only 10 ft, which is much smaller than the NPSH (available), the pump will *not* cavitate.

2.9 PIPE NETWORK

Networks of interconnected pipes are used not only in domestic water supply but also in petroleum product pipelines across the nation. Analysis of the flow distribution in a network is complicated, and it requires the solution of a set of simultaneous algebraic equations derived from the continuity and energy equations. A computer can be used to solve the simultaneous equations, using either commercially available programs, or a program written specifically for the network to be analyzed.

The network equations can be derived from the following considerations:

1. The algebraic sum of the pressure drops around each circuit (loop) in the network is zero, namely,

$$\sum h_L = 0 \text{ (around each loop)} \qquad (2.54)$$

2. Continuity equation must be satisfied at each junction (node). This means all flows going into a junction must equal to flows leaving the junction, or there is no net inflow or outflow at each junction or node, namely,

$$\sum Q = 0 \text{ (at each node)} \qquad (2.55)$$

3. Headloss across each pipe can be determined from the Darcy-Weisbach equation. Fitting losses for each pipe can be neglected when valves in the pipe are fully open and when the pipe is relatively long — say at least 500 times the pipe diameter.

In the analysis of networks of water pipes, to simplify computation (i.e., to reduce iteration and computational time), the Darcy-Weisbach equation is often replaced by the Hazen-Williams equation (Equation 2.41). Results obtained from using the latter are less accurate than from the former. Alternatively, the Darcy-Weisbach equation can be used with the friction factor f treated as a constant for each pipe, determined from the Moody diagram by assuming a velocity in the practical range, say 2 m/s. Using such an approach, the headloss across each pipe can be calculated from

$$h_L = C_o Q^2 \tag{2.56}$$

where C_o is a constant for each pipe equal to

$$C_o = \frac{8fL}{g\pi^2 D^5} \tag{2.57}$$

Once the problem is solved, one should determine the value of f for each pipe from the Moody diagram, and repeat the calculation once more to determine accurate results.

Before a network problem can be solved, one must determine the number of loops and nodes in the network. For example, the network shown in Figure 2.18 has 13 nodes and 7 loops. Therefore, there will be 13 node equations of the form of Equation 2.55, and 7 loop equations of the form

$$\sum h_L = \sum C_o Q^2 = 0 \tag{2.58}$$

This means a total of 20 simultaneous equations must be solved for this problem.

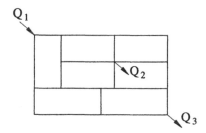

FIGURE 2.18 Network of pipes — an example.

Another method to solve pipe network, without having to solve simultaneous equations, is the **Hardy Cross method** described as follows:

1. A flow (including discharge and direction) is assumed in each pipe of the network such that $\Sigma Q = 0$ at each node (junction).
2. Calculate the headloss for each loop in both the clockwise and counterclockwise directions and compare them. They would be equal to each other in magnitude only by a miracle. If the headloss is greater in the clockwise direction, it indicates that the assumed discharge(s) in the clockwise direction is (are) too large. Therefore, subtract a correction ΔQ from the clockwise discharge(s), and add ΔQ to the counterclockwise discharge(s). If the Darcy-Weisbach formula is used, it can be proved that

$$\Delta Q = \frac{\left(\sum_c h_L - \sum_{cc} h_L \right)}{2 \left[\sum_c (h_L / Q) + \sum_{cc} (h_L / Q) \right]} \tag{2.59}$$

or

$$\Delta Q = \frac{\left(\sum_c C_o Q^2 - \sum_{cc} C_o U^2 \right)}{2 \left[\sum_c C_o Q + \sum_{cc} C_o Q \right]} \tag{2.60}$$

where c and cc beneath the summation sign represent the clockwise and counterclockwise directions, respectively.
3. Repeat step 2 using the corrected discharges. Iterate until accurate results are obtained.

The foregoing method, invented by Hardy Cross in 1936, converges quickly. It was the standard method used for solving pipe network problems prior to the advent of high-speed digital computers. Although the Hardy Cross method can be programmed on computers, in recent years many other numerical methods have been introduced to solve pipe network simultaneous equations. Commercial programs are available nowadays for solving complicated pipe networks.

2.10 UNSTEADY FLOW IN PIPE

When the discharge or velocity of the flow in a pipe varies with time, the flow is said to be **unsteady**. Two types of unsteady flow are treated herein, those that vary slowly with time and those that vary rapidly with time. The former, the **quasi-steady flow,** can be treated with negligible error by using steady incompressible flow

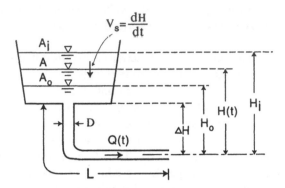

FIGURE 2.19 Pipeline drainage of liquid from a reservoir.

solutions at any given time. The latter must be treated as **truly unsteady flow** with due regard to compressibility effect.

2.10.1 QUASI-STEADY SOLUTION

Slowly varying flows through a pipe, as occurring when a valve is being slowly* opened, or slowly closed, or when draining liquid from a large reservoir with the water level in the reservoir falling slowly, can be treated as quasi-steady incompressible flow without consideration of the compressibility effect of the flow. In what follows, three practical types of such flow are treated. They include (a) drainage of a reservoir or pipe, (b) fluid acceleration due to valve opening, and (c) flow oscillations in interconnected reservoirs. They are discussed in the following sections.

2.10.1.1 Drainage of a Reservoir or Pipe

Figure 2.19 depicts water being drained by gravity from a reservoir. The water surface at time $t = 0$ is at initial height H_i above the pipe outlet. One wishes to know how much time t_o is required to drain the reservoir to the level H_o. The pipeline length is L and the pipe diameter is D.

To solve this problem, we use the steady-state energy equation (Equation 2.19) at any given time t, when the water height is H. Using Equation 2.19 between the water surface (point 1) and the pipe outlet (point 2) yields

$$H = \left(1 + \sum K + f \frac{L}{D}\right) \frac{V^2}{2g} \tag{2.61}$$

or

---

* How slow is "slowly" depends on pipe length. Long pipes require a longer valve closure or opening time without causing significant unsteady effect. This aspect is discussed further in the treatment of pressure surges (water hammer) in the next subsection.

$$V = \sqrt{\frac{2gH}{1 + \sum K + f\frac{L}{D}}} \quad (2.62)$$

The speed at which the water level in the reservoir is falling at time t is $V_s = -dH/dt$. The discharge is

$$Q = VA = -A_s \frac{dH}{dt} \quad (2.63)$$

where A is the cross-sectional area of the pipe, and A_s is the reservoir area at water level H.

Combination of Equations 2.63 with 2.62 yields

$$\frac{A_s dH}{\sqrt{H}} = -A\sqrt{\frac{2g}{1 + \sum K + f\frac{L}{D}}} \, dt \quad (2.64)$$

Integrating the above equation and using the conditions that $H = H_i$ when $t = 0$ and $H = H_o$ when $t = t_o$, the following result is obtained:

$$t_o = \frac{\sqrt{1 + \sum K + f\frac{L}{D}}}{A\sqrt{2g}} \int_{H_o}^{H_i} \frac{A_s}{\sqrt{H}} dH \quad (2.65)$$

which gives the time to drain the reservoir by gravity from initial height H_i to final height H_o. The above integral can be solved if the variation of A_s with H is known. For instance, if A_s does not vary with H, Equation 2.65 reduces to

$$t_o = \frac{2A_s\sqrt{1 + \sum K + f\frac{L}{D}}}{A\sqrt{2g}} \left(H_i^{1/2} - H_o^{1/2}\right) \quad (2.66)$$

Another similar example is to drain a nonhorizontal pipe by gravity. If the pipe diameter is D, the pipe length is L, and the elevation difference between the inlet and outlet is ΔH, the time required to drain this pipe is

$$t_o = \frac{f^{1/2} L^{3/2}}{\sqrt{2gD\Delta H}} \quad (2.67)$$

Example 2.11 A swimming pool has a surface area of 25 m × 50 m. The depth of the water in the pool is 1.5 m. The bottom of the pool is connected to an 8-inch-diameter steel pipe for drainage into a gravity-flow sewer located 3 m below the bottom of the pool. The pipe is 50 m long. A globe valve is used to control the flow. (a) How long does it take to drain the water completely from the pool? (b) After the water is drained from the pool, how long does it take to drain the water completely from the pipe?

[Solution] (a) The time to drain the pool can be calculated from Equation 2.66. The values of the quantities in this equation are

$A_s = 50 \times 25 = 1250$ m².
$\Sigma K = K_1 + K_2 = 0.5 + 5 = 5.5$, where K_1 is the entrance headloss coefficient, and K_2 is the headloss coefficient for the globe valve when it is open.
$L = 50$ m, $D = 0.203$ m, and $L/D = 246$.
$A = \pi D^2/4 = 0.0325$ m².
$g = 9.81$ m/s².
$H_i = 1.5 + 3 = 4.5$ m.
$H_o = 3$ m.

The value of f can be determined as follows:

When the water surface in the reservoir is at height $H_i = 4.5$ m (i.e., at the beginning of the drainage), Equation 2.62 yields

$$V = \sqrt{\frac{2 \times 9.81 \times 4.5}{1 + 5.5 + 246f}} = \sqrt{\frac{88.29}{6.5 + 246f}} \tag{a}$$

In the first iteration, assume that $f = 0.015$. Equation a yields $V = 2.94$ m/s. Assuming that the kinematic viscosity of water at 19°C is 1×10^{-6} m²/s, the Reynolds number is $\mathfrak{R} = 2.94 \times 0.203/10^{-6} = 6 \times 10^5$, and $e/D = 0.00025$. From the Moody diagram, $f = 0.0155$. Then, from Equation a, $V = 2.93$ m/s. $\mathfrak{R} = 2.93 \times 0.203/10^{-6} = 6 \times 10^5$, $f = 0.0155$. Therefore, $f = 0.0155$ when the water starts draining.

When $H = H_o = 3.0$ m (i.e., at the end of the pool drainage),

$$V = \sqrt{\frac{2 \times 9.81 \times 3}{1 + 5.5 + 246f}} = \sqrt{\frac{58.86}{6.5 + 246f}} \tag{b}$$

Following the same iteration procedure used for solving Equation a, Equation b can be solved to yield $V = 2.38$ m/s and $f = 0.0159$. Thus, the value of f to be used in Equation 2.66 must be between 0.0155 and 0.0159.

Using the average value of $f = 0.0157$, Equation 2.66 yields

$$t_o = \frac{2 \times 1250\sqrt{1+5.5+0.0157 \times 246}}{0.0325 \times \sqrt{2 \times 9.81}}(4.5^{1/2} - 3.0^{1/2})$$

$$= 21,750 \text{ s} = 362 \text{ min} = 6.04 \text{ hr}$$

Thus, it takes about 6 hr to drain the swimming pool.

(b) To calculate the additional time required to drain the pipe completely, Equation 2.67 is used and yields

$$t_o = \frac{0.02^{1/2} \times 50^{3/2}}{\sqrt{2 \times 9.81 \times 0.203 \times 3}} = 14.5 \text{ s}$$

Note that the value of $f = 0.02$ used in the above calculation is only an approximate average value. More accurate calculation can be done by determining the different values of f for different stages of pipe drainage. However, because the t_o calculated for pipe drainage is so much smaller than the t_o for draining the pool, a more accurate determination of the pipe drainage time is unwarranted.

2.10.1.2 Flow Establishment (Fluid Acceleration Due to Sudden Valve Opening)

A pipe of length L and diameter D is connected to a large reservoir as shown in Figure 2.20. Initially, the valve at the end of the pipe is closed and there is no flow through the pipe. However, at $t = 0$, the valve is rapidly opened, and the liquid starts to flow through the pipe. Initially (at $t = 0$), the mean flow velocity in the pipe, V, is zero. As time progresses, V increases until it reaches a steady state valve V_o. Find the time that it takes for V to reach 99% of V_o.

To solve this problem, we first calculate the steady-state velocity V_o. By using the energy equation for steady, incompressible flow (i.e., Equation 2.19),

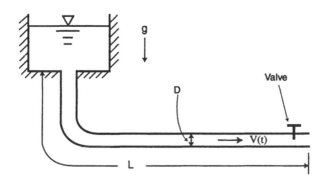

FIGURE 2.20 Flow establishment in pipe following sudden opening of valve.

$$V_o = \sqrt{\frac{2gH}{\sum K + f\frac{L}{D}}} \tag{2.68}$$

where H is the height of the water in the reservoir above the pipe exit, and ΣK includes all the local losses generated by pipe entrance, exit, and all fittings.

When the pipe is long and when local losses are negligible, Equation 2.68 reduces to

$$V_o = \sqrt{\frac{2gHD}{fL}} \tag{2.69}$$

At any time t after the valve is quickly opened, the velocity of the fluid in the pipe is V, and the acceleration of the fluid is dV/dt. Application of Newton's second law, $\vec{F} = m\vec{a}$, to the fluid in the pipe yields

$$(p_2 - p_3)A + AL\rho g\frac{\Delta z}{L} - \pi DL\tau_o = (AL\rho)\frac{dV}{dt} \tag{2.70}$$

Substituting the values $p_2 = \rho g H_1$, $p_3 = 0$, $H_1 + \Delta z = H$ and $\tau_o = \rho f V^2/8$ into the above equation yields

$$\frac{dV}{dt} + \frac{f}{2D}V^2 - \frac{gH}{L} = 0 \tag{2.71}$$

which can be rearranged as

$$dt = \frac{2D}{f}\frac{dV}{(V_o^2 - V^2)} \tag{2.72}$$

where V_o is given by Equation 2.69.

Integrating Equation 2.72 by using the conditions $V = 0$ at $t = 0$ yields

$$t = \frac{LV_o}{2gH} \ell n\left(\frac{V_o + V}{V_o - V}\right) \tag{2.73}$$

This shows that the velocity V approaches V_o asymptotically, which means that it will take infinite time to reach V_o. Assume that it takes a time t_o for the velocity V to reach 99% of the steady-state value V_o. From Equation 2.73,

$$t_o = 2.65\frac{LV_o}{gH} \tag{2.74}$$

For practical purposes, t_o can be regarded as the time to establish steady flow velocity V_o.

In a homework problem (Problem 2.19), the reader will be asked to prove that Equations 2.73 and 2.74, which are derived from Equation 2.69, also hold for the more general case of Equation 2.68.

Example 2.12 A 10-km-long steel pipe of 0.6-m diameter is connected to a reservoir having a water surface elevation 30 m above the outlet of the pipe. There is a valve at the end of the pipe that is initially closed. When this valve is suddenly opened, how long does it take for the flow to reach 99% of its final (steady-state) velocity? Assume that all local losses are negligible.

[Solution] The steady-state velocity can be calculated from Equation 2.69 as follows:

$$V_o = \sqrt{\frac{2 \times 9.81 \times 30 \times 0.6}{0.013 \times 10000}} = 1.648 \text{ m / s}$$

This yields a Reynolds number of $\Re \approx 1 \times 10^6$ and $e/D = 0.000076$. From the Moody diagram, $f = 0.013$, which shows that the assumed f value is approximately correct.

Next, from Equation 2.74,

$$t_o = 2.65 \times \frac{10000 \times 1.648}{9.81 \times 30} = 148 \text{ s} = 2.47 \text{ min}$$

This shows that it takes approximately 2.5 min for steady flow to be established.

2.10.1.3 Flow Oscillations in Interconnected Tanks

Consider two tanks connected by a pipe or U-tube as shown in Figure 2.21. Assume that initially the liquid level in tank 1 is higher than in tank 2 by an amount equal to ΔH, and there is no flow in the system due to the closure of a valve in the pipe or U-tube. Suppose that at time $t = 0$ the valve is suddenly opened. As soon as the valve is opened, the liquid will flow from tank 1 to tank 2 at a velocity V in the pipe varying with time.

Assume that at any time t the liquid surface in tank 1 is at a height z_1 above the final equilibrium position, the liquid surface in tank 2 is at a height z_2 below the equilibrium position, and a particle of the liquid in the pipe is displaced a distance ℓ along the pipe from its equilibrium position. From the continuity equation of incompressible flow,

$$\ell A = z_1 A_1 = z_2 A_2 \qquad (2.75)$$

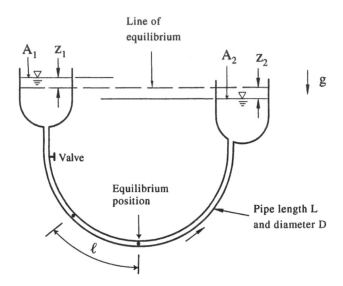

FIGURE 2.21 Flow oscillation in interconnected tanks.

in which A is the cross-sectional area of the pipe, and A_1 and A_2 are the liquid surface areas in tanks 1 and 2, respectively. For simplicity, it is assumed that A_1 and A_2 are both constant (i.e., they do not vary with height).

Applying Newton's second law to the fluid in the pipe or U-tube yields

$$\rho g A(z_1 + z_2) - \pi DL \frac{f\rho V^2}{8} - \sum K \frac{\rho V^2}{2} A = LA\rho \frac{dV}{dt} \qquad (2.76)$$

Substituting Equation 2.75 into Equation 2.76 and using the relationship $V = d\ell/dt$, Equation 2.76 reduces to

$$\frac{d^2\ell}{dt^2} \pm \frac{1}{2L}\left(\sum K + \frac{fL}{D}\right)\left(\frac{d\ell}{dt}\right)^2 + \frac{gA}{L}\left(\frac{1}{A_1} + \frac{1}{A_2}\right)\ell = 0 \qquad (2.77)$$

The sign in front of the second term in the above equation is minus for diminishing values of ℓ ($d\ell/dt < 0$), and plus for increasing ℓ ($d\ell/dt > 0$). Equation 2.77 can be solved numerically by using the initial conditions $\ell = \ell_o$ and $d\ell/dt = 0$ at $t = 0$. Once the variation of ℓ with time is found from Equation 2.77, the variation of the water levels, z_1 and z_2, can be found from Equation 2.75. This is illustrated in the following example.

Example 2.13 Two water tanks of 4-ft diameter are connected by a 1-ft-diameter steel pipe of 100-ft length. Initially, the water level in tank 1 is 8 ft above that in tank 2, and no flow exists because a ball valve in the pipe is closed. At $t = 0$, the ball valve is instantaneously opened, and the water starts to oscillate in the two tanks. Determine the oscillation of the water levels in the tanks as a function of time, and plot the results.

[Solution] Assume that the entrance, the exit and the ball valve (when open) generate headloss coefficients equal to 0.5, 1.0, and 0.1, respectively, and the bends in the pipe generate another headloss coefficient equal to 0.6. Therefore, $\Sigma K = 0.5 + 1.0 + 0.1 + 0.6 = 2.2$. In addition, $D = 1$ ft, $L = 100$ ft, $g = 32.2$ ft/s^2, $A_1/A = (4/1)^2 = 16$, $A_2/A = (4/1)^2 = 16$, and f is assumed to be a constant equal to 0.014. Based on these values, Equation 2.77 reduces to

$$\frac{d^2\ell}{dt^2} \pm 0.018\left(\frac{d\ell}{dt}\right)^2 + 0.0403\ell = 0 \tag{a}$$

From Equation 2.75, $\ell = 16z_1 = 16z_2 = 16z$, where z represents both z_1 and z_2. Thus, Equation a can be written as

$$\frac{d^2z}{dt^2} \pm 0.288\left(\frac{dz}{dt}\right)^2 + 0.0403z = 0 \tag{b}$$

The initial condition for Equation b is $z = 4$ and $dz/dt = 0$ at $t = 0$. Equation b can be solved numerically, and the solution can be plotted in the sketch below. Note that the amplitude of the oscillations decreases with time due to energy dissipation or headloss.

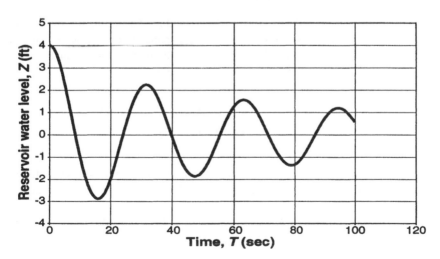

2.10.2 UNSTEADY SOLUTION: WATER HAMMER

Water hammer is the pressure wave created by sudden flow changes generated by rapid valve switching or unexpected pump shutdown such as encountered at power failure. It is a rapidly varying unsteady flow (hydraulic transient) that cannot be treated as quasi-steady. It must be analyzed as truly unsteady flow. The pressure waves (surges) in water hammer often are of such high amplitudes that they cause damage to pipes, pumps, valves, and other fittings. Transient low pressure can also be generated by water hammer to cause vapor columns and cavities that can be damaging to pipelines. A good understanding of the characteristics of water hammer is essential to the design of a safe and reliable pipeline system.

2.10.2.1 Propagation of Small Pressure Disturbances in Pipes

Pressure waves of small amplitudes (i.e., sudden pressure changes created by small disturbances) propagate in a pipe at the following celerity (i.e., wave speed in stationary fluid):

$$C = \sqrt{\frac{E}{\rho}} \tag{2.78}$$

where E is the bulk modulus of elasticity of the fluid; and ρ is the fluid density. When the mean fluid velocity in the pipe is V, the pressure wave propagates upstream at a speed of $(C-V)$, and it moves downstream at speed $(C+V)$. Usually, the celerity $C = \sqrt{E/\rho}$ is more than 100 times larger than the velocity of flow, V. Consequently, in most analyses of water hammer, it is assumed that the pressure wave propagates both upstream and downstream at the same wave speed equal to the celerity.

2.10.2.2 Celerity of Water Hammer Waves

For water hammer waves such as generated by a sudden valve closure, the pressure rise is usually very high. It causes significant expansion of the pipe wall, and Equation 2.78 is no longer applicable. From a consideration of the elastic expansion of pipe due to pressure rise, it can be proved that [2,3]

$$C = \frac{\sqrt{E/\rho}}{\sqrt{1+(E/E_p)(D/\delta)\varepsilon}} \tag{2.79}$$

where E_p is the Young's modulus of the pipe material; δ is the thickness of the pipe wall; and ε is a dimensionless factor equal to 1.0 when the pipe wall is thin (i.e., when $D/\delta > 25$).

For thick-walled pipes ($D/\delta < 25$), the factor ε differs for different conditions, as follows:

Case 1: For a pipeline anchored at upstream,

$$\varepsilon = \frac{2\delta}{D}(1+\mu_p) + \frac{D}{(D+\delta)}\left(1-\frac{\mu_p}{2}\right) \tag{2.80}$$

Case 2: For a pipeline anchored against longitudinal movement,

$$\varepsilon = \frac{2\delta}{D}\left(1+\mu_p\right) + \frac{D}{(D+\delta)}\left(1-\frac{\mu_p^2}{2}\right) \tag{2.81}$$

Case 3: For a pipeline with expansion joints throughout its length,

$$\varepsilon = \frac{2\delta}{D}\left(1+\mu_p\right) + \frac{D}{D+\delta} \tag{2.82}$$

In the above equations, μ_p is the Poisson's ratio of the pipe material, which is approximately 0.3 for steel. Values of μ_p and E_p for various solids are listed in Table C.8 in Appendix C.

2.10.2.3 Rise and Drop of Pressure in Pipe Due to Sudden Valve Closure

Assume the celerity of the water hammer to be C, and the mean velocity (discharge velocity) in the pipe before valve closure to be V. It can be proved that due to a sudden (instantaneous) valve closure, the maximum rise of pressure in the pipe on the upstream side of the valve is [2,3]

$$\Delta p = \rho C V \tag{2.83}$$

From the above equation, the rise of pressure head is

$$\Delta H = \frac{CV}{g} \tag{2.84}$$

On the downstream side of the closed valve, a pressure drop equal to that given by Equation 2.83 will be experienced, provided that the pressure there is above the vapor pressure. Otherwise, column separation will happen and the pressure behind the valve will be vapor pressure. **Column separation** is a phenomenon that often accompanies water hammer. It happens when a portion of the pipe is subject to very low pressure, which causes the liquid to vaporize and form a column of vapor (steam) separated from the liquid water column in the pipe. Column separation can disrupt the operation of pipelines and hence needs to be carefully evaluated or prevented from happening through proper design and operation.

Note that Equations 2.83 and 2.84 are derived for the idealized case in which the valve is closed instantaneously, and that there is no static pressure present after the valve is closed. In real situations, the upstream end of the pipe is either connected to a reservoir or a pump, which imposes a static pressure on the fluid in the pipe upstream of the valve after the valve is closed. This static pressure is the same as the hydrostatic pressure generated by the liquid in the upstream reservoir, or by the pump when running it at zero discharge. The static pressure must be added to the Δp in Equation 2.83 when evaluating the maximum pressure upstream of the valve. Therefore, for the real case, the maximum water pressure in the pipe upstream of the valve caused by instantaneous valve closure is

$$p_1 = p_s + \rho CV \tag{2.85}$$

and

$$H_1 = H_s + \frac{CV}{g} \tag{2.86}$$

where p_1 is the pressure upstream of the valve, H_1 is the pressure head upstream of the valve, p_s is the static pressure generated by an upstream reservoir or pump, and H_s is the corresponding static pressure head.

Equations 2.85 and 2.86 can be used to determine the maximum pressure in a pipe on the upstream side of a suddenly closed valve. They hold regardless of whether the valve is located in the middle of the pipe or near the pipe exit. Nonetheless, one should realize that when the valve is in a sloped pipe of declining elevation, the static pressure p_s and the static head H_s are less when the valve is located in the pipe middle than at the exit. Thus, p_1 and H_1 will also be less when the valve is in the middle instead of the exit of the pipe.

The situation is more complicated for the pressure on the downstream side of the valve, p_2. When the valve is at or near the end (exit) of the pipe where the flow is discharged under atmospheric pressure, p_2 is always atmospheric ($p_2 = p_a$). On the other hand, when the valve is in the middle of the pipe or near the pipe inlet, as soon as the valve is closed, a drop of pressure occurs downstream of the valve. Before valve closure, the pressure at the valve location is p_o, which can be calculated from the energy equation. Due to frictional loss along the pipe, p_o is smaller than p_s.

Immediately after valve closure, p_2 will be either of an amount Δp below p_o, or be the same as p_v (vapor pressure), whichever is greater. Therefore,

$$p_2 = p_o - \rho CV \text{ (when } p_2 > p_v) \tag{2.87}$$

or

$$p_2 = p_v \text{ (when } p_2 \leq p_v) \tag{2.88}$$

2.10.2.4 Water Hammer Force on Valve

The thrust (force) acting on a valve due to water hammer is

$$F_v = A(p_1 - p_2) \tag{2.89}$$

where A is the cross-sectional area of the pipe; and p_1 and p_2 are, respectively, the pressures immediately upstream and downstream of the valve.

Using Equations 2.85 through 2.89, the maximum thrust on a valve due to water hammer caused by sudden valve closure is as follows.

For a valve in the middle of a pipe:

$$F_v = A(p_s - p_o + 2\rho VC) \text{ (when } p_2 > p_v)$$ (2.90)

$$F_v = A(p_s - p_v + \rho VC) \text{ (when } p_2 \le p_v)$$ (2.91)

For a valve at or near the pipe exit:

$$F_v = A (p_s - p_a + \rho VC)$$ (2.92)

In using Equations 2.87 through 2.92, the values of p_1, p_2, p_a, p_o, and p_s must all be in **absolute pressure** because p_v is usually given in absolute pressure.

2.10.2.5 Water Hammer Wave Propagation Due to Sudden Valve Closure

Referring to Figure 2.22, a horizontal pipe is connected to a reservoir with a valve on the other end. The water level in the reservoir is H_s above the pipe, and the velocity in the pipe is V. At $t = 0$, the valve is suddenly (instantaneously) closed. As soon as the valve is closed, a water hammer wave of head rise $\Delta H = CV/g$ above the piezometric head h_p is generated. If the flow is assumed to be frictionless, the headloss along the pipe is zero, and $h_p = H_s$. Thus, a pressure wave of a constant amplitude ΔH above H_s moves upstream at celerity C as shown in Figure 2.22a. The pressure heads upstream and downstream of the wave front are H_s and $(H_s + \Delta H)$ respectively. The velocity of the flow in the pipe upstream and downstream of the wave front is V and 0 (zero) respectively. It takes a time L/C for the wave to reach the reservoir. At this time, V becomes zero throughout the pipe, and the higher pressure in the pipe than in the reservoir causes the flow in the pipe to enter the reservoir. As soon as this happens, the positive pressure wave of magnitude ΔH (above H_s) travels downstream as shown in Figure 2.22b. Behind the wave (i.e., upstream of the wave front*), the pressure head is simply H_s. The velocity of the flow upstream and downstream of the wave front becomes $-V$ and 0 (zero), respectively. It takes another period of L/C (or $t = 2L/C$ since valve closure) for this wave to reach the downstream valve. When the wave reaches the valve, the entire pipe has a pressure head H_s and a reverse flow of velocity $-V$. Then during the next period L/C, a negative pressure wave of amplitude ΔH below H_s moves back toward the reservoir as shown in Figure 2.22c. The wave reaches the reservoir at $t = 3L/C$ after valve closure. Upon reaching the reservoir, the

---

* Because the flow is from left to right, the region to the left of the wave front is considered upstream, and the region to its right is considered downstream.

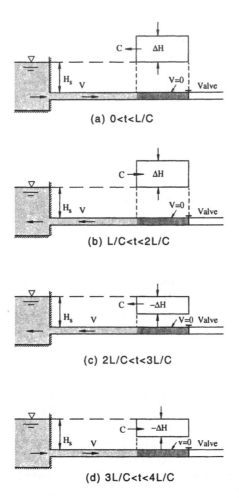

(a) 0<t<L/C

(b) L/C<t<2L/C

(c) 2L/C<t<3L/C

(d) 3L/C<t<4L/C

FIGURE 2.22 Propagation of water hammer waves generated by sudden valve closure.

negative wave is reflected back again toward the valve as shown in Figure 2.22d. During this period, the velocity on the upstream and downstream sides of the wave front is V and 0, respectively. The wave reaches the valve at $t = 4L/C$. This completes one full cycle of the water hammer wave. If frictional loss can be neglected, the wave will repeat in this manner indefinitely without change of amplitude or frequency. In reality, there is damping and the amplitude of the water hammer wave decreases as time progresses.

From the above description of the water hammer wave propagation, it can also be seen that it takes a time of $2L/C$ for the wave to travel upstream to the reservoir and then return to the valve, and it takes a time of $4L/C$ to complete one cycle of the pressure wave. At the valve location, the variation of the water hammer wave as a function of time is shown in Figure 2.23. Note that the regular rectangular wave in Figure 2.23 is for the idealized case of water hammer without energy dissipation (frictionless flow). In reality, energy dissipation always occurs, causing the pressure

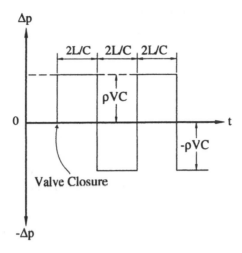

FIGURE 2.23 Variation of pressure with time at valve location due to sudden valve closure.

wave to become damped — to decrease in amplitudes with time cycle after cycle — and to have a rising rather than a leveled top. Moreover, strong turbulence occurs, which causes pressure fluctuations superimposed on the rectangular waves.

From the foregoing description, it can also be seen that during half of each cycle of water hammer (4 $L/C > t > 2L/C$), at least a portion of the pipe encounters a low pressure equal to $\gamma (H_s - \Delta H)$, where γ is the specific weight of the liquid, and $\Delta H = \rho VC$. If this low pressure is less than the vapor pressure of the liquid, p_v, then the pressure should be p_v instead of $\gamma (H_s - \Delta H)$, and column separation and cavitation occur. This shows that column separation and cavitation can occur not only on the downstream side of a rapidly closed valve but also on the upstream side, during each cycle of the water hammer wave when the pressure is sufficiently low.

2.10.2.6 Water Hammer Caused by Partial Closure of Valve

A valve is closed partially such that the velocity in the pipe is decreased from V to V'. If this partial closure is conducted rapidly (i.e., in a time $t_c < 2L/C$), the rise of water pressure head in the pipe due to the water hammer will be

$$\Delta H' = \frac{C(V - V')}{g} \tag{2.93}$$

When a valve is completely closed, V' becomes zero, and Equation 2.93 reduces to Equations 2.83 and 2.84. This means that the last two equations are a special case of Equation 2.93. Much of the earlier discussion on water hammer waves generated by complete closure of a valve also holds for partial closure of a valve, except that the latter does not cause as severe a problem as the former does.

2.10.2.7 Water Hammer with Finite Closure Time

All valves take a finite time to close either partially or fully. When the closure time t_c is less than $2L/C$, it is termed **rapid closure** or **fast closure,** and when t_c is longer than $2L/C$, it is termed **slow closure.** They are discussed separately, as follows.

Rapid closure ($t_c < 2L/C$): When $t_c < 2L/C$, the maximum pressure rise at the valve is the same as due to sudden closure. This makes equations such as 2.83 and 2.85, which are derived for sudden (instantaneous) valve closure, also applicable to rapid closure of finite time. However, for rapid closure this maximum rise will occur over only a portion of the pipe, and over only a part of the time $2L/C$. The part of the pipe subjected to peak pressure is within the distance X upstream from the valve, where

$$X = L - \frac{Ct_c}{2} \tag{2.94}$$

Slow closure ($t_c > 2L/C$): For $t_c > 2L/C$ (slow closure), the maximum pressure at the valve location is

$$\Delta p_s = \frac{2L/C}{t_c} \Delta p = \frac{2L\rho V}{t_c} \tag{2.95}$$

where Δp is the maximum pressure developed corresponding to an instantaneous valve closure as given in Equation 2.83 (for full closure) and Equation 2.93 (for partial closure).

From Equation 2.95, the maximum pressure head generated from the full closure of a valve slowly (in time $t_c > 2L/C$) is

$$\Delta H_s = \frac{2V_o L}{g t_c} \tag{2.96}$$

Equations 2.95 and 2.96 are reasonably accurate and err on the side of safety. The maximum force generated on the valve for slow closure can be calculated from Equations 2.89 through 2.92 by using $2L\rho V/t_c$ in lieu of ρVC.

Example 2.14 A steel pipe of 2-inch diameter and 400-ft length is connected to a reservoir having a water level 20 ft above the pipe outlet. The pipe wall thickness is 1/8 inch, and the pipe has expansion joints throughout its length. A gate valve located near the exit is closed in 3 s. (a) Determine the celerity of the water hammer wave generated. (b) Determine the maximum pressure generated in the pipe due to water hammer. (c) Determine the force on the gate valve caused by the valve closure. (d) Rework parts (b) and (c) if the valve is closed in 0.1 s.

[**Solution**] Assume that local headlosses due to entrance, exit, and fittings are negligible. Then, Equation 2.69 is applicable, and it yields

$$V_o = \frac{0.733}{\sqrt{f}} \qquad (a)$$

By iteration and using the Moody diagram, $f = 0.025$ and $V_o = 4.64$ fps. The cross-sectional area of the pipe is $A = \pi D^2/4 = 0.0218$ ft².

(a) The celerity C can be determined from Equation 2.79. For this pipeline, $D/\delta = 16$, and Equation 2.82 yields $\varepsilon = 1.104$. From Equation 2.79,

$$C = \frac{\sqrt{3 \times 10^5 \times 144 / 1.94}}{\sqrt{1 + \left(\dfrac{3 \times 10^5}{3 \times 10^7}\right) \times 16 \times 1.104}} = \frac{4719}{1.085} = 4350 \text{ fps}$$

(b) $2L/C = 2 \times 400/4350 = 0.184$ s. Because $t_c = 3$ s, which is longer than $2L/C$, the valve closure is considered **slow**, and the maximum water hammer pressure is

$$\Delta p_s = \frac{2L\rho V}{t_c} = \frac{2 \times 400 \times 1.94 \times 4.64}{3} = 2400 \text{ psf} = 16.7 \text{ psi}$$

(c) Because the valve is located near the pipe exit, the pressure p_2 on the downstream side of the valve is approximately atmospheric (0 psig), and the pressure acting on the upstream face of the valve is Δp_s, which is 16.7 psig. Thus, the water hammer force generated on the valve is approximately $F = \Delta p_s A = 2400 \times 0.0218 = 52.4$ lb.

(d) If the valve is closed in 0.1 s instead of 3 s, $2\ L/C$ being 0.184 s longer than t_c, the valve closure is considered **fast.** The maximum water hammer pressure for this case is $\Delta p = \rho VC = 1.94 \times 4.64 \times 4350 = 39,160$ psf $= 272$ psi. This is 16 times greater than the pressure generated by the longer closure time of 3 s. The force on the valve for the 0.1-s closure is $F = \Delta p A = 39,160 \times 0.0218 = 854$ lb, which is about 16 times greater than the force generated from 3-s closure time. This illustrates the great importance of valve closure time in affecting water hammer and the safety of pipeline systems. Whenever practical, valves should be closed slowly (i.e., $t_c \gg 2L/C$).

2.10.2.8 Characteristic Method

The variation of the velocity V and the pressure p (or head H) with time t and distance x along a pipe due to the water hammer can be predicted by solving a set of partial differential equations using a special technique called the **characteristic method.**

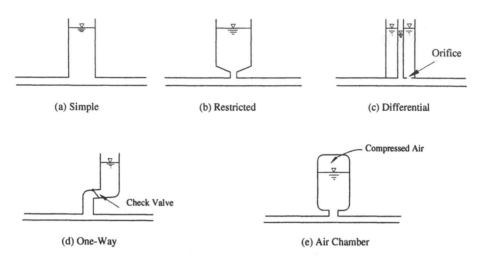

(a) Simple　　　　　　(b) Restricted　　　　　　(c) Differential

(d) One-Way　　　　　　　　　(e) Air Chamber

FIGURE 2.24 Various types of surge tanks.

It is beyond the scope of this book to discuss the characteristic method. It suffices to mention that this is a standard method used for solving unsteady flow problems in pipelines and open channels. Readers interested in the characteristic method should consult References 1, 3, and 4.

2.10.3 SURGE TANKS

A surge tank is a tank (column of liquid) connected to a pipeline for reducing the high pressure generated by the water hammer. When pressure in the pipe rises, the liquid in the pipe enters the surge tank. When the pressure in the pipe drops, the liquid in the pipe leaves the surge tank. Such actions damp out the pressure fluctuations in the pipe caused by the water hammer. Surge tanks are often used in penstocks (the large pipe that conveys water to a turbine for hydropower) and other pipelines. They are placed near turbines, at pumping stations on the discharge side of the pumps, in building water supply systems in which quick acting valves are installed, and at the end of a long pipeline upstream of a valve.

There are many different configurations of surge tanks. They include (a) simple surge tank, (b) restricted inlet tank, (c) differential tank, (d) one-way surge tank, and (e) air chamber. They are illustrated in Figure 2.24.

Consider a simple surge tank placed near the end of a long pipeline (penstock) as shown in Figure 2.25. The surge tank protects the pipe when the downstream valve is closed suddenly, as in the case of an emergency.

Prior to valve closure, the water surface in the surge tank is at an elevation h_L below the water level in the upstream reservoir, where h_L is the headloss along the pipe. For simplicity, assume that the valve is closed instantaneously at time $t = 0$. This causes the water in the pipe to rush into the surge tank, which in turn causes the water level in the tank to rise at the initial velocity $V_s = V_o A/A_s$, where V_o is the mean flow velocity in the pipe at and before $t = 0$. The water level in the tank will rise and surpass the water level in the upstream reservoir until it reaches a maximum

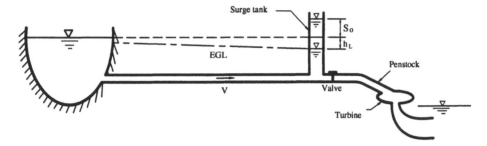

FIGURE 2.25 Maximum surge height generated by sudden valve closure.

surge height S_o above the water level in the reservoir. Then it will oscillate about the equilibrium value. The oscillation of the water in the surge tank is governed by the following differential equation:

$$\frac{d^2S}{dt^2} \pm \left(\frac{f}{2D}\frac{A_s}{A}\right)\left(\frac{dS}{dt}\right)^2 + \left(\frac{g}{L}\frac{A}{A_s}\right)S = 0 \tag{2.97}$$

where S is the surge height, which is the height of water in the surge tank above its equilibrium value.

The sign in front of the second term of the above equation is plus when $dS/dt > 0$, and minus when $dS/dt < 0$. The initial conditions of the above equation are

$$S = -h_L \quad \text{and} \quad \frac{dS}{dt} = \frac{V_o A}{A_s} \quad \text{at} \quad t = 0 \tag{2.98}$$

Equation 2.97 cannot be solved analytically. However, it can be solved by numerical methods for individual cases.

If we can neglect friction loss in the system, Equation 2.97 reduces to

$$\frac{d^2S}{dt^2} + \Phi^2 S = 0 \tag{2.99}$$

$$\Phi^2 = \frac{gA}{LA_s} \tag{2.100}$$

Solving Equation 2.99 yields

$$S = S_o \sin(\Phi t + \Psi) \tag{2.101}$$

which has two arbitrary constants S_o and Ψ, which can be found from using the initial conditions in Equation 2.98. The result is

$$S_o = \sqrt{h_L^2 + \left(\frac{V_o A}{\Phi A_s}\right)^2} \quad \text{and} \quad \Psi = -\tan^{-1}\left(\frac{\Phi A_s h_L}{AV_o}\right) \tag{2.102}$$

The above result yields a sinusoidal oscillation having a period t_o and amplitude (maximum surge) S_o given as follows:

$$t_o = 2\pi \sqrt{\frac{LA_s}{gA}} \tag{2.103}$$

and

$$S_o = \sqrt{h_L^2 + \left(\frac{V_o A}{\Phi A_s}\right)^2} = \sqrt{h_L^2 + \frac{ALV_o^2}{A_s g}} \tag{2.104}$$

When h_L is neglected, the above equation reduces to

$$S_o = V_o \sqrt{\frac{LA}{gA_s}} \tag{2.105}$$

Due to the neglect of frictional losses in their derivation, Equations 2.102 and 2.103 give a conservative estimate of the maximum surge. A more accurate solution of the problem is to solve Equation 2.97 numerically and then plot the result S as a function of t in a manner similar to that in Example 2.13 for water oscillation in interconnected tanks.

The effect of a surge tank in reducing water hammer in pipe is to be discussed next. Due to sudden closure of a valve, the pressure rise due to water hammer in the pipe protected by the surge tank (i.e., the pipe upstream of the surge tank) is

$$\Delta p_t = \frac{\rho C V}{1 + \dfrac{CA_s}{C_t A}} \tag{2.106}$$

where C and C_t are the celerity of pressure waves in the pipe and the surge tank, respectively. From Equations 2.106 and 2.83, the ratio of the water hammer pressures with and without a surge tank (i.e., the **attenuation ratio**) is

$$m = \frac{1}{\dfrac{CA_s}{C_t A} + 1} \tag{2.107}$$

The above equation shows that when $C_t = C$

$$m = \frac{A}{A + A_s} \qquad (2.108)$$

More discussion of surge tanks and unsteady flow in pipe can be found in Reference 4.

Example 2.15 Water flows at 6 fps through the penstock of a hydroelectric power plant. The penstock is made of steel pipe of 4-ft diameter and 0.5-inch thickness. The penstock is protected by a simple surge tank of 12-ft diameter, made of 1-inch-thick steel. The surge tank is located at a distance of 3000 ft from the upstream reservoir, which supplies the water through the penstock. (a) Find the water surface elevation in the surge tank during normal steady-state operation. (b) If a valve downstream of the surge tank is closed in an emergency in 30 s, what is the maximum water hammer pressure generated in the penstock with or without the surge tank? (c) Estimate the maximum surge height and the period of the first cycle of the surge.

[Solution] (a) During steady-state operation, using energy equation between the water surface in the reservoir, point 1, and the water surface in the surge tank, point 2, yields

$$\left(z_1 - z_2\right) = h_L = \left(\sum K + f \frac{L}{D}\right)\frac{V^2}{2g} \qquad (a)$$

Next, $V = 6$ fps, $D = 4$ ft, $L = 3000$ ft, $v = 1 \times 10^{-5}$ ft²/s, $R = VD/v = 2.4 \times 10^6$, $e/D = 0.0000375$. From the Moody diagram, $f = 0.0113$. Thus, $fL/D = 8.48$. Assume $\Sigma K = 2$. Therefore, Equation a yields $h_L = 5.86$ ft. This shows that during normal steady-state operation, the water level in the surge tank is at an elevation 5.86 ft below that of the reservoir.

(b) Using Equation 2.79, the celerity of the water hammer waves in this penstock is calculated to be $C = 3370$ fps. $2L/C = 2 \times 3000/3370 = 1.78$ s, $t_c = 30$ s. Because t_c is much longer than $2L/C$, valve closure in this case is considered *slow*. The maximum water hammer pressure generated in the penstock due to this slow closure is, from Equation 2.95, $\Delta p_s = 2328$ psf = 16.2 psi. This is without surge tank. With the surge tank, Equation 2.79 yields $C_s = 3021$ fps, and Equation 2.107 yields $m = 0.0906$. Thus, with the surge tank the maximum water hammer in the penstock is $\Delta p_t = m\Delta p_s = 0.0906 \times 2328 = 210$ psf = 1.46 psi. This shows that with the surge tank there is about 11 times the reduction in the maximum pressure in the penstock due to the water hammer.

(c) The maximum surge height can be estimated from Equations 2.100 and 2.104, which yield $\Phi = 0.0345$ and $S_o = 20.2$ ft. From Equation 2.103, the period of oscillation in the surge tank is $t_o = 182$ s $= 3.03$ min.

PROBLEMS

2.1   The same experimental setup in Example 2.1 is used to test turbulent flow in pipe, using water as the fluid at 5 fps. Determine whether the 8.06 ft of pipe entrance length determined for laminar flow in Example 2.1 is sufficient to establish fully developed turbulent flow in this pipe.

2.2   The velocity profile of laminar flow in pipe, as described in Equation 2.8, is

$$u = V_c \left(1 - \frac{r^2}{a^2} \right)$$

where V_c is the velocity at the pipe center; r is the radial distance from the pipe center; and a is the pipe radius. Find the energy and momentum correction factors, α and β, for the flow.

2.3   The velocity profile of turbulent flow in a smooth pipe can be approximated by

$$u = V_o \left(\frac{y}{a} \right)^{1/7}$$

where y is radial distance from the wall (i.e., $y = a - r$). Find the value of β for this flow.

2.4   From the Darcy-Weisbach formula in fluid mechanics, determine the manner in which EI varies with pipeline diameter D and the mean-flow velocity, V, in pipe. If V is held constant, how much will the value of EI decrease by doubling D? If D is held constant, how much will EI increase by doubling V?

2.5   For transporting water through a pipeline of 12-inch diameter at a velocity of 3 fps, what is the value of EI in Btu/TM? Assume pump efficiency to be 80%.

2.6   Prove that the headloss for a sudden contraction of pipe is

$$h_L = \left(\frac{1}{C_c} - 1 \right)^2 \frac{V_2^2}{2g}$$

where C_c is the contraction coefficient given in Table 2.1, and V_2 is the mean velocity of the flow in the downstream pipe as shown in the sketch in Table 2.1.

2.7 Find the thrust generated by fluid on a 90° horizontal bend. Determine the magnitude of this thrust if the pipe diameter is 3 ft, the mean velocity is 10 fps, and the fluid is water. Assume the headloss coefficient for the bend to be $K = 1.0$, and fluid discharges into the atmosphere as soon as it goes around the bend.

2.8 If the bend in the previous problem is in the middle of a long pipe, with pressure immediately upstream of the bend equal to 300 psig, what is the thrust generated on the bend? Discuss how your result differs from the previous problem and the implications.

2.9 Prove that Equation 2.35 reduces to Equation 2.36 by changing the natural logarithm to the common logarithm based on ten.

2.10 Show that to change the Manning formula from ft-lb units to SI units without changing the value of n, the constant 1.486 must be changed to 1.0 (exactly 1). This is due to the fact that the Manning formula was originally written in SI units without 1.486.

2.11 Derive the relationship between Manning's n value and the Darcy-Weisbach resistance factor f. Do it in both SI and ft-lb units.

2.12 In the problem stated in Example 2.8, if the pipeline must cross a hill 150 m above the lake water elevation before it can reach the water tower, analyze what problem the hill will produce and what possible solutions exist. The length of the pipe from the water tower to the top of the hill is 1.2 km. Assume that there are one gate valve and three bends located between the lake and the hill top, and one gate valve and two bends located between the hill top and the water tank.

2.13 An 8-inch PVC pipe 1000 ft long is used to convey water by gravity from one reservoir to another having a difference in water surface elevations of 40 ft. Find the discharge through the pipe. If 100 ft of both ends of this pipe is replaced by a 6-inch PVC pipe, what is the new discharge through the pipe?

2.14 Determine the flow into or out of each of the reservoirs shown in the sketch. All pipes are commercial steel. (Hint: Follow the steps given in Section 2.8 for interconnected reservoirs. The only difference here is that no pumps exist.)

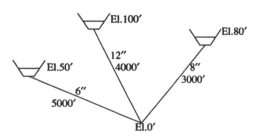

2.15 Solve Example 2.10 by using a digital computer. Program the problem by following the approach used in Example 2.10.

2.16 A total of 14 cfs of water goes through the two parallel pipes in the following sketch. Determine the division of flow and the headloss from A to B.

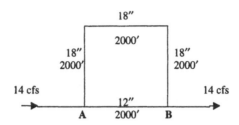

2.17 Find the distribution of flow in the network of pipes as shown in the following sketch. Assume the pipes to be commercial steel, and that the water in the pipe has a kinematic viscosity of $v = 1 \times 10^{-5}$ ft^2/s. Use the Hardy Cross method to solve this problem.

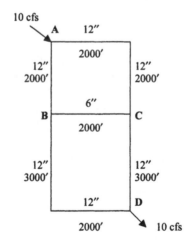

2.18 A cylindrical tank of 6-ft diameter contains 10 ft of water. The bottom of the tank is connected to a straight vertical steel pipe of 4-inch diameter and 5-ft length. The pipe entrance is blunt (sharp-edged), and near the pipe exit there is a threaded gate valve. Initially, the valve is closed and there is no flow. Then, at $t = 0$, the valve is suddenly opened, and the tank is allowed to drain. Determine (a) the time to drain the entire tank, and (b) the time to drain one half of the tank.

2.19 Prove that one can use Equation 2.68 instead of Equation 2.69 to derive Equations 2.73 and 2.74.

2.20 Derive an equation equivalent to Equation 2.74 for the time t_{50} to reach 50% of the final velocity V_o. Then, determine the time t_{50} for the flow in Example 2.12, and compare with t_o found in the example.

2.21 Solve Equation b in Example 2.13 numerically using any method or computer software familiar to you. Then plot the result, z as a function of t, and compare with the graph given in the example.

2.22 (a) Prove that for water hammer waves traveling through a rigid tunnel of diameter D, Equation 2.79 reduces to

$$C = \frac{\sqrt{E/\rho}}{\sqrt{1+2\dfrac{E}{E_t}(1+\mu_t)}}$$

where E_t and μ_t are the modulus of elasticity and the Poisson's ratio of the tunnel material, respectively.

(b) For water moving through a concrete-lined tunnel, what is the celerity of C? Assume that the concrete has a modulus of elasticity of 3×10^6 psi, and a Poisson's ratio of 0.3.

2.23 A 24-inch steel pipe is 0.5 inch thick and 5000 ft long with a control valve located near the outlet. The modulus of elasticity for the steel is 3×10^7 psi, and for water is 3×10^5 psi. The discharge through the pipe is 10 cfs. (a) Find the celerity of the pressure waves generated by valve closure. (b) Find the maximum water hammer pressure in this line due to instantaneous valve closure. (c) What should the valve closure time be in order to reduce this pressure by a factor of two?

2.24 A simple surge tank of 4-ft diameter is placed near the exit of the pipe in the previous problem. Determine the following: (a) The maximum surge in the surge tank due to instantaneous valve closure? (b) How soon after instantaneous valve closure will this surge reach the maximum? (c) How much will the maximum pressure in the pipe be reduced due to the use of the surge tank? Solve this problem by neglecting frictional losses.

2.25 Solve Problem 2.24 without neglecting frictional losses. Plot the surge height as a function of time for the first ten cycles of the oscillation. (Hint: Solve Equation 2.97 numerically by using any method or software familiar to you.)

REFERENCES

1. Streeter, V.L., *Fluid Mechanics*, McGraw-Hill, New York, 1979.
2. Morris, H.M., *Applied Hydraulics in Engineering*, Ronald Press, New York, 1963.
3. Chaudhry, M.H., *Applied Hydraulic Transients*, 2nd ed., Van Nostrand Reinhold, New York, 1987.
4. Roberson, J.A., Cassidy, J.J., and Chaudhry, M.H., *Hydraulic Engineering*, Houghton Mifflin, Boston, 1988, chap. 11, p. 572.

3 Single-Phase Compressible Flow in Pipe

3.1 FLOW ANALYSIS FOR IDEAL GAS

3.1.1 GENERAL ANALYSIS

For steady single-phase compressible flow in a pipe, the continuity, energy, and momentum equations are, respectively,

$$\rho_1 V_1 A_1 - \rho_2 V_2 A_2 = \text{constant} \tag{3.1}$$

$$\frac{V_1^2}{2} + \frac{p_1}{\rho_1} + gz_1 = \frac{V_2^2}{2} + \frac{p_2}{\rho_2} + gz_2 + \left(i_2 - i_1\right) + \frac{dQ_e}{dm} \tag{3.2}$$

$$F_x = \rho_2 V_2^2 A_2 \cos\theta - \rho_1 V_1^2 A_1 \tag{3.3}$$

$$F_y = \rho_2 V_2^2 A_2 \sin\theta \tag{3.4}$$

where the subscripts 1 and 2 refer to an upstream point and a downstream point, respectively, and subscript x refers to the direction of flow before a bend if there is any. The bend angle is θ, and y is perpendicular to the x-axis as in Figure 2.8. The pipe is in the x–y plane. Equation 3.2 is for a pipe without turbines or pumps (blowers or compressors)* between sections 1 and 2. Also, the energy correction factor α and the momentum correction β are assumed to be unity in the above equations.

When the pipe is straight and of constant diameter, Equations 3.1 and 3.3, reduce to

$$\rho_1 V_1 = \rho_2 V_2 = \text{constant} \tag{3.5}$$

---

* The equivalent of a pump in a gas pipeline is a blower (for a low head) and a compressor (for a high head).

83

$$F = \rho_2 V_2^2 A - \rho_1 V_1^2 A \tag{3.6}$$

where the subscript x has been dropped from F for simplicity.

Due to frictional loss, the pressure along a pipeline decreases in the flow direction, causing the density of the gas to decrease along the pipe. From Equation 3.5, a decrease in density causes an increase in velocity in the flow direction. Note that in long pipelines such as those used for transporting natural gas, there can be great changes in the density and the velocity of gas between compressor stations.

Application of Equation 3.6 to a control volume (C.∀.) between two sections at an infinitesimal distance dx apart as shown in Figure 3.1 yields

$$F = (\rho + d\rho)(V + dV)^2 A - \rho V^2 A$$
$$= A(2\rho V dV + V^2 d\rho) \tag{3.7}$$

But,

$$F = pA - (p + dp)A - \tau_o \pi D dx$$
$$= -A dp - \pi D \tau_o dx \tag{3.8}$$

Equating Equations 3.7 and 3.8 yields

$$2\rho V dV + V^2 d\rho + dp + \frac{4\tau_o dx}{D} = 0 \tag{3.9}$$

From Chapter 2, the shear stress at the wall of a straight pipe is

$$\tau_o = \frac{f \rho V^2}{8} \tag{3.10}$$

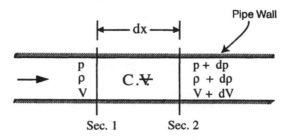

FIGURE 3.1 Variation of flow properties along a pipe in one-dimensional analysis.

Substituting Equation 3.10 into 3.9 yields

$$2VdV + V^2 \frac{d\rho}{\rho} + \frac{dp}{\rho} + \frac{fV^2 dx}{2D} = 0 \tag{3.11}$$

Next, from Equation 3.5

$$\frac{dV}{V} + \frac{d\rho}{\rho} = 0 \tag{3.12}$$

Eliminating $d\rho/\rho$ from the last two equations yields

$$\frac{dV}{V} + \frac{dp}{\rho V^2} + \frac{fdx}{2D} = 0 \tag{3.13}$$

From its derivation, it can be seen that the above equation holds for any compressible flow (adiabatic, isothermal, flow with or without friction, etc.), as long as the pipe diameter remains constant throughout the length of the pipe. Although the equation was derived for a horizontal pipe, it can also be used for sloped pipes because for gas flow the gravity effect is usually negligible unless a large elevation difference exists along the pipe. Because Equation 3.13 has three variables, V, p, and ρ, two more equations are needed to solve the equation: the equation of state that relates p to ρ, and another thermodynamic equation that describes the heat transfer process, such as the isothermal process or the adiabatic process.

At moderate pressure (not much higher than the atmospheric pressure), the relation between p and ρ is given by the equation of state of ideal gas as follows:

$$\frac{p}{\rho} = RT \tag{3.14}$$

where

$$R = nB \tag{3.14a}$$

In the above equations, p is the absolute pressure, ρ is the density, R is the engineering gas constant (gas constant, for short), T is the absolute temperature; n is the number of moles per unit mass, and B is the Boltzmann's constant. Unlike the gas constant R, which varies with the type of gas, the Boltzmann's constant B is a universal constant for all gases. The value of B is 1.379×10^{-16} dyne-cm/K or 1545 lb-ft/°R.

For air, the molecular weight is 29 lbm/mole, and hence $n = 1/29$ mole/lbm. From Equation 3.14a,

$$R = \frac{1}{29} \frac{\text{mole}}{\text{lbm}} \times 1545 \frac{\text{lb-ft}}{\text{°R}} = 53.3 \frac{\text{lb-ft}}{\text{lbm °R}} = 1715 \frac{\text{lb-ft}}{\text{slug °R}} \qquad (3.14b)$$

Note that while the pressure p_1 and p_2 in Equation 3.2 can be either relative (gage) pressure, or absolute pressure, the p in Equation 3.14 and in any equation derived from Equation 3.14 must be absolute pressure. To be consistent in notation, all equations for compressible flow should use absolute pressure.

3.1.2 ISOTHERMAL COMPRESSIBLE PIPE FLOW WITH FRICTION

When the temperature along a pipe is constant (i.e., isothermal), the equation of state of ideal gas yields $p = C\rho$, where $C = RT = $ constant. Consequently,

$$\frac{dp}{p} = \frac{d\rho}{\rho} \qquad (3.15)$$

From Equations 3.12 and 3.15,

$$\frac{dV}{V} = -\frac{dp}{p} \qquad (3.16)$$

Substituting Equation 3.16 into Equation 3.13 yields

$$\frac{dp}{dx} = \frac{\dfrac{fp}{2D}}{\left[1 - \dfrac{p}{\rho V^2}\right]} \qquad (3.17)$$

Because the pressure p in the pipe of constant diameter must decrease in the flow direction, the quantity dp/dx in Equation 3.17 is negative. This shows that $p/\rho V^2$ must be greater than one, or $p > \rho V^2$. The limiting value (minimum value) of p is

$$p_o = \rho_o V_o^2 \qquad (3.18)$$

where the subscript o indicates limiting conditions.

By definition, the Mach number is

$$M = \frac{V}{\sqrt{kRT}} = \frac{V}{\sqrt{k\dfrac{p}{\rho}}} \qquad (3.19)$$

Substituting Equation 3.19 into Equation 3.17 yields

$$\frac{dp}{dx} = \left[\frac{\dfrac{fp}{2D}}{1 - \dfrac{1}{kM^2}} \right] \tag{3.20}$$

Equation 3.20 shows that the limiting value of p is reached when $M = 1/\sqrt{k}$. For air, $k = 1.4$ and $M_o = 0.845$. This means for air in a pipe where the pressure decreases in the flow direction, the Mach number must be below 0.845. This limiting condition is called **choking** because it is not possible to have constant temperature flow through pipe at a Mach number greater than M_o. Figure 3.2 shows the variation of p with x along a pipe of constant diameter and constant temperature. The solid line is for subsonic flow. Note that point B, the limiting point where $p = p_o$ and $M = M_o$, can occur only at the pipeline exit. For gases other than air, the value of k can be found from Tables C.5 and C.6 in Appendix C.

Equations 3.15 and 3.16 show that when the pressure p in a pipe of constant diameter and constant temperature decreases in the flow direction, the density ρ decreases and the velocity V increases in the flow direction. Then, from Equation 3.19, the Mach number M must also increase in the flow direction.

For constant temperature along the pipe, Equation 3.13 can be integrated as follows. For constant temperature and constant pipe diameter, both ρV and the dynamic viscosity μ are constant along the pipe. This means the Reynolds number $\Re$ and hence the resistance factor f must be constant along the pipe. The fact that f is constant makes the third term of Equation 3.13 directly integrable. To be able to integrate the second term, the following relations must be used:

$$\frac{p}{\rho} = \frac{p_1}{\rho_1} = \text{constant}, \quad \text{or} \quad \frac{\rho_1}{\rho} = \frac{p_1}{p} \text{ (for isothermal flow)} \tag{3.21}$$

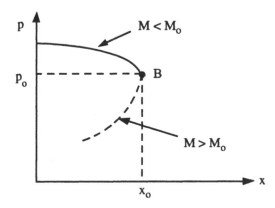

FIGURE 3.2 Variation of pressure along a pipe of constant diameter and constant temperature (i.e., under isothermal condition).

$$\rho V = \rho_1 V_1 = \text{constant}, \quad \text{or} \quad \frac{\rho_1}{\rho} = \frac{V}{V_1} \text{ (for constant diameter pipe)} \quad (3.22)$$

where the subscript 1 refers to a fixed point upstream.

From the two above equations,

$$\frac{\rho_1}{\rho} = \frac{V}{V_1} \quad \text{or} \quad pV = p_1 V_1 = \text{constant} \quad (3.23)$$

From Equation 3.22,

$$\rho V^2 = \frac{\rho_1^2 V_1^2}{\rho} \quad (3.24)$$

Therefore,

$$\frac{dp}{\rho V^2} = \frac{\rho dp}{\rho_1^2 V_1^2} = \frac{p dp}{(\rho_1 V_1)(p_1 V_1)} \quad (3.25)$$

Substituting Equation 3.25 into Equation 3.13 and realizing that the terms $\rho_1 V_1$ and $p_1 V_1$ in Equation 3.25 are both constant, Equation 3.13 can be integrated to yield

$$p_1^2 - p_2^2 = \rho_1 p_1 V_1^2 \left[2\ln\frac{V_2}{V_1} + \frac{fL}{D} \right] \quad (3.26)$$

where the subscripts 1 and 2 refer, respectively, to an upstream and a downstream location, and L is the distance between the two locations.

Equation 3.26 can be rearranged to yield

$$\frac{fL}{D} = \frac{1}{kM_1^2} \left[1 - \frac{p_2^2}{p_1^2} \right] + 2\ln\frac{p_2}{p_1} \quad (3.27)$$

The above equation can be used to calculate p_2 when the upstream conditions such as p_1 and M_1 are known.

In a similar manner, it can be proved that

$$\frac{fL}{D} = \frac{1}{kM_2^2} \left[\frac{p_1^2}{p_2^2} - 1 \right] - 2\ln\frac{p_1}{p_2} \quad (3.28)$$

which can be used to calculate the upstream pressure p_1 when the downstream conditions such as p_2 and M_2 are known.

Example 3.1 Carbon dioxide is pumped through a long underground 6-inch-diameter steel pipeline to an oil field for use in underground oil recovery. Due to the great length of the pipeline, several booster pumps are needed along the pipeline at intervals of 50 mi. The pressure of the gas in the pipe leaving any booster pump station is 2000 psig, and before the gas reaches the next booster station the pressure drops to 100 psig. The temperature of gas in the pipe is approximately constant at 60°F. Find the Mach number of the flow at both ends of the pipe between two neighboring booster stations, and determine whether the limiting condition (choking) has been reached in the pipe. Also, determine the density and the velocity of the flow at both ends of the pipe.

[Solution] The known quantities in this case are p_1 = 2000 psig = 2014.7 psia = 290,117 psfa, p_2 = 100 psig = 114.7 psia = 16,517 psfa, T = 60°F = 520°R, k = 1.29 (from Table C.5 in Appendix C), R = 35.1 ft-lbf/1bm/°R = 1130 ft-lbf/slug/°R, L = 50 mi = 264,000 ft, D = 6 inch = 0.5 ft, and f = 0.012 (assumed). Substituting these values into Equation 3.27 yields M_1 = 0.0111, and M_2 = 0.194. The limiting Mach number for this case (carbon dioxide) is M_o = $1/\sqrt{k}$ = 0.880. Since M_2 is much smaller than M_o, the limiting condition is not approached, and there is no choking of the flow. From the equation of state, $\rho_1 = p_1/RT$ = 0.494 slug/ft³ and ρ_2 = 0.0281 slug/ft³. The velocity of the gas upstream is $V_1 = M_1/\sqrt{kRT}$ = 9.66 fps.

Note that the foregoing calculation is incomplete because the values of M_1 and V_1 are based on an assumed value of f = 0.012. Now that the density and the velocity of the gas have been calculated, a more accurate value of f can be determined from the Moody diagram in Chapter 2, as follows.

From Table C.5, the viscosity of carbon dioxide at 60°F is approximately 3.0 × 10⁻⁷ lb-s/ft². Thus, the Reynolds number is $\Re = \rho_1 V_1 D/\mu$ = 8.0 × 106. The relative roughness is e/D = 0.0003. From the Moody diagram, f = 0.0148. This is a more accurate value of f than the one assumed previously. Based on this new value of f, repeating the foregoing calculations yields M_1 = 0.00994 and V_1 = 8.65 fps. This shows how to calculate M_1, V_1, and f through iteration. Depending on the accuracy of the assumed value of f, two to four iterations are normally needed to reach accurate values.

Now that V_1, ρ_1, and ρ_2 have been determined, Equation 3.5 can be used to yield V_2 = 152 fps. From the foregoing calculation, it can be seen that the velocity of the gas in this case increases by approximately 18 times along the pipeline over a distance of 50 mi, whereas the density of gas decreases by the same amount (approximately 18 times). The product ρV remains constant along the pipe, which is what should be expected for steady flow along a pipe of constant diameter (see Equation 3.22).

3.1.3 ADIABATIC COMPRESSIBLE PIPE FLOW WITH FRICTION

For adiabatic flow along a horizontal pipe, dQ_c/dm is zero, and $z_1 = z_2$. Consequently, Equation 3.2 reduces to

$$\left(\frac{p_1}{\rho_1} - \frac{p_2}{\rho_2}\right) + c_v(T_1 - T_2) = \frac{V_2^2 - V_1^2}{2} \tag{3.29}$$

But, for ideal gas,

$$c_v = \frac{R}{k-1}, \quad p_1 = \rho_1 RT_1 \quad \text{and} \quad p_2 = \rho_2 RT_2 \tag{3.30}$$

Substituting Equation 3.30 into Equation 3.29 and rearranging terms yields

$$\frac{(k-1)}{k}\frac{V_1^2}{2} + \frac{p_1}{\rho_1} = \frac{(k-1)}{k}\frac{V_2^2}{2} + \frac{p_2}{\rho_2} \tag{3.31}$$

$$\frac{(k-1)}{k}\frac{V^2}{2} + \frac{p}{\rho} = \text{constant} \tag{3.32}$$

or,

$$\left[1 + \frac{(k-1)}{2}M^2\right]\frac{p}{\rho} = \text{constant} \tag{3.33}$$

Differentiating Equation 3.32 and using Equation 3.12 yields

$$\frac{d\rho}{\rho} = \frac{dp}{p + \left(\frac{k-1}{k}\right)\rho V^2} \tag{3.34}$$

Substituting Equation 3.24 into Equation 3.34 yields

$$\frac{d\rho}{\rho} = \frac{dp}{p + (\Gamma/\rho)} \tag{3.35}$$

where

$$\Gamma = \frac{(k-1)}{k}\rho_1^2 V_1^2 \tag{3.36}$$

Therefore, Equation 3.13 becomes

$$\frac{dp}{dx} = -\frac{fV^2}{2D}\left[\frac{1+(k-1)M^2}{1-M^2}\right] \tag{3.37}$$

Conclusions:

1. From Equation 3.37, $dp/dx < 0$ when $M < 1$ or vice versa. This means in any subsonic adiabatic flow through a pipe of constant diameter, the pressure should decrease in the flow direction. Conversely, if the pressure of an adiabatic flow in a pipe of constant diameter decreases in the flow direction, the flow must be subsonic.
2. The limiting pressure p_o, limiting density ρ_o and limiting velocity V_o are reached when the Mach number is unity. This exists at the outlet of the pipe if the pipe is sufficiently long. The minimum length to produce the limiting condition is $x = x_o$.

Writing Equation 3.33 between an upstream point 1 and a downstream point 0 corresponding to limiting condition (i.e., $M_o = 1$) yields

$$\frac{p_1}{\rho_1}\left[1+\frac{(k-1)}{2}M_1^2\right] = \frac{p_o}{\rho_o}\left[1+\frac{(k-1)}{2}\right] \tag{3.38}$$

The celerities of pressure waves at 1 and 0 are, respectively,

$$C_1 = \sqrt{kp_1/\rho_1}, \quad C_o = \sqrt{kp_o/\rho_o} \tag{3.39}$$

From Equation 3.39,

$$\frac{p_1}{\rho_1} = \frac{C_1^2}{k}, \quad \frac{p_o}{\rho_o} = \frac{C_o^2}{k} \tag{3.40}$$

Substituting Equation 3.40 into Equation 3.38 yields

$$\frac{C_o}{C_1} = \left[\frac{2+(k-1)M_1^2}{k+1}\right]^{1/2} \tag{3.41}$$

From $C_o = V_o/M_o = V_o$ and $C_1 = V_1/M_1$, we have

$$\frac{C_o}{C_1} = \frac{V_o}{V_1}M_1 = \frac{\rho_1}{\rho_o}M_1 \tag{3.42}$$

Substituting Equation 3.42 into Equation 3.41 yields

$$\frac{p_o}{p_1} = \left[\frac{(k+1)M_1^2}{2+(k-1)M_1^2} \right]^{1/2} \tag{3.43}$$

From Equation 3.38 and Equation 3.43,

$$\frac{p_o}{p_1} = M_1 \left\{ \frac{2+(k-1)M_1^2}{k+1} \right\}^{1/2} \tag{3.44}$$

Finally, it can be proved that

$$\frac{fL}{D} = \frac{(k-1)}{kB}\left[1 - \left(\frac{p_2}{p_1}\right)^2 - \frac{p_2}{p_1}\sqrt{\frac{p_2^2}{p_1^2}+B}+\sqrt{1+B} \right]$$

$$+ \frac{(3k-1)}{k}\ln\left[\frac{\frac{p_2}{p_1}+\sqrt{\frac{p_2^2}{p_1^2}+B}}{1+\sqrt{1+B}} \right] \tag{3.45}$$

where

$$B = (k-1)M_1^2\left[2+(k-1)M_1^2 \right] \tag{3.46}$$

To obtain Equation 3.45, the value of f had to be assumed constant, which is not true for adiabatic flow. However, since the change of the f value is relatively small, Equation 3.45 is approximately correct. The value of f used should be the average of the upstream and downstream values.

Note that Equation 3.45 is written in a form for calculating p_2 once the flow properties upstream, such as p_1 and M_1, are known. To use the same equation to determine the upstream flow properties such as p_1 from the downstream conditions such as p_2, one can first assume the value of M_1. Then, one can determine p_1 from Equation 3.45, and determine ρ_1 from Equation 3.33. Then, V_1 can be calculated from the continuity equation ρV = constant along the pipe. Once we know $V_1, p_1,$ and ρ_1, we can calculate M_1 from Equation 3.19 and compare with the assumed value of M_1. This offers an iteration process to calculate p_1 and other upstream flow conditions from known downstream conditions by using Equation 3.45 for adiabatic pipe flow with friction.

A comparison of the values of p_2/p_1 for the isothermal, adiabatic, and incompressible cases are given in Figure 3.3 a and b. This figure provides a simple graphical way to solve Equations 3.27 (isothermal case) and 3.45 (adiabatic case).

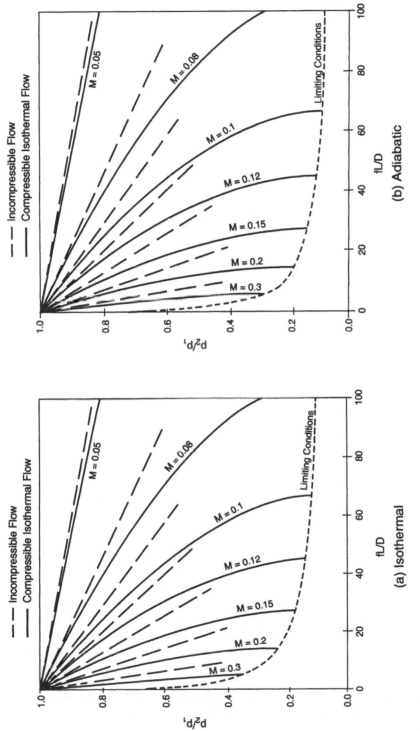

FIGURE 3.3 Variation of p_2/p_1 with fL/D for isothermal and adiabatic flows.

Example 3.2 Air enters a horizontal wrought iron pipe of 6-inch diameter at a pressure of 85 psia, a temperature of 68°F, and a velocity of 101 fps. The pipe length is 700 ft. Determine the pressure drop in the pipe by assuming three different conditions: (a) incompressible flow, (b) compressible isothermal flow, and (c) compressible adiabatic flow.

[Solution] p_1 = 85 psia = 12,240 psfa, T_1 = 68° + 460° = 528°R, k = 1.4, R = 1715 ft-lb/slug/°R, e/D = 0.0003, L/D = 1400, $\rho_1 = p_1/RT_1$ = 0.0135 slug/ft³, $M_1 = V_1/\sqrt{kRT_1}$ = 0.0897, D = 0.5 ft, L = 700 ft.

(a) Incompressible flow solution:

The dynamic viscosity of the air in the pipe is μ = 4.0 × 10⁻⁷ lb-s/ft². Thus, the Reynolds number at the pipe entrance is $\Re = \rho_1 V_1 D/\mu$ = 1.7 × 10⁶. The relative roughness of the pipe is e/D = 0.0003. From the Moody diagram, f = 0.0153. Then, the Darcy-Weisbach formula yields

$$\Delta p = f\frac{L}{D}\frac{\rho_1 V_1^2}{2} = 1477 \text{ psf} = 10.26 \text{ psi}$$

(b) Compressible isothermal flow solution:

From Equation 3.27, $p_2/p_1 = 0.870, p_2 = 0.870\, p_1 = 74.0$ psia, $p_1 - p_2 = 11.05$ psi.

(c) Compressible adiabatic flow solution:

From Equation 3.45, B = 0.006447, p_2/p_1 = 0.869, p_2 = 0.869 p_1 = 73.9 psia, $p_1 - p_2$ = 11.1 psi.

The foregoing results show that while the pressure drops for compressible isothermal flow and for compressible adiabatic flow are almost identical, they are both higher than that predicted from incompressible flow model by about 8%. Their values would be more different from each other when the pipe length is increased, or when the velocity of the air is higher than in this example. This shows that if the pipe is short and velocity is low, gas flow can be treated as incompressible flow with little error. On the other hand, when the pipe is long and/or when the velocity is high, the three models (incompressible, isothermal compressible, and adiabatic compressible) will yield very different results. In such a case, it is important to determine in each application which of the three models best represents the process. Generally, long pipelines maintain the same temperature as that of their environment and hence are constant temperature, whereas insulated pipes of short to moderate lengths cause little heat transfer through pipe wall and hence are adiabatic.

3.1.4 ISENTROPIC (ADIABATIC FRICTIONLESS) PIPE FLOW

Equation 3.31 holds for adiabatic flow in general — with or without friction. Therefore, for isentropic (i.e., adiabatic frictionless) flow,

$$V_2^2 - V_1^2 = \frac{2k}{(k-1)}\left(\frac{p_1}{\rho_1} - \frac{p_2}{\rho_2}\right) \qquad (3.47)$$

From thermodynamics, the relation between pressure p and density ρ for any isentropic ideal gas is

$$\frac{p_1}{\rho_1^k} = \frac{p_2}{\rho_2^k} \quad \text{or} \quad \rho_2 = \rho_1\left(\frac{p_2}{p_1}\right)^{1/k} \qquad (3.48)$$

From the continuity equation of steady flow, Equation 3.1, and Equation 3.48,

$$V_1 = \frac{\rho_2 V_2 A_2}{\rho_1 A_1} = \left(\frac{p_2}{p_1}\right)^{1/k} \frac{V_2 A_2}{A_1} \qquad (3.49)$$

Substituting Equation 3.48 and Equation 3.49 into Equation 3.47 yields

$$V_2 = \sqrt{\frac{2k}{(k-1)}\left(\frac{p_1}{\rho_1}\right)\frac{\left[1-\left(\dfrac{p_2}{p_1}\right)^{1-1/k}\right]}{\left[1-\left(\dfrac{p_2}{p_1}\right)^{2/k}\left(\dfrac{A_2^2}{A_1^2}\right)\right]}} \qquad (3.50)$$

Finally, the mass flow rate, $\dot{m}$, and the weight flow rate, $\dot{w}$, can be calculated from

$$\dot{m} = \rho_2 A_2 V_2 = \rho_1\left(\frac{p_2}{p_1}\right)^{1/k} A_2 V_2 \qquad (3.51)$$

$$\dot{w} = g\rho_2 A_2 V_2 = g\rho_1\left(\frac{p_2}{p_1}\right)^{1/k} A_2 V_2 \qquad (3.52)$$

where V_2 is given by Equation 3.50.

Equations 3.48 to 3.50 can be used to determine the change of p, V, and ρ along a pipe caused by the variation of pipe diameter and/or flow cross section, as encountered in a Venturi or orifice flowmeter. Once such changes are calculated from these equations, they can be substituted into Equations 3.51 and 3.52 to find the mass flow rate $\dot{m}$ and weight flow rate $\dot{w}$. This yields the basic equations of flow metering for compressible flow. However, the aforementioned equations are strictly correct only for truly isentropic flow, which is an idealized condition. In real flows, there is always some frictional losses between sections 1 and 2, and so Equations 3.50, 3.51, and 3.52 overpredict the values of V_2, $\dot{m}$, and $\dot{w}$. To correct for such errors, it is customary to apply a correction factor C_d, called the **discharge coefficient**, to the right side of Equations 3.51 and 3.52. The magnitude of C_d is always less than 1.0. As in the case of incompressible flow, the value of C_d depends on both the Reynolds number and the type and geometry of the flowmeter. When the Reynolds number in the pipe is greater than 10^5, C_d is approximately 0.98. Smaller values of C_d should be used if the Reynolds number is less than 10^5.

Substituting Equation 3.50 into Equation 3.51 and using the discharge coefficient C_d yields

$$\dot{m} = C_p C_d A_2 \rho_1 \sqrt{\frac{2(p_1 - p_2)/\rho_1}{1-(A_2/A_1)^2}} \tag{3.53}$$

where

$$C_p = \sqrt{\frac{k(p_2/p_1)^{2/k}\left[1-(p_2/p_1)^{(k-1)/k}\right]\left[1-(A_2/A_1)^2\right]}{(k-1)\left[1-(A_2/A_1)^2(p_2/p_1)^{2/k}\right]\left[1-(p_2/p_1)\right]}} \tag{3.54}$$

Note that Equation 3.53 is identical to its counterpart for incompressible flow except for the factor C_p. For this reason, C_p is commonly referred to as the **compressibility factor**. In this book, C_p will be referred to as the **compressibility coefficient** instead of the compressibility factor, because the latter term will be needed in Section 3.2 to define something totally different. Values of C_p for air ($k = 1.4$) can be found graphically from Figure 3.4.

Example 3.3 Air in a 6-inch pipe flows through a 6×2 inch Venturi flowmeter (i.e., 6-inch diameter upstream and 2-inch diameter at the throat). The pressure and temperature of the air upstream of the Venturi are 40 psig and 60°F, respectively. The pressure at the throat is measured to be 20 psig. Determine: (a) the mass flow rate $\dot{m}$, (b) the density of the air both upstream and at the throat, (c) the velocity of air upstream and at the throat, and (d) the Mach number of the flow both upstream and at the throat.

[Solution] (a) $k = 1.4$, $R = 1715$ lb-ft/slug/°R, $T = 60°F = 520°R$, $p_1 = 40$ psig $= 54.7$ psia, $= 7877$ psfa, $p_2 = 20$ psig $= 34.7$ psia $= 4977$ psfa, $p_2/p_1 = 0.6344$, $A_2/A_1 = (D_2/D_1)^2 = 4/36 = 1/9$, and $(A_2/A_1)^2 = 1/8 = 0.012346$. Thus, from Equation 3.54, $C_p = 0.778$. The density upstream is $\rho_1 = p_1/RT = 0.008833$

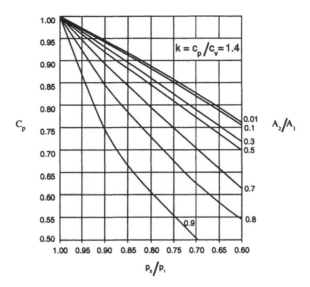

FIGURE 3.4 Compressibility coefficient, C_p, as a function of pressure ratio, p_2/p_1, for $k = 1.4$.

slug/ft³, $A_2 = 0.02182$ ft². Assuming that $C_d = 0.98$, Equation 3.53 yields $\dot{m} = 0.1194$ slug/s.

(b) The density of the air upstream was already found to be 0.008833 slug/ft³. The density of the air at the throat of the Venturi is $\rho_2 = p_2/RT = 0.00560$ slug/ft³. This shows that the density of the air decreases by 1.6 times when reaching the throat of the Venturi.

(c) From continuity, $\dot{m} = \rho_1 V_1 A_1$. Knowing that $A_1 = 0.19635$ ft², $\dot{m} = 0.1194$ slug/s, and $\rho_1 = 0.008833$ slug/ft³, $V_1 = \dot{m}/\rho_1 A_1 = 68.8$ fps. Likewise, $V_2 = \dot{m}/\rho_2 A_2 = 977$ fps.

(d) The Mach number upstream is $M_1 = V_1/\sqrt{kRT} = 0.062$. The Mach number at the throat is $M_2 = V_2/\sqrt{kRT} = 0.87$. Since the calculated Mach number at the throat is less than 1.0, the assumption of subsonic flow throughout is correct, and the calculated results are also correct. Had the calculated value of M_2 been greater than 1.0, the assumption of subsonic flow would be violated, and the result would be meaningless.

More about compressible flow of ideal gas through pipe for both subsonic and supersonic flows can be found in fluid mechanics texts such as Reference 1.

3.2 FLOW ANALYSIS FOR REAL (NONIDEAL) GAS

3.2.1 EQUATION OF STATE

The foregoing derivations are based on the equation of state of a perfect gas (Equation 3.14), which is correct only for gases under relatively low pressure (not more than a few atmospheres) and away from critical conditions. When a gas is compressed to many times that of the atmospheric pressure, the distance between neighboring molecules becomes so small that the molecules are under the influence of one another's force fields, and the gas no longer follows the ideal gas law given by Equation 3.14. In such a case, the equation of state can be modified as

$$\frac{p}{\rho} = zRT \tag{3.55}$$

where z is the **compressibility factor** of gas. In the mechanical engineering literature, z is sometimes referred to as the **supercompressibility factor**, in order to distinguish from C_p, which is mentioned in the same literature as the **compressibility factor**.

Usually, z is a function of the **reduced pressure, P_r,** and the **reduced temperature, T_r,** namely,

$$z = f(P_r, T_r) \tag{3.56}$$

By definition, the reduced pressure is the ratio between the actual pressure of the gas, p, and the critical pressure of the gas, p_c. Likewise, the reduced temperature is the ratio between the actual temperature and the critical temperature of the gas, namely,

$$P_r = \frac{p}{p_c} \quad \text{and} \quad T_r = \frac{T}{T_c} \tag{3.57}$$

Note that the **critical temperature** is the temperature beyond which a gas cannot be compressed into a liquid, and the **critical pressure** is the minimum pressure required to compress a gas into liquid at the critical temperature.

Many equations have been proposed to relate z to P_r and T_r. One of the most useful is by Redlich and Kwong [2]. It yields the curves given in Figure 3.5. Note that the curves do not yield accurate values of z when the critical condition is approached. The values of z for various pure gases under critical condition ($p = p_c$ and $T = T_c$) are listed in Table 3.1. They range from 0.230 for steam to 0.304 for hydrogen. Under critical conditions, the z value for the gases represented by Figure 3.5 is 0.28. When treating gases near the critical conditions, one must make sure that the gases have not changed to the liquid phase. Otherwise, the material is a liquid, and it is not governed by either Equation 3.55 or 3.14.

For multi-component gases such as air or natural gas, the foregoing approach to determine the compressibility factor z is still valid if the values of p_c and T_c used are the molal averages determined from the molal composition of the gases as follows:

$$\bar{P}_c = \sum_{i=1}^{N} y_i P_{ci} \tag{3.58}$$

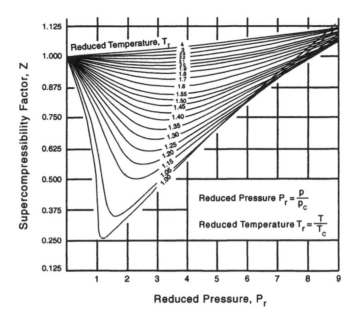

FIGURE 3.5 Supercompressibility factor, z, as a function of reduced pressure, P_r, and reduced temperature, T_r.

TABLE 3.1
Molecular Weight and Critical Properties of Certain Gases

Gas	Molecular Weight, w (lbm/mole)	Critical Temperature, T_c (°R)	Critical Pressure, p_c (psia)	Compressibility Factor, z_c
Methane	16.04	343.3	673.1	0.290
Ethane	30.07	549.8	708.3	0.285
Propane	44.09	666.0	617.4	0.277
Isobutane	58.12	734.7	529.1	0.283
n-Butane	58.12	765.3	550.7	0.274
Hydrogen	2.02	60	189	0.304
n-Pentane	72.15	845.6	489.5	0.269
Air	29.0	238.4	547	—
Nitrogen	28.02	226.9	492	0.291
Oxygen	32.0	277.9	730	0.292
Carbon dioxide	44.01	547.7	1073	0.274

$$\bar{T}_c = \sum_{i=1}^{N} y_i T_{ci} \tag{3.59}$$

where y_i is the mole fraction of component i in the gas, and p_{ci} and T_{ci} are, respectively, the critical pressure and the critical temperature of component i. N is the number of different gases in the multi-component gas under study.

Note that in the literature the molal average critical pressure $\bar{p}_c$, and the molal average critical temperature, $\bar{T}_c$, are called the **pseudocritical pressure** and **pseudocritical temperature,** respectively.

Example 3.4 Calculate the pseudocritical pressure and the pseudocritical temperature of air, and then determine the density of air at 100 atmospheres and 80°F.

[**Solution**] For practical purposes, air can be regarded as a two-component gas mixture with 25% oxygen and 75% nitrogen, in terms of mole fraction. Thus, $y_1 = 0.25$ and $y_2 = 0.75$. The critical pressure and critical temperature of the two components are (see Table 3.1) $p_{c1} = 730$ psia, $T_{c1} = 278°R$ for oxygen and $p_{c2} = 492$ psia, $T_{c2} = 227°R$ for nitrogen.

From Equations 3.58 and 3.59,

$$\bar{p}_c = y_1 p_{c1} + y_2 p_{c2} = 0.25 \times 730 + 0.75 \times 492 = 182.5 + 369.0 = 552$$

$$\bar{T}_c = y_1 T_{c_1} + y_2 T_{c_2} = 0.25 \times 278 + 0.75 \times 227 = 69.5 + 170.3 = 240$$

In this example, the air is at a pressure of 100 atmospheres (1470 psia) and a temperature of 80°F (540°R). Therefore, Equation 3.55 yields

$$P_r = \frac{p}{\bar{p}_c} = \frac{1470}{552} = 2.66$$

$$T_r = \frac{T}{\bar{T}_c} = \frac{540}{240} = 2.25$$

Using these values of P_r and T_r in Figure 3.6 yields $z = 0.97$. Therefore, from Equation 3.55,

$$\rho = \frac{p}{zRT} = \frac{1470 \times 144}{0.97 \times 1715 \times 540} = 0.236 \text{ slugs/ft}^3$$

If the air were assumed to be a perfect gas, Equation 3.14 would yield $\rho = 0.229$ slugs/ft³, which is different from that calculated above by only 3%. Note

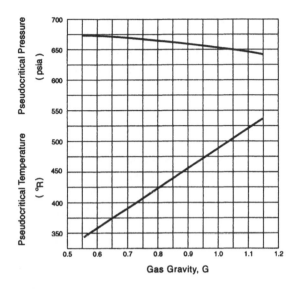

FIGURE 3.6 Pseudocritical properties of a typical natural gas.

that a much larger error would result if the temperature of the air were closer to the critical temperature. For instance, if in the foregoing example the air temperature is cooled to −200°F (260°R) and the other conditions remain the same, then $T_r = 260/240 = 1.08$, and from Figure 3.5, $z = 0.392$. This yields a density 61% greater than that predicted from using $z = 1.0$. This shows the importance of using the correct value of z when the temperature of a gas approaches the critical temperature.

3.2.2 GAS GRAVITY

By definition, the **gas gravity**, G, of any gas is the molecular weight of the gas divided by the molecular weight of air. From this definition and from Equation 3.14a and b, the engineering gas constant R for any gas having a gas gravity G and molecular weight m is

$$R = nB = \frac{1}{m}B = \frac{B}{29G} = \frac{53.3}{G}\frac{\text{lb-ft}}{\text{lbm °R}} = \frac{1715}{G}\frac{\text{lb-ft}}{\text{slug °R}} \qquad (3.14c)$$

The gas gravity reflects the composition of a gas mixture, and it can be calculated once the composition of the gas mixture is known (see Example 3.5). For instance, natural gas is a mixture containing mostly methane (CH_4) and several other less dominant components such as ethane (C_2H_6), propane (C_3H_8), butane (C_4H_{10}), and carbon dioxide (CO_2). For different natural gases produced from the same gas field or for natural gases having less than 3% of nonhydrocarbons, the pseudocritical temperature and the pseudocritical pressure can be predicted from G alone. This is illustrated in Figure 3.6.

Example 3.5 A natural gas has 90% methane, 5% ethane, 3% propane, 1% n-butane and 1% carbon dioxide — by number of molecules or mole fraction. The temperature is 100° F and the pressure is 1000 psia. Find the pseudocritical temperature, pseudocritical pressure, compressibility factor, and gas gravity of the gas.

[**Solution**] The following table is constructed for the calculations:

Gas Constituent	Mole Fraction, y	Molecular Weight, w	Fractional Molecular Weight, yw	Critical Temperature (°R)		Critical Pressure (psi)	
				T_c	yT_c	p_c	yp_c
Methane	0.90	16	14.4	343	308.7	673	605.7
Ethane	0.05	30	1.5	550	27.5	708	35.4
Propane	0.03	44	1.32	666	20.0	617	18.5
n-Butane	0.01	58	0.58	765	7.7	551	5.5
CO_2	0.01	44	0.44	548	5.5	1073	10.7
	1.00		18.24		$\bar{T}_c = 369.4$		$\bar{p}_c = 675.8$

From the above table, we have $T_r = T / \bar{T}_c = 560/369.4 = 1.52$, $P_r = p / \bar{p}_c = 1000/675.8 = 1.48$, and $G = 18.24/29 = 0.629$. Then, from Figure 3.5, $z = 0.875$. An alternative way to determine the pseudocritical properties of the gas mixture and z is to use the gas gravity $G = 0.629$. From Figure 3.6, $T_c = 366$ and $P_c = 670$. They yield $T_r = 1.53$, $P_r = 1.49$ and $z = 0.875$, approximately.

The above example shows that there are two ways to determine $\bar{T}_c$, $\bar{p}_c$, and z for any gas mixture — from the gas constituents of the mixture, and from the gas gravity, G. Both yield approximately the same results. Further discussion of the compressibility effect of natural gas and other gas mixtures can be found in Reference 3.

3.2.3 VISCOSITY OF GAS MIXTURE

At relatively low pressure (i.e., at any pressure not much higher than one atmosphere or 14.7 psia), the dynamic viscosity of a gas such as oxygen is a function of the temperature and not a function of the pressure of the gas. Therefore, the dynamic viscosity of the gas at relatively low pressure can be assumed to be the same as that of the gas at standard atmospheric pressure (14.7 psia), as given for various gases in Tables C.5 to C.7 in Appendix C. At relatively low pressure, the dynamic viscosity, μ_m, of a gas mixture can be calculated from the dynamic viscosity of each (ith) component, μ_i, by using the following equation:

$$\mu_m = \frac{\sum_{i=1}^{N} \mu_i y_i \sqrt{w_i}}{\sum_{i=1}^{N} y_i \sqrt{w_i}} \tag{3.60}$$

where y_i is the mole fraction of component i, and w_i is the molecular weight of component i.

Instead of using Equation 3.60, the dynamic viscosity of any natural gas at one atmospheric pressure can also be determined from Figure 3.7 if the gas gravity or the molecular weight of the mixture gas is known, and if the mole fraction of the nonhydrocarbon gases such as H_2S, N_2, and CO_2 are known [3,4].

When the pressure of a gas is much higher than the atmospheric pressure, the gas can no longer be considered an ideal gas, and the dynamic viscosity varies not only with temperature but also with pressure. In such a situation, the viscosity of the gas, μ, can be calculated from

$$\frac{\mu}{\mu_1} = f(T_r, P_r) = f(G) \tag{3.61}$$

where μ_1 is the dynamic viscosity of the gas at 1 atmospheric pressure. For any gas mixture, μ_1 is the same as the μ_m calculated from Equation 3.60. Equation 3.61 holds not only for pure nonpolar gases but also for gas mixtures such as natural gases, provided that the pseudocritical properties of the mixture are used, such as using μ_m

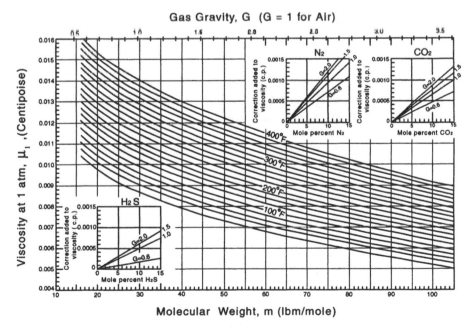

FIGURE 3.7 Dynamic viscosity of gases at atmospheric pressure as a function of gas gravity, temperature, etc. (From Carr, N.L., Kobayashi, R., and Burrows, C.B., Viscosity of hydrocarbon gases under pressure, *Transactions AIME*, 201, 264, 1954. With permission from AIME.)

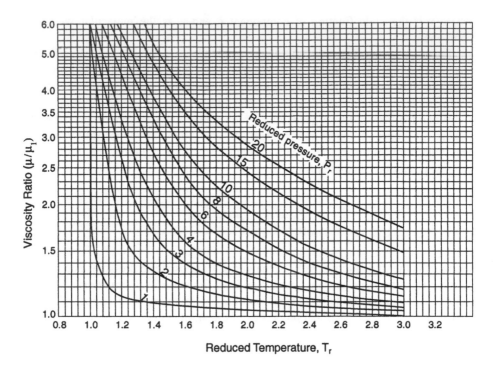

Reduced Temperature, T_r

FIGURE 3.8 Viscosity ratio μ/μ_1 as a function of reduced temperature T/T_c. (From Carr, N.L., Kobayashi, R., and Burrows, C.B., Viscosity of hydrocarbon gases under pressure, *Transactions AIME*, 201, 264, 1954. With permission from AIME.)

for μ_1, and using $\overline{T}_c$ and $\overline{P}_c$ instead of T_r and P_r. The functional form of Equation 3.61 is given graphically in Figure 3.8 [3,4].

Example 3.6 Find the viscosity of a natural gas at 195°F and 1815 psia. The gas gravity is 0.702.

[Solution] From Figure 3.6, $\overline{T}_c = 391°$ and $\overline{P}_c = 667$ psia when $G = 0.702$. Therefore,

$$T_r = \frac{195+460}{391} = 1.68 \quad \text{and} \quad P_r = \frac{1815}{667} = 2.72$$

From Figure 3.8, $\mu/\mu_1 = 1.30$. But, from Figure 3.7, $\mu_1 = 0.0122$ centipoise. Therefore, $\mu = 1.30 \times 0.0122 = 0.0159$ centipoise.

3.2.4 FLOW EQUATIONS

For gases under high pressure and constant temperature along a pipe, the counterpart of Equation 3.27 is

$$\frac{fL}{2D} = \frac{z_1^2 p_c^2}{p_1^2 k M_1^2} \int_2^1 \frac{P_r dP_r}{z} + \ln\left(\frac{z_1 p_2}{z_2 p_1}\right) \tag{3.62}$$

where the pressure p has been changed to the reduced pressure P_r and the critical pressure p_c to facilitate the evaluation of the integral in the equation. Because z is a function of P_r and T_r, the integral can be found numerically from

$$\int_2^1 \frac{P_r dP_r}{z} = \int_0^{P_{r1}} \frac{P_r dP_r}{z} - \int_0^{P_{r2}} \frac{P_r dP_r}{z} \tag{3.63}$$

where each of the two integrals on the right of Equation 3.63 can be obtained from Table 3.2.

Equation 3.62 can be used to calculate the length of a pipe, L, when D, T, p_1, p_2, V_1, and the gas composition are known, and to calculate the downstream pressure p_2 when L, D, T, p_1, V_1, and the gas composition are known.

Example 3.7 A natural gas that consists of 75% methane, 21% ethane, and 4% propane flows through a 100-mile-long steel pipe of 13.375-inch inner diameter. The temperature of the gas in the pipe is constant at 40°F. The pressure at upstream and downstream ends of the pipe are 1300 and 300 psia, respectively. Find the flow rate through the pipe. Due to the large change of pressure encountered, do not treat the gas as a perfect gas.

[Solution] The pseudocritical properties of the gas are first analyzed in the following table:

Gas Component	Mole Fraction, y	Molecular Weight, w	Fractional Molecular Weight, yw	Critical Temperature (°R) T_c	yT_c	Critical Pressure (psi) P_c	yP_c
Methane	0.75	16	12	343	257	673	505
Ethane	0.21	30	6.3	550	115	708	148
Propane	0.04	44	1.76	666	27	617	25
	1.00		20.1		$\bar{T}_c = 399$		$\bar{P}_c = 678$

From the foregoing calculations, $\bar{P}_c = 678$ psi, and $\bar{T}_c = 399°$R. But, $p_1 = 1300$ psia, $p_2 = 300$ psia, and $T = 40° + 460° = 500°$R. Therefore, $P_{r1} = p_1 / \bar{P}_c = 1.917$, $P_{r2} = p_2 / \bar{P}_c = 0.442$, and $T_r = T / \bar{T}_c = 1.253$. Furthermore, $G = 20.1/29 = 0.693$, and $R = 1715/G = 2475$. From Figure 3.5, $z_1 = 0.695$ and $z_2 = 0.935$. Assuming that $f = 0.014$ and $k = 1.31$ (same as for methane only), Equation 3.62 yields

$$3316 = \frac{0.1003}{M_1^2} \int_2^1 \frac{P_r dP_r}{z} - 1.763 \tag{a}$$

TABLE 3.2

Values of $100 \int_0^{P_r} \dfrac{P_r}{z}\, dP_r$

Pseudo-Reduced Pressure, P_r	Pseudo-Reduced Temperature, T_r														
	1.10	1.20	1.30	1.40	1.50	1.60	1.70	1.80	1.90	2.00	2.20	2.40	2.60	2.80	3.00
0.10	0	0	0	0	0	0	0	0	0	0	0	0	0	0	0
0.20	1	1	1	1	1	1	1	1	1	1	1	1	1	1	1
0.30	4	4	4	4	4	4	4	4	4	4	4	4	4	4	4
0.40	8	8	7	7	7	7	7	7	7	7	7	7	7	7	7
0.50	13	13	12	12	12	12	12	12	12	12	12	12	12	12	12
0.60	19	19	18	18	18	18	17	17	17	17	17	17	17	17	17
0.70	27	26	25	25	25	24	24	24	24	24	24	23	23	23	23
0.80	37	35	34	33	33	32	32	32	31	31	31	31	31	31	31
0.90	48	45	43	42	42	41	41	40	40	40	40	39	39	39	39
1.00	62	57	55	53	52	51	51	50	50	50	49	49	49	49	48
1.10	78	71	67	65	64	63	62	61	61	60	60	59	59	59	59
1.20	97	86	81	78	77	75	74	73	73	72	71	71	70	70	70
1.30	120	104	97	93	91	89	87	86	86	85	84	84	83	83	82
1.40	146	123	114	109	106	104	102	101	100	99	98	97	96	96	96
1.50	178	145	133	127	123	120	117	116	115	114	113	112	111	111	110
1.60	216	169	153	146	141	137	134	132	131	130	128	127	126	126	125
1.70	259	195	176	166	160	156	152	150	149	147	145	144	143	142	141
1.80	305	224	200	188	181	176	172	169	167	165	163	162	160	160	159
1.90	355	256	226	212	203	197	192	189	187	185	182	180	179	178	177
2.00	408	290	254	237	227	219	214	210	208	205	202	200	198	197	196
2.10	463	328	284	264	251	243	237	232	229	227	223	221	219	218	216

2.20	237	239	240	243	245	250	252	256	261	268	278	292	316	368	521
2.30	259	261	263	266	269	273	277	281	286	295	306	323	350	411	580
2.40	282	284	286	290	293	298	302	306	313	322	335	354	387	456	641
2.50	306	308	311	315	318	324	328	333	341	351	365	388	425	503	704
2.60	331	334	336	340	345	351	356	361	370	382	397	423	465	552	768
2.70	357	360	363	367	372	379	384	391	400	413	431	460	507	602	833
2.80	384	387	390	395	401	408	414	421	432	446	466	498	550	655	899
2.90	412	415	419	424	430	439	445	453	454	480	502	538	596	709	966
3.00	440	444	448	454	461	470	477	486	438	516	540	580	643	764	1034
3.10	470	474	479	485	492	502	510	519	533	553	579	623	692	821	1103
3.20	501	505	510	517	525	536	544	555	570	591	620	667	742	879	1171
3.30	532	537	543	550	559	571	580	591	608	631	662	713	794	938	1241
3.40	565	570	576	584	593	606	616	628	647	671	705	761	847	998	1310
3.50	598	604	611	619	629	643	654	667	657	714	750	810	901	1059	1380
3.60	632	639	646	655	666	681	692	707	728	757	796	860	957	1122	1450
3.70	668	674	682	692	704	720	732	747	770	802	843	912	1014	1185	1520

or,

$$M_1^2 = 3.02 \times 10^{-5} \int_2^1 \frac{P_r dP_r}{z} \qquad \text{(b)}$$

From Equation 3.63 and Table 3.2,

$$\int_2^1 \frac{P_r dP_r}{z} = 2.44 - 0.1 = 2.34 \qquad \text{(c)}$$

Substituting c into b yields $M_1^2 = 7.07 \times 10^{-5}$, and $M_1 = 0.00841$.

The density of the gas upstream is $\rho_1 = p_1/(z_1 RT) = 0.218$ slug/ft^3. Therefore, $V_1 = M_1\sqrt{kRT} = 10.7$ fps. Finally, the cross-sectional area of the pipe is $A = 0.976$ ft^2, and the mass flow rate is $\dot{m} = \rho_1 V_1 A = 2.28$ slug/s. The volumetric flow rate (discharge) upstream in the pipe is $Q_1 = V_1 A = 10.4$ cfs. This is the discharge of the gas under the high pressure of 1300 psia upstream. The discharge of this gas under standard atmospheric condition would be much higher due to the expansion of gas following pressure drop. It can be calculated by using the equation of state.

Note that the above analysis is incomplete because the f value of 0.014 was assumed. Now that the density and the velocity of the gas based on this assumed value of f have been found, the Reynolds number can be calculated and used to obtain more accurate values of f, M_1, V_1, and Q_1. This shows how Equation 3.62 can be used to calculate the flow rate and the velocity of super-compressed gas flowing through pipe. The reader should complete the problem as an exercise.

3.2.5 APPROXIMATE FLOW EQUATIONS

The more exact equations given in the previous section for real (nonideal) gas flow are often cumbersome to use. Consequently, simplifying assumptions are often introduced to reduce the foregoing equations to simpler forms, which are easier to use. This is discussed next.

If it is assumed that the velocity of the flow is relatively small, the first term in Equation 3.13 becomes much smaller than the other two terms, and thus it can be dropped from the equation. In addition, if the flow is isothermal and real gas, it can be proved that Equation 3.13 reduces to the following equation:

$$Q = 32.5 \frac{T_o}{P_o} \sqrt{\frac{(p_1^2 - p_2^2)D^5}{GTLfz_a}} \qquad \text{(3.64)}$$

where Q is the volumetric discharge of the gas in cfs under standard atmospheric condition; T_o is the temperature under standard condition ($T_o = 60°F = 520°R$); p_o is the pressure of the standard atmosphere ($p = 14.7$ psia $= 2117$ psfa); p_1 and p_2 are the gas flow pressure in psfa, at points 1 and 2, respectively; D is the pipe diameter in ft; G is gas gravity; T is the temperature of the gas in the pipe (same as the ambient temperature) in °R; L is the pipe length in ft; f is the Darcy-Weisbach friction factor; and z_a is the average value of the compressibility factor of the gas, namely, $z_a = (z_1 + z_2)/2$. Because the equation is homogeneous in units, it can also be used in SI units.

Assuming that $f = 0.032/D^{1/3}$, and changing the units of the quantities in Equation 3.64 (D to inches, L in miles, p in psia, and Q in ft³/hr), the equation can be rewritten as

$$Q = 18.06 \frac{T_o}{p_o} \sqrt{\frac{(p_1^2 - p_2^2)D^{16/3}}{GTLz_a}} \tag{3.65}$$

which is known as the **Weymouth equation**.

On the other hand, if it is assumed that $1/f = 52(GQ/D)^{0.1461}$, Equation 3.64 reduces to:

$$Q = 436E\left(\frac{T_o}{p_o}\right)^{1.079}\left(\frac{p_1^2 - p_2^2}{L}\right)^{0.539}\left(\frac{1}{T}\right)^{0.539}\left(\frac{1}{G}\right)^{0.461}D^{2.62} \tag{3.66}$$

where E is an **efficiency factor** equal to 0.92 approximately; Q is in ft³/day; and the other quantities have the same units as Equation 3.65. Equation 3.66 is a form of the Panhandle equation, developed from data of large natural gas pipelines [5]. As before, the equations given in this section are applicable to all gases, including natural gas. A discussion of these and other formulas for computing the pressure drop of hydrocarbon gases and liquids are given in Reference 5.

3.3 WORK, ENERGY, AND POWER REQUIRED FOR COMPRESSION OF GAS

3.3.1 GENERAL RELATIONSHIPS

The reversible work required to compress a unit mass of gas by a compressor from pressure p_1 to pressure p_2 is

$$w = \int_1^2 \frac{dp}{\rho} \tag{3.67}$$

where the limits of the integral, 1 and 2, represent the intake and discharge sides, respectively, of the compressor.

The energy required to compress a gas of unit mass is the same as the work, w, calculated from Equation 3.67, divided by the efficiency of the compressor, η_c. Therefore, the power required to compress a gas of mass flow rate $\dot{m}$ is

$$P = \dot{m}w / \eta_c \tag{3.68}$$

In English units, P is the power in **ft-lb/s**; $\dot{m}$ is the mass flow rate in **slugs/s**; and w is the work per unit mass in **ft-lb/slug**. To convert power to horsepower and kilowatts, one needs to know that one horsepower is equal to 550 ft-lb/s, and one kilowatt is equal to 737 ft-lb/s. In SI units, P is in N-m/s, which is the same as watt used by electrical engineers; $\dot{m}$ is in kg/s; and w is in N-m/kg.

3.3.2 ISOTHERMAL COMPRESSION OF IDEAL GAS

Under relatively low pressure, the gas can be regarded as an ideal gas. Substituting the equation of ideal gas into Equation 3.67 and assuming an isothermal process yields

$$w = RT \int_1^2 \frac{dp}{p} = RT \ln \frac{p_2}{p_1} = \frac{1715T}{G} \ln \frac{p_2}{p_1} \tag{3.69}$$

Note that the constant 1715 is for English units only, and when w is the work in **ft-lb** done on each unit mass in **slug**. The constant should be changed to 53.2 if **lbm** is used for mass.

3.3.3 ISOTHERMAL COMPRESSION OF REAL GAS

When the gas is under high pressure, the equation of state of real gas, with the factor z, must be used. Substituting Equation 3.55 into Equation 3.67 yields

$$w = RT \int_1^2 \frac{z\,dp}{p} = \frac{1715}{G} T \left[\int_{0.2}^{P_{r2}} \frac{z\,dp_r}{p_r} - \int_{0.2}^{P_{r1}} \frac{z\,dp_r}{p_r} \right] \tag{3.70}$$

Note that the lower limits of the integrals in the above equation have been chosen to be 0.2, because normally p_r is greater than 0.2 in the range of practical interest. The use of 0.2 instead of 0 (zero) reduces the size of the table needed for calculating the integral. Difference in the two integrals in Equation 3.70 does not depend on the value of the lower limits. Table 3.3 gives the values of the integrals.

3.3.4 ISENTROPIC COMPRESSION OF IDEAL GAS

If the compression is adiabatic rather than isothermal, and if the gas can be regarded as an ideal gas, then the equation for isentropic gas, Equation 3.48, holds. Substituting it into Equation 3.67 yields

$$w = \frac{p_1^{1/k}}{\rho_1} \int_1^2 \frac{dp}{p^{1/k}} = \frac{k}{(k-1)} \frac{p_1^{1/k}}{\rho_1} \left[p_2^{(k-1)/k} - p_1^{(k-1)/k} \right]$$

$$= \frac{k}{(k-1)} RT_1 \left[\left(\frac{p_2}{p_1} \right)^{\frac{(k-1)}{k}} - 1 \right] \tag{3.71}$$

$$= \frac{k}{(k-1)} \frac{1715}{G} T_1 \left[\left(\frac{p_2}{p_1} \right)^{\frac{(k-1)}{k}} - 1 \right]$$

TABLE 3.3

Values of $\int_o^{P_r} \frac{z}{P_r} dP_r$

Pseudo-Reduced Temperature, T_r

Pseudo-Reduced Pressure, P_r	1.10	1.20	1.30	1.40	1.50	1.60	1.70	1.80	1.90	2.00	2.20	2.40	2.60	2.80	3.00
0.2	0	0	0	0	0	0	0	0	0	0	0	0	0	0	0
0.3	0.350	0.350	0.350	0.350	0.350	0.350	0.350	0.350	0.350	0.350	0.350	0.350	0.350	0.350	0.350
0.4	0.619	0.626	0.630	0.633	0.635	0.636	0.637	0.638	0.639	0.639	0.640	0.640	0.640	0.640	0.640
0.5	0.816	0.834	0.844	0.851	0.856	0.860	0.852	0.864	0.866	0.867	0.868	0.869	0.869	0.869	0.869
0.6	0.971	0.998	1.022	1.040	1.048	1.049	1.049	1.050	1.050	1.050	1.051	1.051	1.052	1.052	1.052
0.7	1.100	1.145	1.178	1.199	1.207	1.210	1.211	1.213	1.214	1.216	1.218	1.219	1.220	1.220	1.220
0.8	1.207	1.264	1.300	1.322	1.340	1.347	1.352	1.357	1.359	1.360	1.363	1.364	1.364	1.364	1.364
0.9	1.300	1.365	1.403	1.429	1.450	1.462	1.472	1.480	1.485	1.489	1.492	1.494	1.495	1.495	1.495
1.0	1.375	1.455	1.500	1.530	1.551	1.568	1.530	1.590	1.598	1.602	1.607	1.608	1.609	1.610	1.610
1.1	1.438	1.528	1.573	1.606	1.631	1.653	1.657	1.676	1.684	1.691	1.699	1.702	1.706	1.709	1.711
1.2	1.500	1.600	1.645	1.682	1.710	1.737	1.753	1.761	1.770	1.780	1.790	1.795	1.802	1.808	1.812
1.3	1.545	1.057	1.709	1.746	1.779	1.810	1.828	1.836	1.845	1.858	1.868	1.875	1.883	1.890	1.896
1.4	1.590	1.713	1.772	1.810	1.847	1.882	1.903	1.911	1.920	1.935	1.945	1.954	1.964	1.972	1.980
1.5	1.620	1.757	1.824	1.867	1.906	1.938	1.962	1.973	1.984	1.997	2.010	2.019	2.027	2.036	2.045
1.6	1.649	1.800	1.875	1.923	1.964	1.993	2.021	2.035	2.047	2.059	2.074	2.083	2.090	2.100	2.110
1.7	1.670	1.834	1.917	1.969	2.012	2.043	2.072	2.089	2.102	2.116	2.131	2.141	2.148	2.159	2.169
1.8	1.690	1.867	1.958	2.014	2.060	2.093	2.123	2.142	2.157	2.172	2.188	2.198	2.205	2.217	2.227
1.9	1.708	1.896	1.993	2.054	2.100	2.136	2.165	2.187	2.204	2.219	2.237	2.247	2.256	2.267	2.279
2.0	1.725	1.924	2.027	2.093	2.140	2.178	2.207	2.231	2.250	2.265	2.285	2.295	2.307	2.317	2.330
2.1	1.743	1.947	2.057	2.126	2.176	2.215	2.248	2.272	2.292	2.307	2.326	2.337	2.350	2.361	2.375
2.2	1.761	1.971	2.086	2.160	2.212	2.252	2.288	2.313	2.334	2.349	2.366	2.380	2.394	2.404	2.420

TABLE 3.4 *(Continued)*

Values of $\int_o^{P_r} \frac{z}{P_r} dP_r$

Pseudo-Reduced Pressure, P_r	Pseudo-Reduced Temperature, T_r																		
	1.10	1.20	1.30	1.40	1.50	1.60	1.70	1.80	1.90	2.00	2.20	2.40	2.60	2.80	3.00
2.3	1.779	1.994	2.116	2.193	2.249	2.288	2.329	2.354	2.375	2.391	2.407	2.422	2.437	2.448	2.465
2.4	1.797	2.018	2.145	2.227	2.285	2.325	2.369	2.395	2.417	2.433	2.447	2.465	2.481	2.491	2.510
2.5	1.815	2.041	2.175	2.260	2.321	2.362	2.410	2.436	2.459	2.475	2.488	2.507	2.524	2.535	2.555
2.6	1.830	2.061	2.198	2.288	2.350	2.392	2.442	2.469	2.492	2.508	2.523	2.544	2.562	2.574	2.593
2.7	1.845	2.081	2.221	2.316	2.379	2.423	2.474	2.502	2.525	2.541	2.559	2.581	2.599	2.612	2.630
2.8	1.860	2.101	2.245	2.344	2.407	2.453	2.506	2.534	2.557	2.575	2.594	2.617	2.637	2.651	2.568
2.9	1.875	2.121	2.268	2.372	2.436	2.484	2.538	2.567	2.590	2.608	2.630	2.654	2.674	2.689	2.705
3.0	1.890	2.140	2.291	2.400	2.465	2.514	2.570	2.600	2.623	2.641	2.665	2.691	2.712	2.728	2.743
3.1	1.904	2.157	2.311	2.423	2.489	2.540	2.597	2.628	2.652	2.670	2.694	2.722	2.744	2.759	2.775
3.2	1.918	2.175	2.331	2.446	2.512	2.565	2.623	2.657	2.681	2.700	2.723	2.753	2.775	2.790	2.806
3.3	1.932	2.102	2.350	2.469	2.536	2.591	2.650	2.685	2.709	2.729	2.752	2.783	2.807	2.821	2.838
3.4	1.946	2.210	2.370	2.492	2.559	2.616	2.676	2.714	2.738	2.759	2.781	2.814	2.838	2.852	2.869
3.5	1.960	2.227	2.390	2.515	2.583	2.642	2.703	2.742	2.767	2.788	2.810	2.845	2.870	2.883	2.901
3.6	1.974	2.243	2.407	2.535	2.603	2.664	2.726	2.766	2.792	2.813	2.836	2.872	2.910	2.911	2.929
3.7	1.988	2.259	2.424	2.556	2.624	2.686	2.748	2.791	2.817	2.839	2.862	2.899	2.950	2.938	2.957
3.8	2.002	2.275	2.440	2.576	2.644	2.708	2.771	2.815	2.843	2.864	2.888	2.925	2.990	2.966	2.984
3.9	2.016	2.291	2.457	2.597	2.665	2.730	2.793	2.840	2.868	2.890	2.914	2.952	3.030	2.993	3.012
4.0	2.030	2.306	2.474	2.617	2.685	2.752	2.816	2.864	2.893	2.915	2.940	2.979	3.070	3.021	3.040

3.3.5 ISENTROPIC COMPRESSION OF REAL GAS

For real gas under high pressure or near critical temperature, the above equation changes to

$$w = \frac{k}{(k-1)} \frac{1715}{G} T_1 \left[\left(\frac{p_2}{p_1} \right)^{\frac{(k-1)}{k}} - 1 \right] (1+\lambda) \tag{3.72}$$

where

$$\lambda = \left(\frac{z_2}{z_1} - 1 \right) \left\{ \frac{k + r^{(k-1)/k} [r(k-1) - 2k + 1]}{(2k-1)(r-1) [r^{(k-1)/k} - 1]} \right\} \tag{3.73}$$

In the above two equations, z_2 and z_1 are the z values at the compressor discharge and intake, respectively, and $r = p_2/p_1$ is the **compression ratio**. Note that Equation 3.72 differs from the ideal gas case, Equation 3.71, only by the factor $(1 + \lambda)$. When z_1 equals z_2, λ becomes zero and $(1 + \lambda)$ becomes unity.

PROBLEMS

3.1 Prove that Equation 3.27 reduces to the Darcy-Weisbach equation for incompressible flow when the pressure difference $p_1 - p_2$ is small.

3.2 Derive Equation 3.28, which can be used to calculate p_1 from the downstream conditions p_2 and M_2.

3.3 Air is transported through a steel pipe of 6-inch diameter at a temperature of 68°F, a pressure of 200 psia, and a velocity of 200 ft/s at the compressor outlet, which is near the pipeline inlet. Find the pressure in the pipe at a distance of 500 ft downstream by assuming (a) incompressible flow and (b) compressible isothermal flow.

3.4 Work problem 3.3 for compressible adiabatic flow and compare your result with the previous solutions for incompressible flow and isothermal flow. Discuss results briefly.

3.5 Show how the limiting conditions in Figure 3.3a and 3.3b are determined.

3.6 Plot p_2/p_1 versus fL/D for air assuming (a) incompressible flow, (b) compressible isothermal flow, and (c) compressible adiabatic flow. Do it for Mach number $M_1 = 0.05, 0.08, 0.1, 0.12, 0.15, 0.2,$ and 0.3, and present all your results in a single graph to facilitate comparison. Use a computer to do your computation and attach your computer program and result to the graph.

3.7 A 4×1-inch Venturi is used in a 4-inch pipe to determine the airflow rate going through the pipe. The pressure and temperature of the air at the Venturi entrance are 120 psia and 60°F, respectively. The pressure at the throat of the Venturi is 100 psia. (a) Find the velocity of the air at the throat of the Venturi. (b) Find the velocity of the air in the pipe at the Venturi entrance. (c) Find the mass and weight flow rates of the air through the pipe. Assume that the discharge coefficient is 0.96. (d) Find the Mach number of the flow at the throat of the Venturi. (Note: The Mach number at the throat should not be greater than one. Otherwise, something must be wrong with the problem.) (e) How much error would one make in the determination of the flow rate if the flow is assumed to be incompressible?

3.8 Work problem 3.3b by using Equation 3.62. Compare with previous results.

3.9 In Example 3.7, the discharge of the gas at 1300 psia and 40°F was found to be $Q = 10.4$ cfs. Convert this discharge to that under standard atmospheric conditions.

3.10 For the same problem stated in Example 3.7, find the discharge Q by using the Weymouth formula and the Panhandle formula. Compare the Q values found from these two methods with that found in the previous problem, and discuss the results.

3.11 A compressor compresses air at 0°F and 14.7 psia to 2000 psia at a rate of 50 lbm/min. If the compressor efficiency is 90%, determine the power required assuming (a) isothermal ideal gas, (b) isentropic ideal gas, and (c) isentropic real gas. Compare the results of the three cases, and explain the differences.

REFERENCES

1. Olson, R.M. and Wright, S.J., *Essentials of Engineering Fluid Mechanics*, 5th ed., Harper & Row, New York, 1990, chap. 10.
2. Redlich, O. and Kwong, J.N.S., On the thermodynamics of solutions: V: an equation of state, fugacities of gaseous solution, *Chemical Reviews*, 44, 1949.
3. Katz, D.L., Ed., *Handbook of Natural Gas Engineering*, McGraw-Hill, New York, 1959.
4. Carr, N.L., Kobayashi, R., and Burrows, C.B., Viscosity of hydrocarbon gases under pressure, *Transactions AIME*, 201, 264, 1954.
5. Pipeline Design for Hydrocarbon Gases and Liquids, Report of the Task Committee on Engineering Practice in the Design of Pipelines, Pipeline Division, American Society of Civil Engineers (ASCE), 1975.

4 Non-Newtonian Fluids

4.1 INTRODUCTION

Fluids can be classified into two broad categories: **Newtonian** and **non-Newtonian**. A Newtonian fluid is a fluid that has a constant viscosity independent of the magnitude of the shear stress. For a parallel flow of a Newtonian fluid with velocity u varying in the y direction as shown in Figure 4.1, Newton's law of viscosity holds, as follows:

$$\tau = \mu \frac{du}{dy} \tag{4.1}$$

where y is the distance perpendicular to the wall and the flow, measured from the wall; u is the velocity of the fluid at y; τ is the shear stress at y; and μ is the viscosity (more specifically, the dynamic viscosity) of the fluid. When τ is plotted against du/dy, the result for a Newtonian fluid is a straight line going through the origin of the coordinates as shown in Figure 4.2. Note that graphs plotting τ versus du/dy are called *rheograms*. The quantity du/dy is often referred to as the **shear rate**, which is a misnomer. Physically, du/dy is the velocity gradient or the rate of angular deformation of the fluid.

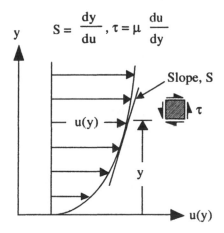

FIGURE 4.1 Definition of viscosity for Newtonian fluid.

115

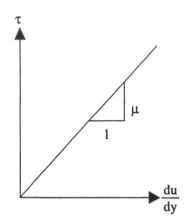

FIGURE 4.2 Rheogram of a Newtonian fluid.

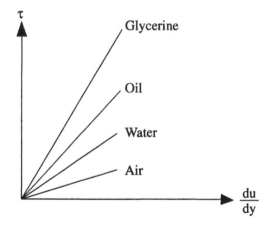

FIGURE 4.3 Rheogram of various Newtonian fluids.

The slope of the straight line in any rheogram of a Newtonian fluid represents the viscosity (or more specifically, the dynamic viscosity) of the fluid. The higher the viscosity of a fluid is, the steeper the slope in the rheogram becomes. Figure 4.3 gives the rheogram of several Newtonian fluids of quite different values of viscosity.

4.2 CLASSIFICATION OF NON-NEWTONIAN FLUIDS

For non-Newtonian fluids, the line in the rheogram is either curved, does not pass through the origin, or both. Figure 4.4 shows various types of non-Newtonian fluids. Note that while the rheogram of a pseudoplastic fluid curves downward (i.e., having decreasing slope with increased shear), for a dilatant fluid the rheogram curves upward (i.e., having increasing slope with increased shear). Otherwise, the two are similar: they both pass through the origin of the rheogram, as is the case with

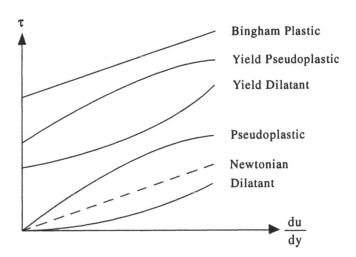

FIGURE 4.4 Rheogram of non-Newtonian fluids.

Newtonian fluids. As shown in Figure 4.4, a Bingham plastic (or simply *Bingham*) fluid is represented by a straight line in rheograms, but the line does not pass through the origin. It takes a certain minimum shear stress, called the **yield stress**, τ_y, to cause a Bingham fluid to behave like a fluid. For τ less than τ_y, a Bingham plastic fluid behaves like a solid rather than a fluid. When τ becomes greater than τ_y, it behaves like a Newtonian fluid. Finally, a yield-pseudoplastic fluid is similar to a pseudoplastic fluid except that it requires a minimum shear (yield stress) to behave like a fluid, and a yield-dilatant fluid is similar to a dilatant fluid but requires a yield stress.

Pure fluids such as water or air are Newtonian fluids. Non-Newtonian fluids are solutions or suspensions of particulates (i.e., large molecules or fine solid particles suspended in a pure fluid). Whether a non-Newtonian fluid is pseudoplastic, dilatant, or another type depends on not only the kind but also the concentration of the suspended particles. In some cases, at low concentration of particulates the fluid is pseudoplastic. It changes to Bingham plastic when the concentration is moderate, and then changes to dilatant when the concentration is high. At very low concentration of solids, all fluids behave like a Newtonian fluid, with increasing viscosity as the concentration of solids increases. The reason that a pseudoplastic fluid has a decreasing viscosity when the shear increases is believed to be a reversible breakdown of loosely bonded aggregates by the shearing action of the flow. Examples of pseudoplastic fluids include aqueous suspension of limestone, aqueous and nonaqueous suspension of certain polymers, hydrocarbon greases, etc.

The reason that a dilatant fluid has an increasing viscosity when shear increases is believed to be due to the shift, under shear, of a closely packed particulate system to a more open arrangement, which entraps some of the liquid. Examples are aqueous suspensions of magnetite, galena, and ferro-silicons. Examples of Bingham plastic fluids include water suspensions of clay, fly ash, sewage sludge, paint, and fine minerals such as coal slurry. The yield stress, τ_y, for a Bingham fluid may be very small (less than 0.1 dynes/cm^2 for some sewage sludge), or very large (more than 10^{10} dyne/cm^2 for some asphalts and bitumens).

Finally, some clay–water suspensions at intermediate level of concentration exhibit yield-pseudoplastic properties.

The aforementioned non-Newtonian fluids are time-independent. This means their viscosity under shear does not change with time. Some other non-Newtonian fluids, however, have time-dependent rheological properties. For instance, a **thixotropic fluid** is a pseudoplastic fluid whose viscosity under constant shear decreases with time. This is due to particle agglomeration. Water suspension of bentonitic clay — the drilling fluid used by the petroleum industry — is a thixotropic fluid. Crude oil at low temperature, such as the oil from the Pembina Field in Canada, is another example of thixotropic fluid. Another type of time-dependent non-Newtonian fluid is the **rheopectic fluid**. It exhibits negative thixotropic behavior (i.e., the viscosity of the fluid under shear increases with time).

4.3 RHEOLOGICAL PROPERTIES AND LAWS OF NON-NEWTONIAN FLUIDS

While Newtonian fluids have only one rheological property — the viscosity defined by Newton's law of viscosity (Equation 4.1) — non-Newtonian fluids often have two or three rheological properties, defined by the following laws.

4.3.1 POWER-LAW FLUIDS

The relationship between shear, τ, and the velocity gradient, du/dy, for certain non-Newtonian fluids can be expressed satisfactorily with the following power laws:

$$\tau = K\left(\frac{du}{dy}\right)^n \tag{4.2}$$

from which

$$\mu = K\left(\frac{du}{dy}\right)^{n-1} \tag{4.3}$$

Equation 4.2 is applicable to pseudoplastic fluids when $n < 1$, dilatant fluids when $n > 1$, and Newtonian fluids when $n = 1$. From Equation 4.2, the two rheological properties of pseudoplastic and dilatant fluids that can be represented by the equation are the coefficient K and the power n. The constant K is usually referred to as the **consistency index** or **power-law coefficient,** whereas the constant n is referred to as the **flow-behavior index,** or **power-law exponent.** The constant μ in Equation 4.3 is the **apparent viscosity,** which reduces to the dynamic viscosity when the fluid is Newtonian ($n = 1$).

4.3.2 BINGHAM FLUIDS

For any Bingham plastic fluid (or **Bingham fluids**, for short), the following law holds:

$$\tau = \tau_y + \mu \frac{du}{dy} \tag{4.4}$$

where τ_y is the **yield stress**; and μ is the **coefficient of rigidity,** or simply the **rigidity** of the fluid.

4.3.3 YIELD FLUIDS

For yield-pseudoplastic fluids and yield-dilatant fluids, the following law can be used:

$$\tau = \tau_y + K\left(\frac{du}{dy}\right)^n \tag{4.5}$$

which is a combination of Equations 4.2 and 4.4. The exponent n in Equation 4.5 is greater than one for yield-dilatant fluids, and less than one for yield-pseudoplastic fluids. When $n = 1$, Equation 4.5 reduces to Equation 4.4, which is for Bingham fluids.

4.3.4 OTHER NON-NEWTONIAN FLUIDS

There are many other laws proposed in the literature for various types of non-Newtonian fluids. They will not be discussed here. Readers interested in learning more about non-Newtonian fluids should see References 1 and 2.

4.4 NON-NEWTONIAN PIPE FLOW: LAMINAR

Consider a cylindrical element of fluid of length dx and radius r in a horizontal pipe of radius a as shown in Figure 4.5. Application of the momentum equation to the element yields

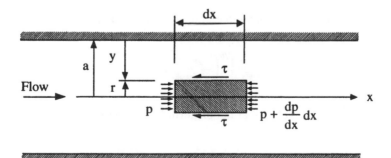

FIGURE 4.5 Force balance on a cylindrical fluid element of radius r and length dx.

$$\tau = -\frac{r}{2}\frac{dp}{dx} = \frac{r}{2}\frac{\Delta p}{L} \qquad (4.6)$$

where τ is the shear on the surface of the element; p is the pressure; x is the coordinate in the direction of the flow; dp/dx is the rate of increase in pressure in the flow direction, which is a negative quantity; L is the length of the pipe or a finite segment of it; and Δp is the decrease in pressure in the flow direction over the distance L.

From Equation 4.6, the shear stress on the pipe wall, τ_o, is

$$\tau_o = \frac{a}{2}\frac{\Delta p}{L} \qquad (4.6a)$$

Equations 4.6 and 4.6a are rather general — applicable to Newtonian as well as non-Newtonian fluids, for either laminar or turbulent flows. Further solution of Equation 4.6 requires a knowledge of the relationship between τ and the velocity gradient du/dy, which depends on the type of non-Newtonian fluid and the rheological law used. Two important cases are to be considered: power-law fluid and Bingham plastic fluid. They are discussed next.

4.4.1 POWER-LAW FLUIDS

Because $y = (a - r)$ and $dy = -dr$, the power law given by Equation 4.2 can be rewritten as

$$\tau = K\left(-\frac{du}{dr}\right)^n \qquad (4.7)$$

Combination of Equations 4.6. and 4.7 yields

$$-\frac{du}{dr} = \left(\frac{r}{2K}\frac{\Delta p}{L}\right)^{\frac{1}{n}} \qquad (4.8)$$

Integrating Equation 4.8 and using the boundary condition $u = 0$ at $r = a$ yields

$$u = \left(\frac{\Delta p}{2KL}\right)^{1/n}\left(\frac{n}{n+1}\right)\left(a^{\frac{n+1}{n}} - r^{\frac{n+1}{n}}\right) \qquad (4.9)$$

From Equation 4.9, the velocity at the centerline of the pipe is

$$V_c = \left(\frac{\Delta p}{2KL}\right)^{\frac{1}{n}}\left(\frac{n}{n+1}\right)a^{\frac{n+1}{n}} \qquad (4.10)$$

The discharge in the pipe is

$$Q = 2\pi \int_0^a urdr \tag{4.11}$$

Substituting Equation 4.9 into Equation 4.11 and integrating yields

$$Q = \left(\frac{n\pi}{3n+1}\right)\left(\frac{\Delta p}{2KL}\right)^{\frac{1}{n}} a^{\frac{3n+1}{n}} \tag{4.12}$$

The mean velocity in the pipe is

$$V = \frac{Q}{\pi a^2} = \left(\frac{n}{3n+1}\right)\left(\frac{\Delta p}{2KL}\right)^{\frac{1}{n}} a^{\frac{n+1}{n}} \tag{4.13}$$

From Equation 4.10 and 4.13,

$$\frac{V_c}{V} = \frac{3n+1}{n+1} \tag{4.14}$$

Next from Equation 4.13,

$$\frac{\Delta p}{L} = \frac{2K(3n+1)^n}{n^n} \cdot \frac{V^n}{a^{n+1}} \tag{4.15}$$

If the same Darcy-Weisbach equation for Newtonian fluids is used herein for the non-Newtonian fluid under discussion,

$$f = \frac{4a\Delta p}{L\rho V^2} \tag{4.16}$$

Substituting Equation 4.15 into Equation 4.16 yields

$$f = \frac{8}{\mathfrak{R}_1}\left(\frac{6n+2}{n}\right)^n \tag{4.17}$$

where

$$\mathfrak{R}_1 = \frac{\rho D^n V^{2-n}}{K} \tag{4.18}$$

Note that $\mathfrak{R}_1$ is a generalized Reynolds number for the power-law non-Newtonian fluid. Another Reynolds number, $\mathfrak{R}_2$, may be defined as

$$\mathfrak{R}_2 = 8\left(\frac{n}{6n+2}\right)^n \mathfrak{R}_1 \qquad (4.19)$$

Substituting Equation 4.19 into Equation 4.17 yields

$$f = \frac{64}{\mathfrak{R}_2} \qquad (4.20)$$

which is a relation analogous to that for a Newtonian fluid of laminar flow.

From the foregoing, the relationship between the resistance factor f and the Reynolds number $\mathfrak{R}_2$ for any power-law fluids is the same as that between f and $\mathfrak{R} = \rho DV/\mu$ for Newtonian fluid, as long as the flow is laminar. This means if we plot f versus $\mathfrak{R}_2$, the result in the laminar regime will be identical to that shown in an ordinary Moody diagram between f and $\mathfrak{R}$. One may also expect that transition from laminar to turbulent flow will take place at a value of $\mathfrak{R}_2$ that approaches 2100 approximately when $n = 1$, which is the case for Newtonian fluids based on $\mathfrak{R}$. This will be discussed later.

All the foregoing equations for non-Newtonian, power-law fluids reduce to those for Newtonian fluids when $n = 1$. For instance, when $n = 1$, Equation 4.9 becomes

$$u = \frac{\Delta p}{4\mu L}(a^2 - r^2) \qquad (4.21)$$

Likewise, Equations 4.10, 4.13, 4.14, 4.15, and 4.20 become, respectively,

$$V_c = \frac{\Delta p}{4\mu L}a^2 \qquad (4.22)$$

$$V = \frac{\Delta p}{8\mu L}a^2 \qquad (4.23)$$

$$\frac{V_c}{V} = 2 \qquad (4.24)$$

$$\frac{\Delta p}{L} = \frac{8\mu V}{a^2} \qquad (4.25)$$

$$f = \frac{64}{\mathfrak{R}} \qquad (4.26)$$

Based on a stability analysis made by Ryan and Johnson [3], it can be shown that transition from laminar to turbulent flow occurs at or above a critical value of $\Re_2$ approximately equal to

$$(\Re_2)_t = \frac{6464n}{(1+3n)^2 \left(\dfrac{1}{2+n}\right)^{\frac{2+n}{1+n}}} \tag{4.27}$$

where the subscript t indicates transition value. The above equation is consistent with the transition Reynolds number $\Re_t = 2100$ for Newtonian fluids ($n = 1.0$).

From Equation 4.27 and Equation 4.20, the value of f at transition from laminar to turbulent is

$$f_t = \frac{(1+3n)^2}{101n} \left(\frac{1}{2+n}\right)^{\frac{2+n}{1+n}} \tag{4.28}$$

4.4.2 BINGHAM FLUIDS

For Bingham plastic fluids, we have

$$\frac{du}{dr} = 0 \quad \text{when } \tau \leq \tau_y \tag{4.29}$$

$$\frac{du}{dr} = \frac{1}{\mu}(\tau_y - \tau) \quad \text{when } \tau > \tau_y \tag{4.30}$$

Equation 4.29 yields a constant velocity profile in the region where the shear is equal to or less than τ_y, namely,

$$u = u_p = \text{constant} \quad \text{for } \tau \leq \tau_y \tag{4.31}$$

This region of constant-velocity is termed the **plug flow**.

Substituting Equation 4.6 into Equation 4.30 yields

$$\frac{du}{dr} = \frac{1}{\mu}\left(\tau_y - \frac{\Delta p}{2L}r\right) \tag{4.32}$$

The radius of the plug flow, i.e., the distance r at which τ decreases to the yield stress τ_y, denoted as r_o, can be found from Equation 4.32 by setting $du/dr = 0$. This yields

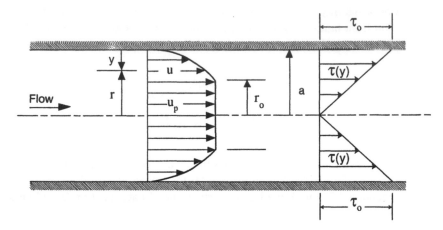

FIGURE 4.6 Velocity profile of a Bingham fluid in a pipe.

$$r_o = \frac{2\tau_y}{\left(\dfrac{\Delta p}{L}\right)} \tag{4.33}$$

When $\tau_y = \tau_o$, Equation 4.33 yields $r_o = a$. This shows that if the yield stress is greater than the wall shear or when $\tau_y > a\Delta p/(2L)$, the fluid does not move in the pipe.

To find the velocity profile in the region $\tau > \tau_y$ or $r > r_o$, we must integrate Equation 4.32, and use the boundary condition $u = 0$ at $r = a$. The result is

$$u = \frac{\Delta p}{4\mu L}(a^2 - r^2) - \frac{\tau_y(a-r)}{\mu} \tag{4.34}$$

At $r = r_o$, Equation 4.34 becomes

$$u = u_p = \frac{1}{\mu}\left(\frac{a^2 \Delta p}{4L} + \frac{\tau_y^2 L}{\Delta p} - a\tau_y\right) \tag{4.35}$$

Equations 4.34 and 4.35 give the velocity distributions in the regions $r > r_o$ and $r \le r_o$, respectively. As shown in Figure 4.6, the velocity distribution across the pipe is characterized by the constant velocity, $u = u_p$ in the region $r < r_o$.

If we define X to be

$$X = \frac{\tau_y}{\tau_o} = \frac{2L\tau_y}{a\Delta p} = \frac{r_o}{a} \tag{4.36}$$

then Equation 4.35 can be rewritten as

$$u_p = \frac{a^2 \Delta p}{4\mu L}(1-X)^2 \qquad (4.37)$$

The discharge of the flow is

$$Q = 2\pi \int_0^{r_o} u_o r dr + 2\pi \int_{r_o}^{a} u r dr \qquad (4.38)$$

Substituting Equations 4.34, 4.36, and 4.37 into Equation 4.38 and integrating yields

$$V = \frac{Q}{\pi a^2} = \frac{a^2 \Delta p}{24\mu L}(3 - 4X + X^4) \qquad (4.39)$$

From the Darcy-Weisbach formula,

$$f = \frac{4a\Delta p}{\rho V^2 L} \qquad (4.40)$$

Substituting Equation 4.39 into Equation 4.40 yields

$$f = \frac{192}{\Re(3 - 4X + X^4)} \qquad (4.41)$$

where $\Re = (VD\rho)/\mu$ is the Reynolds number.

Instead of X, two other dimensionless parameters can be defined: the **yield number, Y,** and the **Hedstrom number, H.**

$$Y \text{ (yield number)} = \frac{D\tau_y}{\mu V} \qquad (4.42)$$

From Equations 4.36, 4.39, and 4.42,

$$Y = \frac{24X}{3 - 4X + X^4} \qquad (4.43)$$

$$H \text{ (Hedstrom number)} = \frac{\rho D^2 \tau_y}{\mu^2} = Y\Re \qquad (4.44)$$

In terms of Y and H,

$$\frac{1}{\mathfrak{R}} = \frac{f}{64} - \frac{Y}{6\mathfrak{R}} + \frac{64Y^4}{3f^3\mathfrak{R}^4} \tag{4.45}$$

$$\frac{1}{\mathfrak{R}} = \frac{f}{64} - \frac{H}{6\mathfrak{R}^2} + \frac{64H^4}{3f^3\mathfrak{R}^8} \tag{4.46}$$

From the foregoing derivation, it can be seen that the friction factor f for the laminar flow of a Bingham plastic fluid is a function of both the Reynolds number and the Hedstrom number (or Yield number). This is illustrated in Figure 4.7, which also includes regions of turbulent flow. Note that the values of f' (Fanning factor), in Figure 4.7, are equal to $f/4$. Figure 4.7 is equivalent to the Fanning's diagram for Newtonian fluids.

Since X is smaller than one, the X^4 term in Equation 4.41 can be neglected to yield

$$f = \frac{192}{\mathfrak{R}(3-4X)} = \frac{64}{\mathfrak{R}\left(1-\frac{4}{3}X\right)} \tag{4.47}$$

The above equation is in the form of $f = 64/\mathfrak{R}$ for laminar flow of Newtonian fluids if the following modified Reynolds number is used:

$$\mathfrak{R}_m = \mathfrak{R}\left(1-\frac{4}{3}X\right) \tag{4.48}$$

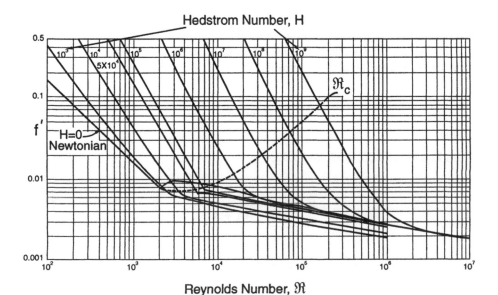

FIGURE 4.7 Fanning friction factor, f', for Bingham plastic fluid in pipe.

For a Bingham plastic fluid, the transition from laminar to turbulent occurs at a modified Reynolds number having approximately the same value as that for a Newtonian fluid based on the ordinary Reynolds number. Considering the transition Reynolds number to be 2100, Equation 4.48 yields

$$\frac{\rho D V_t}{\mu \left(1 + \frac{D \tau_y}{6 \mu V_t}\right)} \simeq 2100 \tag{4.49}$$

where V_t is the critical velocity in pipe corresponding to transition condition.

For tubes of diameter greater than 1 inch, the term $D\tau_y/(6\mu V_t)$ is much greater than one. Therefore,

$$\frac{\rho V_t^2}{\tau_y} \simeq 350 \quad \text{or} \quad V_t = 19\sqrt{\frac{\tau_y}{\rho}} \tag{4.50}$$

4.5 NON-NEWTONIAN PIPE FLOW: TURBULENT

There are many equations proposed in the literature for turbulent pipe flow of non-Newtonian fluids. Only a few will be mentioned here.

4.5.1 TOMITA'S EQUATIONS

For turbulent flow of any power-law fluid in a smooth pipe, Tomita found that [2]

$$\sqrt{\frac{1}{f_m}} = 2\log\left(\Re_m \sqrt{f_m}\right) - 0.80 \tag{4.51}$$

where f_m and $\Re_m$ are the modified Darcy-Weisbach friction factor and the modified Reynolds number, respectively, defined as

$$f_m = \frac{8 D \Delta p (1 + 2n)}{3 L \rho V^2 (1 + 3n)} \tag{4.52}$$

$$\Re_m = \frac{6[(1 + 3n)/n]^{1-n}}{2^n[(1 + 2n)/n]} \frac{D^n V^{2-n} \rho}{K} \tag{4.53}$$

Since Equation 4.51 is the same as Nikuradse's equation for smooth pipe flow of Newtonian fluid, which forms the basis of the curve of smooth pipe in the Moody diagram, the same Moody diagram for Newtonian fluid can be used for power-law non-Newtonian fluids, provided that f_m and $\Re_m$ are used in lieu of f and $\Re$. This is true even for rough pipes. Once f_m is found from Equation 4.51 or the Moody diagram, Equation 4.52 can be used to calculate the pressure gradient $\Delta p/L$.

4.5.2 HANKS-DADIA ANALYSIS

For turbulent flow of Bingham fluids in a smooth pipe, the friction factor f is a function of the Reynolds number, $\mathfrak{R}$, and the Hedstrom number, H. The relation is normally expressed in a graph, as shown in Figure 4.7. The result is based on Hanks and Dadia's study.

Note that Figure 4.7 is for hydraulically smooth pipe. To obtain values of f for hydraulically rough pipe, the value of f for hydraulically smooth pipe, f_s, is first obtained from Figure 4.7. Then, the Moody diagram for Newtonian fluid is used to determine two f values, one for smooth pipe, f_1, and the other for rough pipe, f_2, at the same Reynolds number. Finally, the value of f for rough pipe of non-Newtonian fluid is calculated from

$$f = \left(\frac{f_2}{f_1}\right) f_s \tag{4.54}$$

4.5.3 TORRANCE EQUATION

Torrance has treated the case of turbulent flow in pipe of a yield-pseudoplastic fluid described by Equation 4.5. For the case of rough pipe, he obtained the following equation:

$$\frac{1}{\sqrt{f}} = 2.035 \log\left(\frac{a}{e}\right) + 3.0 - \frac{1.325}{n} \tag{4.55}$$

where e is the absolute roughness of the pipe. The above equation can be used for the prediction of f for fully turbulent flow of power-law yield-pseudoplastics, Bingham, or simple pseudoplastic fluids in rough pipes.

PROBLEMS

4.1 Compare the magnitudes of the two kinds of Reynolds number, $\mathfrak{R}_1$ and $\mathfrak{R}_2$, for power-law fluids. This can be done by calculating the ratio $\mathfrak{R}_2/\mathfrak{R}_1$ at various values of n and plotting the result. Discuss your finding.

4.2 The rheogram of a certain non-Newtonian fluid shows that the fluid follows a power law with $n = 0.7$ and $K = 2 \times 10^{-4}$ in basic English units. The specific gravity of the fluid is 1.2. (a) What kind of non-Newtonian fluid is this? (b) If this fluid is forced to move through a pipe of 1-inch diameter, at what speed of the flow will the transition from laminar to turbulent take place? (c) Determine the pressure drop in the pipe over a distance of 30 ft if the flow is at 0.5 ft/s.

4.3   A certain coal slurry consists of 50% coal and 50% water by weight. The coal particles in the slurry are of –325 mesh. The slurry density is 1.22 grams/cm^3, the yield stress is 20 dynes/cm^2, and the coefficient of rigidity is 28 centipoises. (a) If this slurry is forced through a pipe of 1-inch diameter at a mean velocity of 1 m/s, is the flow expected to be laminar or turbulent? (b) What is the pressure gradient in this pipe? (c) What is the radius of the plug flow in the pipe, r_o? (d) What is the velocity of the plug flow? (e) What is the minimum pressure gradient to cause this slurry to move in the pipe?

4.4   Find the relation between Tomita's resistance factor, f_m, and the Darcy-Weisbach resistance factor, f. Also, find the relation between Tomita's Reynolds number, $\mathfrak{R}_m$, and the Reynolds number $\mathfrak{R}_2$.

4.5   If the same slurry in Problem 3 is pumped through an 18-inch-diameter pipeline 273 mi long at a velocity of 6 ft/s, find (a) the total headloss along the pipeline in feet of slurry and water, (b) the power required in *MW* for pumping, assuming that the pump/motor efficiency is 70%, (c) the tons of coal transported each year by this pipeline, assuming 98% availability, (d) the energy intensiveness for transporting coal by this pipeline, and (e) the cost of energy per ton of coal transported ($/ton), assuming that each kwh of electricity costs 10¢.

REFERENCES

1. Chhabra, R.P. and Richardson, J.F., *Non-Newtonian Flow in the Process Industries: Fundamentals and Engineering Applications*, Butterworth Heinemann, Oxford, U.K., 1999.
2. Govier, G.W. and Aziz, K., *The Flow of Complex Mixture in Pipes*, Van Nostrand Reinhold, New York, 1972.
3. Ryan, N.W. and Johnson, M.M., Transition from laminar to turbulent flow in pipes, *AIChE Journal*, 5, 433–435, 1959.

5 Flow of Solid–Liquid Mixture in Pipe (Slurry Pipelines)

5.1 FLOW REGIMES

This chapter discusses the flow of solid particles transported by a liquid in pipelines, namely, slurry pipelines. For simplicity, the discussion will focus on horizontal or nearly horizontal pipes with particles suspended by the liquid.

It is known from physical chemistry that whenever a solid particle comes into contact with a liquid such as water, there will be differential adsorption of iron at the particle surface causing the solid to become negatively charged and the water around the particle positively charged. Such charged particles are electrically active (i.e., strongly influenced by the surface charges) when the particle size is very small (less than 1 μm). Such small particles suspended in water are called **colloidal particles**. They exhibit strong Brownian motion — random motion of particles in a zigzagged manner. Because these small particles have very small terminal velocities caused by gravity, the Brownian motion alone is sufficient to suspend them in the liquid. No bulk movement of the liquid is needed to cause the particles to remain in suspension. The mixture is said to be homogeneous regardless of fluid motion. Depending on its rheological properties, the mixture is regarded as either a Newtonian fluid or a non-Newtonian fluid. The equations given in previous chapters for single-phase, Newtonian and non-Newtonian fluids are applicable to this case. When the particle concentration is low, say less than 10% by volume, the fluid may be regarded as Newtonian. When the particle concentration is high, say more than 10%, the fluid may be regarded as non-Newtonian.

In general, the ability of fluid in horizontal motion to be able to suspend solid particles depends on the counterbalance of two actions: **gravity**, which causes the particles to fall or settle in the fluid, and an **upward diffusion** of the particles, caused by a concentration gradient of particles (more particles at lower elevations), which in turn is created by gravity. The upward diffusion of particles would not be possible if the particles were not agitated — i.e., in random motions (fluctuations). For solids dissolved in a liquid, the liquid need not have bulk motion to agitate and suspend the solid particles (molecules). Even the Brownian motion of individual liquid molecules can cause the solids to fluctuate and diffuse. However, for large and heavy particles, it may take a strong turbulence, as found in turbulent pipe flow, in order to suspend the particles in a horizontal pipe. Understanding this mechanism of particle suspension helps comprehend what happens to pipe flows of suspended solids — be it for solid–liquid mixture (slurry flow) or solid–gas mixture (pneumotransport).

When the solid particles in a liquid are very fine (say, in the 0.1- to 1-μm range) and significantly denser than the liquid, without liquid motion the particles may settle out by gravity over a long time, but may become uniformly suspended (homogeneous) when the fluid is moving, even when the flow is laminar. Although there is no turbulence in laminar flow, the velocity gradient across the pipe in a laminar flow can cause particles to rotate and travel an irregular path similar to Brownian motion but at a larger scale. Such random motion of very fine particles can cause the particles to be suspended in a homogeneous state in laminar flow of relatively high velocity, but the particles may settle out when the flow is stopped, or when the velocity of the laminar flow is too low.

When the denser-than-fluid particles are in the 1- to 10-μm range, their terminal velocity (settling velocity) may be sufficiently high to cause the particles to settle out of laminar flow, but they become uniformly dispersed or suspended by turbulent flow. Therefore, under turbulent conditions, the mixture is again considered homogeneous, and the equations for single-phase Newtonian or non-Newtonian fluids are again applicable. The term **pseudohomogeneous** is often used in lieu of the term **homogeneous**. Hereafter, a mixture that is homogeneous only in turbulent flow when the velocity is sufficiently high is termed pseudohomogeneous.

When the solid particles are relatively coarse (say, having diameters greater than 10 μm), the turbulence in the pipe must be sufficiently high in order to be able to suspend the solid particles. Depending on the mean velocity of the flow, and the terminal velocity of the solid particles in suspension, the mixture may be in one of the following states or regimes:

Pseudohomogeneous — This occurs when the particle size is relatively small and the velocity of the pipe flow is relatively high, i.e., relatively fine particles in strongly turbulent flow.

Heterogeneous — This occurs when particles are fully suspended but not uniformly distributed — nonhomogeneous. This happens when either the velocity is somewhat smaller or the particle size is somewhat larger than in the previous case.

Moving-Bed Flow — Particles settle out of the flow and form a bed. The particles in the bed move in the flow direction by **sliding, rolling,** or **saltation**. This happens when either the velocity of the flow is less or the particle size is larger than in the previous case. Saltation refers to the phenomenon that some particles on the surface of the bed layer move intermittently in frog leaps.

Stationary-Bed Flow — Particles settle out on the bed and they do not move in the bed. This happens with very coarse particles or very low velocity in pipes.

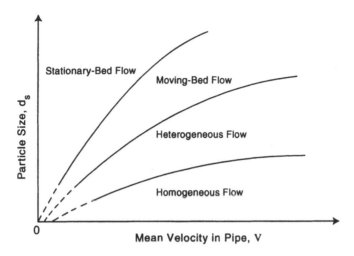

FIGURE 5.1 Four regimes for transport of solids through pipe.

The four states or regimes described above are illustrated in Figure 5.1. Although only two parameters, mixture velocity in pipe and particle size, are used in classifying these four regimes, it should be realized that the density of the solid particles has the same effect as the particle size, namely, a higher density causes the particle to settle more. However, the effect of particle size is more pronounced than the effect of particle density because the former can vary over a much wider range than the latter. The only exception is when the density of the solids approaches that of the liquid in the mixture. In such a case, even a small variation in density has a strong impact on particle settlement. It should be realized that whether a mixture is in one regime or another depends not only on the particle properties (size and density) but also on the flow velocity and the Reynolds number. The tendency to settle is always reduced when the velocity or Reynolds number is high. Therefore, a given solid–water mixture containing coarse particles may be pseudohomogeneous at high velocity, heterogeneous at medium velocity, moving-bed flow at low velocity, and stationary-bed flow at very low velocity. One should realize that fluid density also affects suspension of solids. A dense liquid creates large buoyancy, which has the same effect as low density of suspended particles, for it is the density difference between the particle and the liquid that determines whether a particle will settle or not.

The pressure gradients of the four regimes of flow are shown in Figure 5.2. In the transport of solids through pipelines, the stationary-bed regime should always be avoided because it results in no transport of solids at all. The moving-bed regime should also be avoided under normal conditions because it requires high EI (energy intensiveness) and runs the risk of pipeline blockage.

Whether a pipe flow of solid–liquid mixture is to be treated as pseudohomogeneous or heterogeneous depends on the variation of solids (sediment) concentration in the pipe. If the solids are uniformly or almost uniformly distributed across the pipe, the flow can be treated as pseudohomogeneous. In contrast, if the concentration distribution varies greatly in the vertical plane (having much higher solids

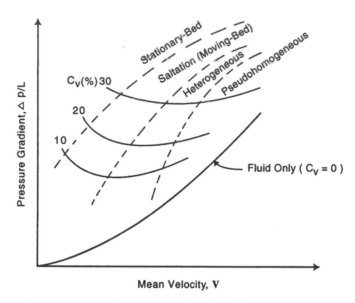

FIGURE 5.2 Variation of pressure gradient with mean-flow velocity at different concentrations of solids.

concentration near pipe bottom than near pipe top), then the flow must be regarded as heterogeneous.

In the study of sediment transport by liquid flow, whether it takes place in a river or in a pipe, the key sediment property that one must deal with is the **settling velocity**, V_s, which is the terminal velocity at which a particle falls (settles) under gravity in the fluid when the fluid is at rest, or when the velocity of the fluid in the pipe, V, is zero. When the fluid is moving in the pipe as turbulent flow, the vertical component of the turbulent velocity fluctuations, v', during its upward movement, must be greater than the settling velocity (namely, $v' > V_s$), before the turbulent flow is capable to suspend the sediment. The sediment is suspended in a horizontal turbulent flow under the balance of two forces: gravity that causes the sediment to fall at V_s, and turbulent diffusion due to the turbulence and the existence of a vertical concentration gradient of the sediment particles in the flow. This balancing act causes sediment particles to be suspended in the flow, with higher concentration of particles toward the lower part of the pipe cross section. Readers interested in predicting such vertical variation of concentration distribution of sediment in pipe or open channel can find such information in Reference 1.

In 1971, Wasp et al. [1] proposed using the following equation for classification of slurry flows:

$$\log \frac{C_T}{C_A} = -1.8 \frac{V_s}{\kappa u_*} \tag{5.1}$$

where log is the common logarithm (10-based); C_A and C_T are the volume concentrations of the solids at the pipe axis (centerline), and the top of the pipe at a location

$0.92D$ from the pipe bottom (or $0.08D$ from the top); V_s is the settling velocity, which is the terminal velocity of solids falling in water (if water is the fluid in the pipe); κ is the von Karman constant, which for slurry flows is slightly less than the value of 0.4 given in Chapter 2 for pure fluids (e.g., for fine coal slurry of 50% concentration by weight, κ is about 0.35); and u_* is the shear velocity, which, according to Equation 2.15, is

$$u_* = \sqrt{\tau_o / \rho} = V\sqrt{f / 8} \qquad (5.2)$$

Equation 5.1 can be used to calculate the concentration ratio C_T/C_A of any slurry pipeline, and the result can be used to classify the flow as follows: **pseudohomogeneous** when $C_T/C_A > 0.8$, **heterogeneous** when $C_T/C_A < 0.1$, and **intermediate** when $0.8 > C_T > 0.1$. The numbers used here for classification, 0.1 and 0.8, are somewhat arbitrary.

The settling velocity V_s in Equation 5.1 can be estimated by assuming that it is the same as that of a sphere of the same material (density), having a diameter d equal to the particle size, d_s, of an irregular-shape particle determined by sieve analysis. This allows calculation of the terminal velocity by using equations in fluid mechanics, such as Stokes law when the Reynolds number of the sphere, $\Re$, based on settling velocity and the particle size, is smaller than 1.0. For $\Re$ greater than 1.0, the terminal velocity can be expressed as a function of the drag coefficient C_D, which, for spheres, is given as a function of $\Re$ in fluid mechanics. However, since particle shape has a strong effect on the terminal velocity, such calculations based on spherical particles represent only rough estimates for particles of irregular shape. More accurate determination of V_s for nonspherical particles can be done through laboratory tests using a settling column (tank), which is a transparent vertical pipe section with an open top and a sealed bottom. The column is filled with water, and solid particles to be tested are dropped into the column. The particles falling through water in the column soon reach their terminal velocity, which is determined from the falling distance divided by time. For accurate measurements, the column must be much longer than the distance required for the falling particles to reach their terminal velocities. The column diameter must be at least 20 times the size of the largest particles to be tested. Needless to say, the test must be conducted with the same solids and the same liquid as to be encountered in the pipeline.

Once the mean velocity of the flow in the pipe, V, is determined or selected, the Moody diagram in Chapter 2 can be used to determine the friction factor f in Equation 5.2 for calculating u_*. Then, Equation 5.1 can be used for calculating the concentration ratio C_T/C_A.

Example 5.1 Sand of specific gravity of 2.4 and size range of 75 to 150 μm is to be transported by water at 20°C through a 6-inch-diameter commercial steel pipe. If the mixture velocity in the pipe is 1.5 m/s, should the slurry flow be classified as pseudohomogeneous, heterogeneous, or intermediate? Suppose that the mean settling velocity of the sand, determined from laboratory tests, is 0.03 m/s.

[**Solution**] For this problem, $V = 1.5$ m/s, $V_s = 0.03$ m/s, $D = 6$ inches $= 0.1524$ m, and the density and the viscosity of water at 20°C are, respectively, $\rho = 998$ kg/m^3 and $\mu = 1.002 \times 10^{-3}$ N-s/m^2. Therefore, the Reynolds number of the flow in the pipe, based on water, is $\Re = \rho \, DV/\mu = 2.28 \times 10^5$. The relative roughness of the pipe is $e/D = 0.0003$. From Moody diagram, $f = 0.0177$. From Equation 5.2, $u_* = 0.0706$ m/s. Finally, assuming that the von Karman constant for the slurry is 0.35, Equation 5.1 yields $\log(C_T/C_A) = -2.19$, or $C_T/C_A = 0.00653$. Since this is well below the limit of 0.1, according to the criterion discussed previously the flow is classified as heterogeneous. The foregoing calculation is inexact because the density and the f value used are based on water instead of slurry flow. Without knowing the concentration of the solids in the flow, the values of ρ and f for the slurry cannot be determined. The concentration will be specified in the next example.

5.2 PSEUDOHOMOGENEOUS FLOWS

A pseudohomogeneous slurry can be treated in the same manner as homogeneous flows (Newtonian and non-Newtonian fluids discussed in Chapters 2 and 4, respectively), except that in the laminar regime and during extended pump shutdown the solids in pseudohomogeneous flow will settle out.

For pseudohomogeneous flows that have a constant ratio between shear, τ, and velocity gradient, du/dy, they can be treated as Newtonian fluids. The pressure gradient in this case can be calculated from

$$i_m = \frac{\Delta P_m}{L} = f_m \frac{1}{D} \frac{\rho_m V^2}{2} \tag{5.3}$$

Note that Equation 5.3 is identical to the Darcy-Weisbach formula for Newtonian fluids, except that the subscript m denotes **mixture**. Thus, Δp_m, f_m, and ρ_m are the pressure drop, Darcy-Weisbach friction factor, and density, respectively, of the pseudohomogeneous mixture. They are different from their counterparts of the water-only flow. The resistance factor f_m can be obtained from the ordinary Moody diagram for Newtonian fluid. However, one should use the viscosity and the density of the mixture, rather than those of water, to calculate the Reynolds number for determining f_m. The mixture (slurry) viscosity can be determined approximately from the following formula given by Thomas [2]:

$$\mu_m = \mu\left[1 + 2.5C_v + 10.05C_v^2 + 0.00273EXP(16.6C_v)\right] \tag{5.4}$$

where μ_m is the dynamic viscosity of the mixture; μ is the dynamic viscosity of the liquid (water); and C_v is the volume concentration of the solids in the slurry, defined as the volume of the solids (particles) in the mixture divided by the mixture (slurry) volume.

Once C_v is known or specified, the mixture density ρ_m can be calculated from

$$\rho_m = \frac{\rho_s}{S + C_w - SC_w} \tag{5.5}$$

where C_w is the weight concentration of the solids in the slurry; and S is the density ratio, which is the density of the solids, ρ_s, divided by the density of the fluid, ρ, namely, $S = \rho_s/\rho$.

The weight concentration, C_w, can be calculated from the volume concentration, C_v, as follows:

$$C_w = \frac{S}{S + (1/C_v) - 1} \tag{5.6}$$

To determine whether Equation 5.3 can be used for any pseudohomogeneous slurry flow, one must conduct a standard rheological test in the laboratory to determine the variation of shear, τ, as a function of the velocity gradient du/dy. Only when the test results show that τ is linearly proportional to du/dy, as shown in Figures 4.2 and 4.3 for Newtonian fluid, can the equation be used. Otherwise, one must compare the rheogram with those listed in Figure 4.4 to determine which non-Newtonian model is applicable. For instance, if the rheogram shows that the slurry is a Bingham plastic fluid, both the yield stress, τ_y, and the coefficient of rigidity, μ, must be found from the rheogram. The equations given in Chapter 4 for Bingham plastic fluid should then be used for calculating the pressure gradient.

The foregoing discussion shows that it is important to determine the rheogram of pseudohomogeneous flow through laboratory tests before the flow can be analyzed. Once the rheological properties of a flow is determined from the rheogram, the flow can be treated either as a Newtonian fluid or a non-Newtonian fluid, whichever is appropriate for the case, in the determination of the pressure drop along the pipe.

Example 5.2 Suppose that the slurry in the previous example has a volume concentration of 27%. Determine C_w, the mixture (slurry) density ρ_m, and the mixture viscosity μ_m. Then, calculate f_m and i_m based on the assumption that Equation 5.3 is applicable to this case. Recalculate C_T/C_A and reclassify the flow regime based on mixture properties to determine if the flow is pseudohomogeneous.

[Solution] Because $S = 2.4$ and $C_v = 0.27$, Equation 5.6 yields $C_w = 0.470$. The solids density is $\rho_s = 2.4 \times 998 = 2395$ kg/m³. Therefore, from Equation 5.5, $\rho_m = 1375$ kg/m³. Because the dynamic viscosity of water is $\mu = 1.002 \times 10^{-3}$ N-s/m², Equation 5.4 yields $\mu_m = 2.65 \times 10^{-3}$ N-s/m². The Reynolds number based on ρ_m and μ_m is $\Re_m = 1.19 \times 10^5$. For the same relative roughness $e/D = 0.0003$, using the Moody diagram yields $f_m = 0.0191$. Then, Equation 5.3 yields $i_m = \Delta p_m/L = 194$ Pa/m. From Equation 5.2, $u_* = 0.0733$ m/s. Finally, from Equation 3.1, $C_T/C_A = 0.00786$. So, the flow still must be classified as heterogeneous. This means the assumption of pseudohomogeneous flow is incorrect, and hence the foregoing calculation of i_m and u_* based on Equation 5.3 is invalid. The correct solution must await the next example, after equations for heterogeneous flow have been introduced.

5.3 HETEROGENEOUS FLOWS

Heterogeneous flow can be subdivided into two categories: **symmetric** or **asymmetric**. Symmetric mixture flow exists when the concentration profile of the solid in the flow is symmetric or approximately symmetric about the centerline of the pipe, though the concentration may not be uniform or homogeneous in the radial direction across the pipe. This happens in vertical pipes when gravity does not affect the concentration profile of the solids, but the higher shear stress near the pipe wall causes solids to rotate more and migrate towards or away from the pipe centerline, according to the Magnus effect of fluid mechanics. On the other hand, when the pipe is horizontal, when the solid particles are relatively large and/or heavy, and when the mean velocity in the pipe is relatively small, gravity causes more solids to concentrate near the pipe bottom than the top, resulting in an asymmetric concentration profile. In this case, the concentration is a function of both radius r and angle θ of the cylindrical coordinates.

5.3.1 LIMIT-DEPOSIT VELOCITY

The **limit-deposit velocity**, V_L, is the minimum velocity required to suspend a solid particle in a pipe. It corresponds to the velocity for transition from the moving-bed regime to the heterogeneous regime. It is also referred to in the literature as the **deposition velocity**, or simply the **deposit velocity**.

Durand [3] conducted extensive tests with pipes ranging in diameter from 38 to 710 mm, using particles of sand and gravel measuring from 0.2 to 25 mm, at volume concentrations between 2 and 23%. Based on this study, Durand and coworkers developed the graph shown in Figure 5.3a for solid particles of uniform or nearly uniform size, and Figure 5.3b for particles of rather nonuniform size distribution. The figures can be used to determine V_L from the particle diameter, d_s, pipe diameter, D, solid specific gravity, S, solid volumetric concentration, C_v, and gravitational acceleration, g. For nonuniform material, the value of d_s is based on the mesh size that passes 85% of the particles. The ordinates of Figure 5.3a and b are the same: $F_L = V_L / \sqrt{2gD(S-1)}$, which is the densimetric Froude number of the slurry flow in pipe corresponding to particle deposition. Once F_L is determined from Figure 5.3, the value of V_L is determined. To make sure that the flow is heterogeneous with fully suspended solids, the pipe should run at a velocity V slightly higher than V_L (say $V = 1.2 \, V_L$).

5.3.2 PRESSURE GRADIENT IN HETEROGENEOUS FLOW

When the solids (sediment) in a pipe are totally suspended by the flowing water (i.e., when the flow is heterogeneous), the frictional loss of the flow along the pipe can be calculated from the **modified Durand's equation** [4]:

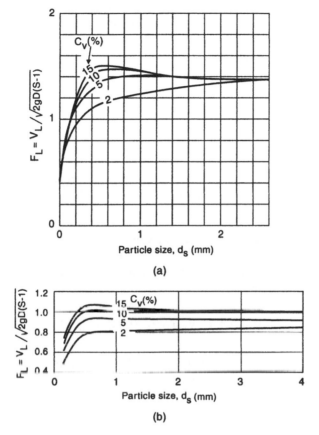

(a)

(b)

FIGURE 5.3 Limit-deposit velocity for solid particles: (a) uniform size particles; (b) non-uniform size particles.

$$\Phi = \frac{i_m - i}{iC_v} = 67 \left(\frac{\sqrt{gD(S-1)}}{V} \right)^3 \left(\frac{V_s}{\sqrt{g(S-1)d_s}} \right)^{3/2} \tag{5.7}$$

where Φ is a dimensionless function; i_m is the pressure gradient of the mixture (slurry) flow, namely, $i_m = \Delta p_m/L$; and i is the pressure gradient of the liquid (water) flow in the same pipe at the same velocity V, namely, $i = \Delta p/L$.

The constant 67 in Equation 5.7 is based on spherical particles. In spite of that, the equation is applicable as an approximation to solid particles of other shapes including irregular shapes as for sand and gravel. Even though Durand only tested sand and gravel in water, Worster included the term $(S - 1)$ in both the numerator and the denominator of Equation 5.7, rendering the equation applicable to other materials and liquids as well [4]. When a liquid other than water is used, S in Equation 5.7 becomes the relative density of the solids in fluid ($S = \rho_s/\rho$), rather than the specific gravity. Equation 5.7 is correct only when the velocity in the pipe is about the same as or higher than V_L. It is not correct when V is much smaller than V_L. Furthermore, the minimum value of pressure gradient in the pipe, i_m, is expected to be at or near the location $V = V_L$.

Equation 5.7 can be rewritten as

$$i_m = C_1 V^2 + \frac{C_2}{V} \tag{5.8}$$

where

$$C_1 = \frac{f\rho}{2D} \quad \text{and} \quad C_2 = 33.5 \frac{f\rho C_v [g(S-1)]^{3/4} D^{1/2} V_s^{3/2}}{d_s^{3/4}} \tag{5.8a}$$

When C_1 and C_2 are constant (i.e., for a given pipe-slurry system at a given value of C_v, and assuming f to be constant, which is approximately correct), the pressure gradient i_m in Equation 5.8 becomes a minimum at a velocity V equal to V_o, given by the following equation:

$$V_o = 3.22[g(S-1)]^{1/4} C_v^{1/3} (DV_s)^{1/2} d_s^{-1/4} \tag{5.9}$$

Many other investigators have proposed their own equations for calculating the pressure gradient of heterogeneous flows. Readers interested in the subject should consult References 4 and 5, which contain detailed and authoritative treatment of the subject.

Example 5.3 (a) Find the deposition velocity of the sand–water flow described in the two previous examples by using Figure 5.3, and compare it with the operation velocity of 1.5 m/s. Determine from the comparison whether the selected operation velocity is adequate to insure heterogeneous flow. (b) Calculate the pressure drop per unit length of the pipe by assuming that the flow at 1.5 m/s is heterogeneous. Compare the result with that obtained in the previous example for pseudohomogeneous flow, and discuss the significance. (c) Increase the operation velocity to the deposition velocity and then calculate the pressure gradient by assuming heterogeneous flow. Compare with the pressure gradient at the lower velocity of 1.5 m/s determined in the previous part. (d) Use the result of part (b) to determine the value of C_T/C_A and the classification of this slurry flow at 1.5 m/s. (e) Use Equation 5.9 to calculate the optimum operation velocity, V_o. Compare this velocity with the limit deposit velocity, V_L, found in part (a), and discuss the result.

[Solution] (a) The sand in this case is rather uniform in size within the narrow range of 75 to 150 μm. Thus, the sand can be considered as being uniform with a mean size of 112.5 μm, which is equal to 0.1125 mm, or 1.125×10^{-4} m. From Figure 5.3a, the densimetric Froude number is 1.05, and hence $V_L =$

$1.05\sqrt{2gD(S-1)}$ = 2.15 m/s. Since this velocity is considerably higher than the operation velocity of 1.5 m/s, the actual flow (at 1.5 m/s) must contain both suspended and moving-bed sediments.

(b) When the velocity V is 1.5 m/s, Equation 5.7 yields $\Phi = (i_m - i)/(iC_v) =$ 40.1. But, for the water-only flow, $e/D = 0.0003$, $\Re = 2.28 \times 10^5$. From the Moody diagram, $f = 0.0177$, and $i = \Delta p/L = fpV^2/(2D) = 130.4$ Pa/m. Therefore, $i_m = 1542$ Pa/m, which is the pressure gradient based on heterogeneous flow assumption. From Example 5.2, the counterpart of pseudohomogeneous flow is $i_m = 194$ Pa/m. This shows that the solution of i_m is very different for pseudohomogeneous than for heterogeneous flow. The latter is almost eight times that of the former.

(c) If the pipe is operated at the deposition velocity of 2.15 m/s, Equation 5.7 yields $\Phi = (i_m - i)/(iC_v) = 13.6$. Thus, $i_m = 5.70i$. But, $e/D = 0.0003$, and $\Re = 3.26 \times 10^5$. From the Moody diagram, $f = 0.0168$. Hence, $i = fpV^2/(2D) = 254$ Pa/m. Finally, $i_m = 4.76i = 1187$ Pa/m. This is significantly smaller than the value found at $V = 1.5$ m/s.

(d) From part (b), for heterogeneous flow $i_m = 1542$ Pa/m, and $\rho_m = 1375$ kg/m³. Thus, from Darcy-Weisbach formula, the friction factor is $f_m = 2Di_m/(\rho_m V^2) = 0.152$. Next, from Equation 5.2, $u_* = 0.207$ m/s. Using this value of u_* in Equation 5.1 yields $C_T/C_A = 0.180$. Since this value is between 0.1 and 0.8, the flow is in the intermediate regime. This means that when V is 1.5 m/s, the true value of i_m must be between the value of 1542 Pa/m for heterogeneous flow and the value of 194 Pa/m for pseudohomogeneous flow. It must be closer to the former (higher) value than the latter (lower) value because $C_T/C_A = 0.180$ is far closer to the lower limit of 0.1 than the higher limit of 0.8.

(e) Equation 5.9 yields $V_o = 2.63$ m/s. From part (a), the limit deposit velocity, V_L, was found to be 2.15 m/s, which is about 18% lower than V_o. To insure that the flow will be heterogeneous without deposition of sand in the pipe, V_o should be used as the operational velocity. Using $V_o = 2.63$ m/s, Equation 5.7 yields $\Phi = 7.44$, $i = 370$ Pa/m, and $i_m = 1113$ Pa/m.

The foregoing examples illustrate two points. First, the use of Equation 5.1 to calculate C_T/C_A depends on the value of u_*, which in turn depends on whether the flow is homogeneous or not. Incorrect assumption of the flow type in the beginning will result in a wrong value of u_*, which yields a value of C_T/C_A in conflict with the initial assumption of the flow type. When that happens, the initial assumption of flow type must be changed to obtain the correct value of C_T/C_A. Several iterations may be needed before the correct value of C_T/C_A can be found from Equation 5.1. Second, the examples show that the velocity V_o is greater than V_L by almost 20%. To ensure proper operation, the pipe should be operated at V_o instead of V_L. Finally, the examples show that large differences exist between the values of i_m for homogeneous flow and for heterogeneous flow. Therefore, there is a strong need for a reliable method to predict the pressure gradient for flows in the intermediate regime — a regime of great practical importance for long-distance transport of fine slurry. This is discussed in the next section.

5.4 INTERMEDIATE FLOW REGIME

When $0.8 > C_T/C_A > 0.1$, the slurry flow is said to be **intermediate**. In this regime, calculations based on pseudohomogeneous model such as Equation 5.3 will underestimate the headloss or pressure gradient, whereas calculations based on heterogeneous flow such as Equation 5.7 will overestimate the headloss or pressure gradient. The true values for the intermediate regime exist somewhere between these two extremes. Wasp et al. [1] proposed an empirical approach to deal with slurries in this regime, assuming that the slurry is rather nonuniform in size (i.e., having a wide range of particle sizes). In the approach, the slurry is first divided into several size fractions. With the use of Equation 5.1, each fraction is subdivided into a homogeneous part (suspended load, or as termed by Wasp, **vehicle load**), and a heterogeneous part (bed load). The homogeneous part is assumed to constitute a fraction of C_T/C_A, whereas the heterogeneous part is assumed to constitute a fraction of $(1 - C_T/C_A)$. The pressure gradient of the heterogeneous part of each size fraction is calculated separately by using Durand's equation. They are then added together to yield the total pressure gradient generated by bed load, i_b. On the other hand, the homogeneous parts (suspended loads) of all fractions are added together to form the total suspended load, which is used to calculate the pressure gradient due to the entire suspended load, using the procedure outlined in Section 5.2 for pseudohomogeneous flow. Finally, the two parts, i_b and i_v, are added together to yield the total pressure gradient of the slurry (mixture) in the intermediate regime, namely, $i_m = i_b + i_v$. A detailed discussion of the Wasp method with an example is given in Reference 6.

Wasp's approach has proved satisfactory for predicting pressure drops along some commercial coal slurry pipelines. This is not surprising because the method was developed from coal slurry pipeline data. Its use for commercial slurry pipelines transporting certain other materials, such as phosphate and iron, have also been substantiated by the designers of such pipelines. However, a recent publication in China compared predictions from Wasp's approach with the measured values of the pressure gradients of five different slurry pipelines in China, for transporting minerals including phosphate concentrates, iron concentrate, and coal. The article reported that Wasp's approach overestimated the pressure gradient of the five pipelines by 16 to 216%. While no explanation was offered for some of the large discrepancies, the reported discrepancies call for caution. The fact that all the predicted values exceeded the measured values makes one wonder whether the slurry particles in the five reported cases, after long distance transport through pipes and after passing through pumps, have broken down into finer particles, which reduce the overall pressure drop along the pipe. In loop tests of slurries, serious attrition of particles is quite possible, due to numerous passages of the particles through pumps. Even for commercial slurry pipelines, particle attrition may still be significant in certain cases. The Wasp approach and most other approaches for predicting the headloss in slurry pipelines assume negligible attrition of particles during pipeline transport. Also, the accuracy of Wasp's approach is expected to depend on the particle size distribution. It appears to be accurate and most useful when the slurry consists of a wide spectrum of particle sizes, as in the case of crushed minerals and coal.

An alternative approach to predict the pressure gradient for intermediate flow regime, developed by Fei in China [7], is based on the following equation:

$$i_{mh} = \frac{i_m}{\rho g} = \frac{\Delta p_m}{\rho g L} = \alpha \frac{fV^2 S_m}{2gD} + 11\eta_s C_v (S - S_m) \frac{V_{sa}}{V} \tag{5.10}$$

where i_{mh} is the headloss gradient in m/m, which is dimensionless; ρ is the water density; f is the Darcy-Weisbach friction factor of the liquid (water) flow; V is the mean velocity of the mixture in the pipe; S_m is the mixture density divided by the liquid (water) density; S is the solids (particles) density divided by the liquid density; g is gravitational acceleration; D is the pipe diameter; η_s is the contact friction coefficient between the particles and the pipe; C_v is the volume concentration of the solids in the slurry; and V_{sa} is the weighted average settling velocity of the solids. The constant α is a correction factor given as follows:

$$\alpha = 1 - 0.4(\log \mu_r) + 0.2(\log \mu_r)^2 \tag{5.11}$$

where log is the common logarithm based on ten; and μ_r is the relative viscosity of the slurry, which is the slurry or mixture viscosity, μ_m, divided by liquid (water) viscosity, μ. The value of μ_m can be determined from Equation 5.4. The first and the second terms on the right of Equation 5.10 represent contributions due to suspended load and bed load, respectively. It was reported in the same recent Chinese article that the predictions based on Equation 5.10 compared within 25% of the measured values for the same five slurry pipelines to which Wasp's approach was compared.

More research is needed to clarify the conditions of applicability and accuracy of each of the two foregoing approaches, or to develop even better methods for predicting pressure gradients in the intermediate regime.

5.5 PRACTICAL CONSIDERATIONS

5.5.1 WEAR OF SLURRY PIPELINES

The solid particles in a slurry pipeline may cause wear (abrasion) to the pipe and to special equipment such as pumps, valves, and bends. Wear is especially serious when the velocity is high, the slurry is coarse, and the solid particles are dense and possess sharp edges. For transporting coarse slurry, wear may be so serious that the entire pipeline needs to be lined by a wear-resistant material such as rubber or urethane. Otherwise, the pipeline will be short-lived. Equipment that is wear-prone must also be protected by coating it with an elastomer (rubber or urethane), hardened metal (such as alloys that contain high chrome or nickel), or ceramics. Due to the centrifugal force generated by centrifugal pumps and pipe bends, the interior wall of the pumps and bends must be coated with wear-resistant materials. The same can also be said about valves; their interior must be protected by coating with a wear-resistant material. Serious wear can occur to a valve if it is partially open. In a slurry pipeline of high pressure, if the valve is not completely closed, the leakage flow through the gap can create a high-speed jet approaching 100 m/s. If that happens in a slurry pipeline, the valve can be seriously damaged by wear within minutes, since such high-speed slurry jet is extremely abrasive. In addition, proper operation and proper design against wear are even more important than using wear-resistant materials or coating. In contrast, fine slurry pipelines, such as the Black Mesa Coal Slurry Pipeline, can run for decades with little wear except in special places such as bends and segments of the pipe that encounter slack flow, which is the free-surface flow created in pipe by steep and long downward slopes, when the elevation drop to the control point exceeds the headloss due to friction. The slack flow can create velocities much higher than that in the rest of the pipe, thus causing severe wear.

5.5.2 CAVITATION

Cavitation is a phenomenon that affects (damages) pipes, pumps, and valves in both liquid-only pipes and slurry pipes. The cause of cavitation and how to design against it is a subject discussed in Section 9.5.2.2, and hence is not repeated here. It is sufficient to mention that proper design of pipeline suction inlet, proper selection of pumps based on NPSH (net positive suction head), maintenance of positive pressure in pipelines, and proper operation of valves are the key to avoid cavitation. In slurry pipelines, cavitation can be aggravated by slurry wear, causing more rapid damage than in ordinary pipelines with liquid only. One should pay no less attention to cavitation in slurry pipelines than in ordinary liquid pipelines.

5.5.3 SLURRY PUMPS, VALVES, AND FLOWMETERS

Slurry pumps are similar to ordinary liquid pumps except for the fact that they are protected against wear and, when used with chemically corrosive slurries, they must possess a lining that is inert to the corrosive slurry. Practically all types of pumps discussed in Chapter 9 are used for slurry pipelines. Positive displacement pumps are used when the head is high, as needed for long-distance slurry pipelines. Centrifugal pumps are used for medium head, and propeller pumps for low head but large discharge. Diaphragm pumps are needed when the slurry is corrosive, or when wear is excessive for piston and plunger pumps. Figure 5.4 is a photograph of a centrifugal slurry pump for pumping coarse solids, such as sand and crushed rock. More information on slurry pumps can be found in books on slurry transport such as Reference 8.

Plug valves, ball valves, gate valves, and check valves are all used in slurry pipelines. The main difference between valves for slurry and for water is that the former must be more rugged and protected against wear. For slurry pipelines that must be pigged (it is a good idea to design slurry pipelines with the ability to pass pigs), full-bore valves must be used to allow free passage of the pigs. This eliminates

FIGURE 5.4 A slurry pump manufactured by Georgia Iron Works (GIW), in Georgia, U.S.A., for transporting coarse slurry, including sand and crushed rock. (Courtesy of GIW.)

plug valves, which cannot have full bore. Full-bore ball valves are always the best for slurry pipelines except that they are more expensive than gate valves. Globe valves and butterfly valves should not be used because they block the passage of slurry and can wear rapidly. Most commercial slurry valves contain mechanisms for flushing the valve with clean water periodically. This keeps the gap between the valve body and the valve seat free of solid particles, thereby minimizing valve wear and optimizing valve operation. Otherwise, the valves will soon be worn out or even be jammed during operation.

Only nonintrusive types of flowmeters, such as magnetic and acoustic flowmeters, should be used. Any intrusive type, such as a turbine flowmeter, an orifice meter, and a Venturi meter, should not be used for slurry pipelines.

5.5.4 APPLICATION OF SLURRY PIPELINES

Slurry pipelines are widely used in many fields and by many industries for transporting bulk solids, which can be exposed to water in a pipe without damage or detrimental effect. Typical examples include short- to medium-distance transport of mineral wastes (tailings) from mineral processing plants to waste disposal sites; transport of mineral concentrates over various distances whenever economical — even short distances are economical when the mineral dressing process is water based; long distance transport of coal; short-distance transport of sand and gravel; dredging applications; short- and medium-distance transport of sludges from municipal and industrial sources, such as water-treatment and sewage-treatment plants; short- and medium-distance transport of wood chips and pulps by paper manufacturers; transporting salt by using brine; and short-distance transport of hazardous

FIGURE 5.5 Valve and choke station #1 of copper–zinc slurry pipeline in northern Peru. (Courtesy of Pipeline Systems Inc.)

wastes, including radioactive waste materials, to reduce the threat to workers handling such waste transport.

A good review of slurry pipeline technology, including a list of major slurry pipelines constructed around the world prior to 1980, is given in Reference 9. Many new slurry pipelines have been constructed worldwide since 1980. They are described in various articles in technical journals, especially in publications of the proceedings of hydrotransport conferences, organized by the British Hydromechanics Research (BHR) group in Cranfield, U.K. Figure 5.5 shows a recently (2001) completed slurry pipeline for transporting 1.5 MT/yr (million tonnes per year) of minerals — three grades of copper and zinc concentrates. The pipeline, 302 km long, is located in northern Peru, with a starting elevation of 4200 m. The pipeline, owned by Minera Antamina, was designed by PSI (Pipeline Systems Inc.). The pipeline is made of steel, with an HDPE lining. It uses pipe diameters of 203 mm (8-inch NPS), and 254 mm (10-inch NPS).

PROBLEMS

5.1 Prove that for a pseudohomogeneous flow that exhibits the property of a Newtonian fluid, the following equation holds:

$$\frac{i_m - i}{iC_v} = S - 1$$

where i is the pressure gradient of the fluid (e.g., water in the case of solid–water mixture); i_m is the pressure gradient of the mixture; S is the density of the solid, ρ_s; and C_v is the volume concentration of the solids in the mixture. (Hint: Use the

Darcy-Weisbach equation, and assume that the friction factor for the fluid, f, is approximately the same as that for the mixture.)

5.2 A clay–water mixture at 70°F contains 20% of clay by weight. The specific gravity of the individual clay particles is 2.6. (a) Find the density and viscosity of the mixture. (b) Find the headloss along a 1-inch-diameter steel pipe that transports this clay–water mixture at 4 fps over a distance of 1500 ft. Assume that the mixture behaves like a Newtonian fluid. (c) Find the pump head and the horsepower required to run this pipe at 4 fps if the pump–motor combined efficiency is 80%. Also, give the input power required in kW.

5.3 Derive Equation 5.6. Hint: From the continuity of incompressible flow, $C_v = Q_s/(Q_s + Q)$, $C_w = \rho_s Q_s/(\rho_s Q_s + \rho Q)$, and $\rho_m = (\rho_s Q_s + \rho Q)/(Q_s + Q)$.

5.4 Show that for very small particles falling slowly in fluid in the Stokes' range, the settling velocity V_s of a sphere of diameter d is $V_s = \rho g(S - 1)d^2/(18\mu)$, where ρ, g, S, and μ are quantities as defined earlier in this chapter. (Hint: Use Stokes' law in fluid mechanics.)

5.5 Show that for particles falling outside the Stokes' range (i.e., when the Reynolds number of the falling sphere, $\Re_s = \rho d V_s/\mu$, is greater than 1.0), the settling velocity can be found from the equation $V_s = \sqrt{4gd(S - 1)/(3C_D)}$, where C_D is the drag coefficient of the particle, which is given as a function of the Reynolds number in fluid mechanics.

5.6 Based on the two previous problems, find from fluid mechanics texts the equations of C_D for spheres in various ranges of Reynolds number. Then, print a computer spreadsheet listing the values of V_s, in m/s, for spheres of different diameter, d, ranging from 0.1 mm to 10 cm. Assume that the sphere has a specific gravity of 2.65, and the fluid is water at 20°C.

5.7 A contractor wants to use a steel pipe with water to transport sand for a construction project at the rate of 80 tons per hour. He plans to transport the sand at a weight concentration of 30%. The sand has a specific gravity of 2.6 and a rather uniform size of 1 mm. (a) What size pipe should be used to achieve the throughput of 80 tons per hour? (b) At what speed should the mixture be pumped through the pipe without deposition in the pipe (i.e., what should be the operational velocity of this pipe)? (Hint: Start the calculation by assuming a 2-inch-diameter pipe. If it proves to be insufficient to transport the required 80 tons per hour at the deposition velocity, increase the pipe size to 4-inch diameter and repeat the calculation. Continue to increase the pipe diameter by 2 inches each time until the right size is found.)

5.8 A 4-inch steel pipe with water at 70°F is used to transport sand and crushed rocks over a 2-mile distance, at the volume concentration of 18%. A sieve analysis of a sample showed that 85% of the solids passed through $1/2$-inch mesh. A settling column test found that the settling velocity for the $1/2$-inch particles is 1.74 fps. Find (a) the limit-deposit velocity V_L; (b) the pressure drop along the entire length of the pipe, assuming that the mixture is run through the pipe at the velocity V_L; (c) the optimum velocity, V_o, to run the system; (d) the pressure drop across the pipeline when operating at V_o; and (e) the electrical power needed to run the system at V_o, assuming that the combined efficiency of the pump–motor system is 78%.

REFERENCES

1. Wasp, E.J., Aude, T.C., Seiter, R.H., and Thompson, T.L., Hetero-homogeneous solids–liquid flow in the turbulent regime, in *Advances in Solid–Liquid Flow in Pipes and Its Application*, I. Zandi, Ed., Pergamon Press, New York, 1971, p. 199.
2. Thomas, D.G., Transport characteristics of suspensions: Part vii. A note on the viscosity of Newtonian suspensions of uniform spherical particles, *Journal of Colloid Science*, 20, 267, 1965.
3. Durand, R., Basic Solids in Pipes — Experimental Research, Proceedings International Hydraulics Conference, Minneapolis, MN, 1953, pp. 89–103.
4. ASCE, Sedimentation Engineering, Task Committee Report, American Society of Civil Engineers, New York, 1975.
5. Govier, G.W. and Aziz, K., *The Flow of Complex Mixture in Pipes*, Van Nostrand Reinhold, New York, 1972.
6. Wasp, E.J., Kenny, J.P., and Gandhi, R.L., *Solid–Liquid Flow Slurry Pipeline Transportation*, Gulf Publishing, Houston, TX, 1979.
7. Fei, X.J., *Hydraulics Transport of Slurry and Particulates*, Tsing Hua University Press, Beijing, China, 1994, p. 380 (in Chinese).
8. Wilson, K.C., Addie, G.R., and Clift, R., *Slurry Transport Using Centrifugal Pumps*, Elsevier Science, New York, 1992.
9. Thompson, T.L. and Aude, T.C., Slurry pipelines: design, research and experience, *Journal of Pipelines*, 1(1), 25–44, 1981.

6 Flow of Solid–Gas Mixture in Pipe (Pneumotransport)

6.1 INTRODUCTION

Pneumatic pipeline, also called **pneumotransport** or **pneumatic conveying,** is the use of air or another gas to transport powdered or granular solids through pipes. It is the counterpart of the slurry pipeline, using a gas instead of a liquid as the medium to transport solids. First used successfully in the 1860s for transporting lightweight materials such as wood shavings, sawdust, and waste papers, the technology of pneumotransport has steadily improved and found increasing use in the last 150 years. Currently, it is used widely in many industries to transport minerals, grain, flour, coal, sand, cement, solid wastes, and hundreds of other products. Due to the high energy intensiveness of pneumotransport, and the abrasion (wear) of the materials transported, such pipelines are for transport over short distances only, usually less than 1 km, most often only a few hundred meters or even shorter. Some of the longest pneumatic pipelines are used to transport cement in major construction projects in remote locations. For instance, in 1933, cement was transported pneumatically over a distance of 1.8 km at the construction site of the Hoover Dam in Nevada, U.S.A. Shortly thereafter, a 2.1-km pneumatic pipeline was built to transport cement in the construction of the Grand Coulee Dam in the State of Washington, U.S.A. The longest systems are used in mining where single pipelines of 3.5 km are reported to be in use in some German coal mines. These mines also feature an extensive network of pneumatic pipelines, with a total length of over 40 km of interconnected pipes. Pneumatic conveying is used most often at harbors, barge terminals, and rail terminals for loading or unloading bulk materials such as grain, cement, and fertilizer to or from ships, barges, and trains. It is also commonly used in chemical plants, paper plants, breweries, cement plants, and food processing plants. Because most such applications involve transportation over short distances only, the technology is often considered as a means for materials handling (conveying) rather than transportation. This explains why it is usually referred to as **pneumatic conveying**. The term conveying implies transportation over short distance.

149

The advantages of pneumatic conveying include the following: (1) economical short-distance transport of bulk materials, (2) dust-free conveyance of powdered materials, (3) automatic and labor-saving, (4) avoid or minimize human contact with the materials being transported, thereby preventing product contamination and enhancing safety and security, (5) occupy less space than belt conveyors, (6) flexibility in routing — the pipeline can be horizontal, vertical, inclined, or a combination thereof, can be placed on the floor, buried, or suspended under or above ceilings, and (7) simultaneous conveyance to and from multiple points or locations by using a single system.

6.2 TYPES OF PNEUMATIC CONVEYING

There are three general types of pneumatic pipelines: **negative-pressure** (or **suction**) systems, **positive pressure** (or **pressure**) systems, and **combined (negative–positive pressure)** systems. They are separately discussed as follows:

6.2.1 NEGATIVE-PRESSURE SYSTEMS

The **negative-pressure systems**, also called **suction systems**, behave like a vacuum cleaner. The prime mover (an air pump) of the system is placed near the exit of the pipe. A vacuum (suction) is created in the pipe by the prime mover to suck or move the solid–air mixture through the pipe. Because the maximum pressure differential across a pipe that can be developed by a suction system is always less than one atmospheric pressure, the suction system can only be used for relatively short distances, normally not more than a few hundred feet. For a suction system of a given negative pressure Δp to transport a given type of solid, longer transport distance can be achieved if the pipe diameter is larger. One cannot use a lower velocity to achieve a longer transport distance because it may cause the solids to settle from the flow and block the pipe.

The suction system may use a common air pump (e.g., blower) at the pipeline outlet for several branches of pipe. This is especially practical when one wishes to transport solids from several locations to a common collection point, as for instance in transporting refuse from individual households to a common trash collection station. Due to its non-polluting nature, the suction system is exclusively used for transporting or conveying toxic or hazardous solids. The system is non-polluting for two reasons: (1) the suction in the pipe provides dust-free feeding of materials into the pipe, and (2) any leakage in the pipe will only draw air into the pipe instead of leaking any gas or solids out the pipe.

The smallest suction systems are vacuum cleaners. The largest systems are those used for unloading ships; they convey thousands of tonnes of solids per hour. An extensive pneumatic conveying suction system is used at Disney World in Orlando, Florida. It consists of an underground network of pipelines for collecting the trash from various buildings to a central station. Such systems are also used widely in Europe and Japan for collecting trash from new apartment complexes. Each system brings trash from individual apartments at various floor levels to a central collection station at the street level. It contributes to the cleanliness and sanitation of the apartments and to the convenience of the residents.

TABLE 6.1
Key Properties of Commonly Used Positive-Pressure and Negative-Pressure Pneumatic Conveying Systems

Type	Negative-Pressure (Suction) Systems	Positive-Pressure Systems	
Phase	dilute	dilute	dense
Operating pressure (bars)	0.4–1.0	1–3	1–9
Pick-up velocity (m/s)	8–41	8–41	0.25–10
Loading ratio (mass of solids/mass of air)	0–30	0–>30	0–>150

Source: Based on data from Williams, O.A., *Pneumatic and Hydraulic Conveying of Solids,* Marcel Dekker, New York, 1983.

6.2.2 POSITIVE-PRESSURE SYSTEMS

The positive-pressure system uses an air pump (blower) located near the inlet of a pipe. It can develop a pressure differential across the pipeline much greater than one atmosphere. The system can transport solids over distances much longer than those of the suction systems, especially if more than one air pump is used — either placed in series at the pipe entrance or spaced at regular intervals along the pipe as booster pumps.* When using booster pumps, one must make sure that the solids to be transported can pass through the booster pumps freely without damage to both the pumps and the solids.

For a positive-pressure system, one can use the same prime mover, located near the inlet, to transport solids from a single source to a number of receiving points through pipe branches. Such transport can be done either sequentially (one branch at a time), or simultaneously (all braches at the same time). If done simultaneously, the branches must use smaller pipes than the main in order to assure that adequate transport velocity is maintained not only in the main but also in each branch.

All positive-pressure systems require special feeders that can feed the solids to be transported into a pressurized pipe. In this case, using gravity or the weight of the solids to feed the solids into the pipe is usually insufficient. The solids must be forcibly fed into the pipe downstream of the air pump. Currently, the highest pressure used for pneumatic transport of solids is approximately ten times the atmospheric pressure. Use of such high-pressure systems requires careful selection of feeders that can feed solids satisfactorily against such pressures.

Table 6.1 lists some key properties of both the negative- and positive-pressure systems of pneumatic conveying. For simplicity, the operation pressures listed in Table 6.1 are all given in the unit of **bar**, which is approximately one standard atmospheric pressure (more precisely, 1 bar = 10^5 Pa = 100 kPa = 14.5 psia = 0.987 standard atmospheric pressure). Note that the **operating pressure** is the pressure difference (i.e., pressure upstream minus pressure downstream) required for driving the solid–air mixture through the pipe used in pneumatic conveying), whether the system is operating in the negative-pressure (less-than-atmospheric pressure) range, or the positive-pressure (above-atmospheric-pressure) range.

---

* The concept of booster pump is explained in Section 9.2.1.4.

6.2.3 COMBINED (NEGATIVE–POSITIVE PRESSURE) SYSTEMS

The combined system, also called the **suck-blow** system, combines the advantages of both the negative and the positive pressure systems into a single system. It uses a negative pressure system with multiple inlets for the upstream part, followed by a positive pressure system with multiple outlets downstream. A single air pump can be used for such a system.

6.2.4 OTHER RELATED SYSTEMS

Three other related pneumatic conveying systems, which are modifications of ordinary suction or pressure systems, should be mentioned. The first is the **gravity-assisted** system. In this case, solids slide down the slope of a pipe or chute to be conveyed to a lower place. Air is often injected through the floor of the inclined pipe or chute to *fluidize* the solids in order to facilitate their movement. Because gravity is used as the prime force to move the solids, the system is very reliable and efficient. It is used in special railroad cars (hopper cars) designed for transporting grain and other granular or powdered products.

Another type is **fluidized dense-phase** pneumatic transport. In such a system, two concentric pipes are used. The inner pipe is perforated and used for conveying the solids, and the outer pipe is for the supply of air for fluidization use. The air in the space between the two pipes is forced through the perforations of the inner pipe to fluidize the solids being conveyed. The conveyance is in **dense-phase**, which means that most of the pipe interior is filled with the solids to be transported, or the solids-to-air weight ratio is very high (greater than 100). In lieu of using two concentric pipes, the system may use a single pipe with nozzles attached to various locations along the pipe for injecting air. Each nozzle in turn is connected by tubing to a common compressor or separate compressors. Such dense-phase systems allow conveyance of large quantities of solids with relatively small pipes at relatively low velocity. It is an efficient mode of pneumatic conveyance.

A third type is the **closed-loop system**. In such a system, the conveying gas is recycled after being cleaned by filters. Since no exhaust gas is released from the system, the closed-loop system is especially suited for conveying toxic or radioactive materials, or for systems that must use an inert gas such as nitrogen, rather than air, for conveying.

6.3 FLOW CHARACTERISTICS

Theoretically, the same four regimes of flow discussed in Chapter 5 for slurry pipelines — homogeneous, heterogeneous, moving-bed, and stationary-bed — exist also for pneumatic pipelines. However, the field of pneumotransport is less developed analytically than the field of hydrotransport. Most equations predicting the flow properties of pneumotransport are empirical and different for different solids. Only a qualitative presentation of the pressure drop along a pneumatic pipeline will be presented here. The discussion is based on the findings from a classic experiment conducted by Gasterstaedt at the Technical High School of Dresden, Germany, using a laboratory pipe loop to transport wheat [1] (see Figures 6.1 and 6.2).

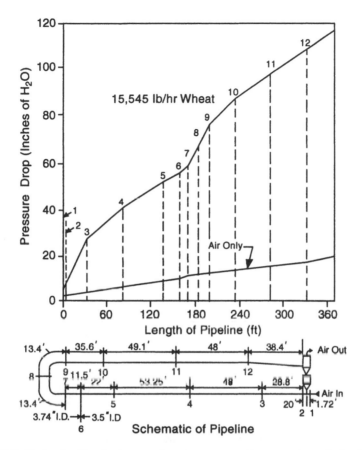

FIGURE 6.1 Gasterstaedt's classic experiment on pneumatic conveying of wheat (Note: 1 m = 3.28 ft; 1 inch = 2.54 cm; 1 kg = 2.2 lb). (From Kraus, M.N., *Pneumatic Conveying of Bulk Materials,* 2nd ed., McGraw-Hill, New York, 1980. With permission.)

Figure 6.1 shows the variation of pressure along Gasterstaedt's test loop as a function of distance from the pipeline intake for a particular case of wheat transport through this loop. Two curves are shown in the figure: the upper curve is for transporting wheat at the throughput of 15,545 lb/hr (7066 kg/hr), and the lower curve is for pumping air alone through the same pipe at the same velocity. The pipeline system (test loop) is shown on the lower part of the figure. The pressure gradient along the pipe in any given region is indicated by the slope of the pressure curve in the region. As can be seen from the upper curve, near the intake of the pipe (between points 0 and 3), the slope of the line is relatively steep, meaning that the pressure gradient in the entrance region is relatively high. This is due to the fact that the solids injected into the pipe must accelerate from zero horizontal velocity to the final fully developed velocity. The acceleration consumes energy, which is supplied by the airflow. From points 3 to 6, the solids in the pipe are not accelerating any more, and hence the pressure gradient is reduced as shown in the curve. After point 6, the flow starts to be affected by the two 90° bends, one after another. The bends cause the wheat

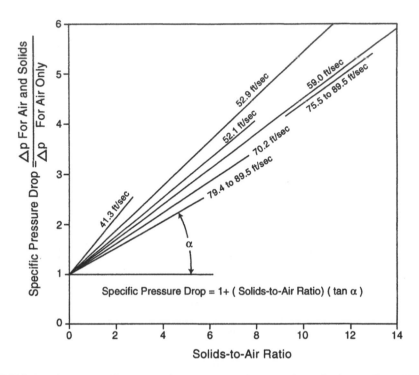

FIGURE 6.2 Gasterstaedt's test results on pneumatic conveying of wheat. (From Kraus, M.N., *Pneumatic Conveying of Bulk Materials,* 2nd ed., McGraw-Hill, New York, 1980. With permission.)

particles to impinge on the outer radius of the bends, causing excessive energy loss. This explains why the pressure drop curve is relatively steep from 6 to 9. After point 9, the retarded solids will again accelerate in the pipe until they reach point 10. This explains why the pressure gradient between points 9 and 2 is slightly greater than that in the fully developed flow regions, such as between 10 and the exit.

Figure 6.2 gives the **specific pressure drop,** N_{sp}, as a function of the solids-to-air ratio (also called **loading ratio**), N_{lr}. The specific pressure drop is the pressure drop of transporting a certain type and amount of solid material pneumatically through a given pipe, divided by the corresponding pressure drop of the air flowing through the same pipe at the same velocity without the solids. For example, based on Figure 6.1, the total pressure head drop along the pipe when transporting wheat is approximately 117 inches of water, whereas when transporting air alone it is only about 20 inches of water. This means the specific pressure drop for this case is 117/20 = 5.9, approximately. On the other hand, the loading ratio is the mass flow rate of the solids transported through the pipe, in kg/s of solids, divided by the mass flow rate of air used in transporting the solids, also in kg/s. From Figure 6.2, it is clear that the specific pressure ratio, N_{sp}, is linearly proportional to the loading ratio, N_{lr}. Written mathematically,

$$N_{sp} = 1 + N_{lr}(\tan \alpha) \tag{6.1}$$

where α is the angle of the lines in Figure 6.2 and tan is the tangent. Note that the lines in Figure 6.2 are obtained for wheat, and hence they should not be used without modification for other types of solids. For other materials, the angle α can be quite different from what are indicated in Figure 6.2. Because more than one line exists in Figure 6.2, each representing a different conveyance velocity, this means even for the same material the angle α depends on the conveyance velocity. Higher velocities correspond to smaller angle α and so on.

Example 6.1 (a) Find the relationship between N_{sp} and the quantities i_m and i defined in Chapter 5. (b) Find the relationship between N_{lr} and C_w; the latter is also defined in Chapter 5. (c) Then, write Equation 6.1 in terms of i_m, i, and C_w, and compare the result with the equation given in Problem 5.1 in Chapter 5 for slurry flow. Discuss the significance.

[Solution] (a) By the definition of specific pressure N_{sp} given in this chapter, and the definition of i and i_m given in Chapter 5 for slurry flow, it can be seen that $N_{sp} = i_m/i$. (b) From the definition of the loading ratio, N_{lr}, and the weight-based solids concentration, C_w, we have $N_{lr} = C_w/(1 - C_w)$. More rigorous proof of this relationship will be given later for Equation 6.7. (c) Substituting the foregoing relationships into Equation 6.1 yields

$$\frac{i_m}{i} = 1 + \frac{C_w}{1 - C_w}(\tan \alpha) \tag{a}$$

Substituting Equation 5.6 into Equation a and rearranging terms yields

$$\frac{i_m - i}{iC_v} = \frac{S}{1 - C_v}(\tan \alpha) \tag{b}$$

If the pneumatic (solid–gas) flow is pseudohomogeneous, then the above equation should reduce to the equation given in Problem 5.1 for pseudohomogeneous slurry flow, namely,

$$\frac{i_m - i}{iC_v} = S - 1 \tag{c}$$

Equating the right-hand sides of Equations b and c yields

$$\tan \alpha = \frac{(S - 1)(1 - C_v)}{S} \tag{d}$$

For pneumotransport, S, which is the density ratio between the solids and the conveying fluid, is very large (of the order of 1000). This means $(S - 1)$ is only slightly smaller than S, and Equation d reduces to $\tan \alpha = (1 - C_v)$. This shows that if the solid–air mixture flow were indeed pseudohomogeneous, the angle α must be constant, and it would not depend on the conveying velocity. However, since the straight lines in Figure 6.2 do depend on conveying velocity (different conveying velocities result in different straight lines), one must conclude that the solid–air mixture flows in Gasterstaedt's experiments on wheat transport are not pseudohomogeneous. They must be in the heterogeneous region, with significant variation in the solids concentration profile along the vertical axis of the pipe cross section. This shows that the theory on slurry flow may be used to interpret or even predict the behavior of pneumatic conveying.

Figure 6.3 is a comparison of the pressure gradient (i.e., pressure drop per unit length along the pipe in the flow direction) between a horizontal and a vertical (upward) pneumatic pipeline. The pressure gradient (the ordinate of the graph) is plotted against the superficial air velocity (the abscissa). The **superficial air velocity** is the volumetric airflow rate through the pipe (in cfs) divided by the pipe cross-sectional area (in ft^2). Both cases are for a constant rate of solids transported. In the horizontal case, initially the pressure gradient along the pipe decreases continuously as the superficial air velocity in the pipe is increased. This corresponds to the sliding-bed regime discussed in the previous chapter for slurry flow. When the velocity has reached a certain critical value (approximately 13 fps in Figure 6.3), a sudden drop in pressure gradient takes place. This corresponds to the state when the solids are being lifted off (fully suspended) by the air, or the sliding-bed regime has ended and changed to a heterogeneous flow regime. Beyond this point, further increase in air velocity will increase the pressure drop due to the increased collisions among the suspended particles caused by the turbulent flow and stronger collisions between the particles and the pipe. For the vertical (upward) flow case, the solids cannot be transported upwards through the pipe unless and until the air velocity is greater than the settling velocity of the particles in air — approximately 13 fps in Figure 6.3. This velocity is called the **choking velocity**. As the air velocity exceeds the choking velocity, the solids move upwards, and the pressure gradient decreases. This is caused by the fluidization of the column of solids in the vertical pipe. Fluidization makes the column of solids loose (less compact), thereby reducing the resistance of airflow passing by the solids. As the velocity increases to a certain higher value (25 fps in Figure 6.3), the pressure gradient reaches a minimum. Further increase in the air velocity will cause increased pressure gradient along the vertical pipe due to increased energy loss caused by strong turbulence. This shows that Figure 6.3 provides the information needed for a good qualitative understanding of the characteristics of the solid–gas flow through pipes in both horizontal and vertical pneumatic conveying systems.

6.4 SYSTEM LAYOUTS

6.4.1 GENERAL SYSTEMS

A typical modern pneumatic conveying system consists of the following main components:

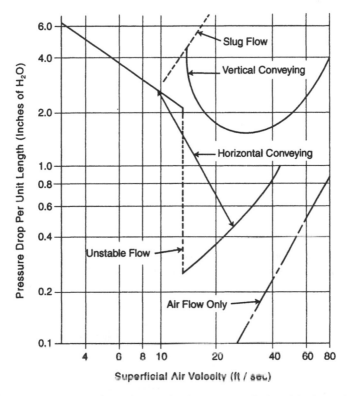

FIGURE 6.3 Variation of pressure drop and velocity in vertical and horizontal pneumatic conveying of granular material. (From Kraus, M.N., *Pneumatic Conveying of Bulk Materials,* 2nd ed., McGraw-Hill, New York, 1980. With permission.)

1. A pipe to convey the solids from point A to point B, over a certain distance.
2. A prime mover (air pump) to provide the energy and the driving force needed to convey the solids through the pipe.
3. An intake that injects the solids into the pipe for transport.
4. A separator that separates the solids from the carrier gas or air at the pipeline outlet in order to recover the solids transported.
5. A filter to remove the dust from the air or gas at the pipeline outlet before the air or gas is released into the atmosphere or recirculated.
6. An automatic control system including a PLC (programmable logic controller), a computer that communicates with the PLC, and sensors, transducers, flowmeters, and relays that interact with the PLC.

In addition, the system may have silencers for noise reduction. For drying gas or air used for conveying, drying equipment such as a dehumidifier may be needed. For systems that must recirculate the air or gas used in conveying, a return pipeline for air with a separate air pump is needed. For more details, refer to Figure 6.4 for a typical positive-pressure system, and to Figure 6.5 for a typical negative-pressure system.

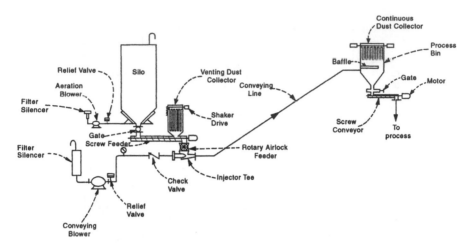

FIGURE 6.4 A typical positive-pressure, pneumatic conveying system. (From Kraus, M.N., *Pneumatic Conveying of Bulk Materials,* 2nd ed., McGraw-Hill, New York, 1980. With permission.)

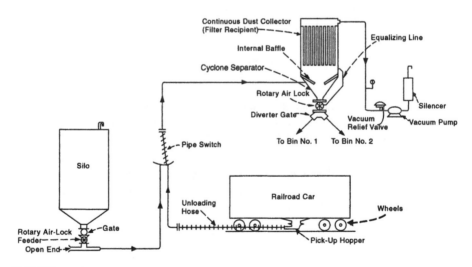

FIGURE 6.5 A typical negative-pressure, pneumatic conveying system. (From Kraus, M.N., *Pneumatic Conveying of Bulk Materials,* 2nd ed., McGraw-Hill, New York, 1980. With permission.)

6.4.2 INTAKES

Figure 6.6 shows the various kinds of solid intakes for a negative-pressure pipeline system [2]. Note that they are grouped into two categories: A and B. The A-type intakes discharge their solids into a horizontal or nearly horizontal pipe with air entering from the upstream part of the pipe rather than from the intake for solids. In fact, part of the intake may be vented (see Figure 6.7). Feeding of solids is controlled by one of the following devices: rotary airlock feeder, butterfly valve, sliding gate, or grid. The most commonly used is the rotary airlock feeder. As its

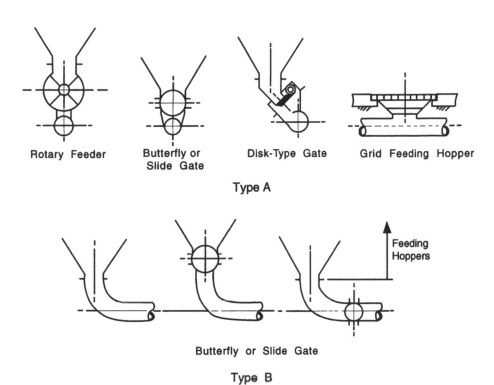

Rotary Feeder　　Butterfly or Slide Gate　　Disk-Type Gate　　Grid Feeding Hopper

Type A

Butterfly or Slide Gate

Type B

FIGURE 6.6 Types of negative-pressure intake systems for pneumatic conveying. (From Williams, O.A., *Pneumatic and Hydraulic Conveying of Solids,* Marcel Dekker, New York, 1983. With permission.)

name suggests, it not only feeds the solids but also keeps a tight seal to prevent air leakage through the feeder. Figure 6.7 is a close-up view of the airlock feeder in operation. As shown in Figure 6.6, type-B feeders discharge their contents (solids) into the pipe inlet via a vertical bend. All the air used for conveying enters through the feeder. Feeding is controlled by a butterfly or slide gate. Some are not controlled by any gate in the feeder. In such a case, control must be obtained with a gate mounted in the pipeline. For positive-pressure systems, many types of feeders are available commercially. They all involve using a pressure tank to hold the solids and applying air or gas pressure to the tank in order to discharge the solids from the tank into the pipe. Figure 6.7 is a special type, called a **blow tank,** widely used to feed solids into a positive-pressure pneumatic pipeline system.

6.4.3 PRIME MOVERS (AIR PUMPS)

The prime mover (air pump) of any pneumatic conveying system may be a fan (for low pressure systems of less than about 0.3 bar), a blower (for medium pressure in the range 0.3 bar to 3 bar, approximate), or a compressor (for pressure above 3 bar). For negative pressure systems that must develop a high vacuum (up to about 0.8 bar), a vacuum pump is used, which is usually a piston pump. Chapter 9 contains

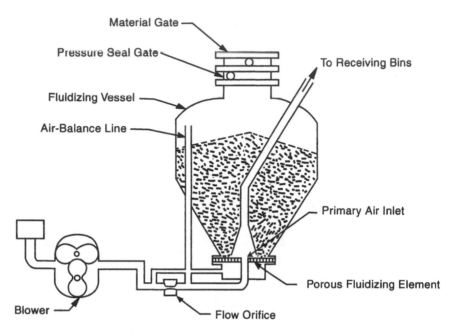

Material Gate

Pressure Seal Gate

To Receiving Bins

Fluidizing Vessel

Air-Balance Line

Primary Air Inlet

Porous Fluidizing Element

Blower

Flow Orifice

FIGURE 6.7 A typical low-pressure blow tank for pneumatic conveying. (Courtesy of Delta-Ducon Company.)

a detailed discussion of pumps (including air pumps), and their selection, operation, and maintenance.

A major difference exists between pneumatic and hydraulic conveying of solids; while the slurry (solid–liquid mixture) is allowed to go through the slurry pumps, in pneumatic conveying the solids are not allowed to pass through the pump. The solids must be either injected downstream of the air pump — for the positive pressure systems, or removed from the air stream before the air enters the pump — for the negative pressure systems. This greatly limits the conveying distance of current pneumatic pipeline systems. The main reasons for not allowing solids to pass through the pump are (1) prevent wear (erosion) to the pump and (2) prevent damage to the product (solid particles) by the pump. Even in situations where damage to the product is of no concern and pump wear can be controlled, there is still a lack of incentive for allowing solids to pass through pumps for long-distance transport. This is due to the high energy-intensiveness of horizontal pneumatic conveying, especially in the dilute-phase transport.

6.4.4 SEPARATOR AND CLEANER

At the end of any negative or positive pneumatic conveying system, a separator is needed to separate the solids from the air. This is most often done by using a cyclone separator (see Figure 6.5). The device efficiently separates the solids from the air by utilizing the centrifugal force generated from the spiral motion of the solids entering the separator in a tangential (circumferential) direction. The mixture enters the separator at a level below the baffle, which prevents the particles from rising and entering the filter zone. The filter above the cyclone is needed to filter out the dust (fine particles) that cannot be removed by the cyclone action. Thus, the cyclone and the filter together remove (separate) the fine and coarse solids from the delivered solid–air mixture. Other simpler systems also exist, such as the one in Figure 6.4, which is done without a cyclone, but still with a baffle plate and a filter. The cleaner (filter) is the dust collector; it is usually a bag made of some fabric material. The largest of such systems is the bag-house used at coal-fired power plants to capture the fly ash from the exhaust gas before it is emitted through the chimney or stack. Combustion exhaust is a special case of dilute-phase pneumatic conveying using positive pressure.

6.5 SYSTEM DESIGN

Due to the lack of sufficient scientific knowledge, and published engineering data on pneumatic conveying, design of any pneumatic conveying system for a new application is always difficult. Unless the equipment vendor has experience from a similar application (the same type of solids to be transported, and the same pipe diameter, length, and fittings), he or she must conduct either a pilot plant test or a full-length test to determine accurately the design parameters, such as what convey-ing speed should be used, and at what loading rate. Also, data must be available so that the pressure drop along the pipe can be accurately predicted. With the pressure drop known, one can then size the air pump and determine the horsepower. The general design procedure involves the following steps:

1. Define the problem. Determine what solids need to be transported, for what distance, in what quantity, in what form (particle size distribution), and so on.
2. Select the conveying gas. In most cases, air can be used. However, in special cases where air may contaminate the product transported or cause fire or explosion, an inert gas such as nitrogen may be used. When air is to be used, determine whether the air must be dried before use.
3. Decide whether to use dilute-phase or dense-phase conveying. The design calculations for the two types are very different.
4. Determine the conveying velocity. If the same material has been tested before for conveying velocity and the critical velocity is known from such tests, then the optimum conveying velocity is known. Otherwise, one must run experiments to determine the critical velocity.
5. Select the solids loading ratio, N_{sl}. Different loading ratios may have to be considered. If the loading ratio is too low, a very large pipe, or more than one pipe, may be needed. If the loading is too high, blockage may occur. Even without blockage, high loading rates may cause excessive pressure drop, which in turn may cause high energy loss.
6. Determine the pipe size to suit the airflow rate and the air velocity. The pipe size should be such that when the pipe cross-sectional area is mul-tiplied by the conveying velocity, the resultant discharge Q must be able to transport the amount of solids to be transported, Q_s.

7. Calculate the pressure drop or headloss across the pipe. This again must be based on past data collected under similar conditions so that they will be applicable. The pressure drop must include not only those produced along fully developed straight pipe, but also those in the entrance region (due to particle acceleration), across bends, and across other fittings.
8. Calculate the power. The power is computed by multiplying the volumetric discharge of the mixture, Q_m, by the pressure drop Δp.
9. Determine the type of air pump needed. Use a fan when the pressure is low (less than 0.3 bar), use a blower for medium pressure (0.3 to 3 bar), and use a compressor for higher pressure (>3 bar).
10. Select the air pump. Similar to the selection of liquid pumps, one must make a selection based on the pump characteristic curves. See Chapter 9 for details on pump selection.
11. Determine the operating pressure and discharge of the pump. This is done by plotting both the system curve and pump curve, and finding the operation point from the intersection, as detailed in Chapter 9.
12. Determine the efficiency and the brake-horsepower of the pump.
13. Select the motor that drives the pump, and determine the required power input to the motor. See Chapter 9 for details.

6.6 SAFETY CONSIDERATIONS

Careless design or misuse of pneumatic conveying systems may not only cause blockage of pipe and frequent disruptions of operation, but also fire and even explosion when the materials transported are combustible, such as grain, flour, coal, or gunpowder. Therefore, safety should be of utmost concern in designing systems of pneumatic pipelines that carry such combustible and/or explosive materials.

The cause of fire and explosion in pneumatic pipelines is the accumulation of electric charge at certain points in the pipe and the subsequent discharge of such charges by sparking or electric arcing generated by the high voltage produced by such charges. Due to impingement of solid particles on pipe walls, bends, or fittings, charges are constantly generated and separated from solids and accumulate in certain places along a pipe, such as at a valve or in the receiving bin or cyclone. If such charges accumulate in an insulated place, they are not discharged to the ground and voltage can build up to a high and dangerous level causing a spark or arc discharge. Then the system may ignite or explode.

Due to their large cumulative surface areas, fine particles carry more charge and are more dangerous than coarse particles. Plastic pipes do not conduct electricity to ground and hence are more dangerous than metallic pipes used in pneumatic conveying. Coating metal pipe and the receiving-bin inner surface by dust also makes pneumatic pipelines more dangerous. To minimize danger and maximize safety, the following precautions should be taken:

1. Use metallic rather than nonmetallic pipe and receiving bins.
2. Ground many places along the pipeline, especially at valves, joints, silos, cyclones, and bins.

3. Avoid transporting fine powder of combustible solids. When this is not possible or practical, pay special attention to safety in the design of such a system. Use the services of a safety expert.
4. Avoid using dry air — 80% relative humidity will avoid most problems. However, the effect of moisture on product quality must be considered.
5. Use an inert gas such as nitrogen for conveying combustible materials.
6. Clean the pipeline interior periodically with water to remove dust coating.
7. Use vibrators to remove dust deposited in bins and cyclones.

6.7 ANALYSES

Even though the design of pneumatic pipeline systems are highly empirical and dependent on test data for individual systems, much can be learned from analysis to help the practitioner understand or gain insight into certain critical features of pneumatic conveying. Some of such analyses are provided next.

6.7.1 PICKUP VELOCITY

The pickup velocity, also called **saltation velocity** in the field of pneumatic conveying, is the minimum velocity needed to cause the solids to be totally suspended by the airflow in a horizontal pipe. The same velocity exists in hydrotransport of solids, i.e., for slurry pipelines, except that those working in the slurry field call it the **limit-deposit velocity,** or **deposition velocity** for short (see Chapter 5). Due to the close similarity between hydraulic and pneumatic transports of solids, one should expect that the same basic conditions and the same forces for causing solids to be lifted (suspended) must be at work in both types of flows. From the theory of slurry transport, it is known that particles become suspended when the vertical component of turbulence (i.e., turbulent velocity fluctuations), v', is greater than the settling velocity, V_s, of the particle in the fluid. This should also be the case for pneumatic transport of solids. This means that the deposition and pickup of particles, in both hydro and pneumo transports, should begin when the turbulence component v' reaches the same magnitude as V_s. However, because v' is directly proportional to the mean flow velocity V, the deposition or pickup velocity for both hydro and pneumo cases must be directly proportional to the settling velocity, as given by the following relationship:

$$V_L' = V_L \frac{V_s'}{V_s'} \tag{6.2}$$

In the above equation, the primed quantities are those for gas flow (pneumotransport), and the unprimed quantities are those for liquid flow (hydrotransport).

Use of Equation 6.2 depends on knowing the settling velocity of the particles (solids) in both the liquid and the gas used for transporting the particles. As discussed in Chapter 5, accurate determination of the settling velocity of particles of arbitrary shapes is possible only by experiments (settling column tests). In the absence of experiments, approximate values of settling velocities can be determined from theory by assuming that each particle is a sphere having a diameter that will yield the same volume of the particle of irregular shape. Such theoretical determination may not be accurate for individual velocities V_s and V_s', but it is sufficiently accurate for determining the ratio V_s'/V_s in Equation 6.2. From fluid mechanics, the settling velocity of any sphere in any Newtonian fluid can be determined from the following equations:

$$V_s = \frac{\rho g(S-1)d_s^2}{18\mu} \quad \text{for } \Re < 1 \tag{6.3}$$

and

$$V_s = \sqrt{\frac{4g(S-1)d_s}{3C_D}} \quad \text{for } \Re > 1 \tag{6.4}$$

where ρ and μ are, respectively, the density and dynamic viscosity of the fluid; g is the gravitational acceleration; d_s is the diameter of the sphere; S is the density ratio, which is the density of the solid particle, ρ_s, divided by the density of the fluid, ρ; C_D is the drag coefficient of the sphere; and $\Re$ is the settling Reynolds number equal to $\rho V_s d_s/\mu$. The values of C_D is a function of the settling Reynolds number, $\Re$, given in graphs and formulas in most fluid mechanics texts.

The following example illustrates the use of the information just described.

Example 6.2 (a) Find the limit-deposit velocity of coarse sand particles of uniform diameter of 2 mm transported by water in a pipe of 10-cm diameter. The specific gravity of the sand is 2.4, and the volume concentration of the sand is 10%. (b) Use the result of part (a) to determine the pickup velocity if the same sand is transported by the same pipe under the same conditions as in part (a), except that air instead of water is used for the transport.

[Solution] (a) From the Durand's graph given in Chapter 5 (Figure 5.3a), for particles of 2 mm size, the densimetric Froude number, F_r, is 1.36. Therefore, $V_L = 1.36\sqrt{2gD(S-1)} = 1.36 \times \sqrt{2 \times 9.81 \times 0.1(2.4-1)} = 2.25$ m/s. This gives the deposition velocity of the sand in water. (b) The S value for the sand in water is 2.4, and the dynamic viscosity of water at 20°C is $\mu = 1.002 \times 10^{-3}$ N-s/m². Using Equation 6.4 and information on C_D given in fluid mechanics, we find the value of C_D to be 0.50, and $V_s = 0.271$ m/s. When air instead of water is used, the S value is now increased greatly: $S = 2400/1.2 = 2000$, approximately. Again, using Equation 6.4 and information on C_D in fluid mechanics, we find that $C_D = 0.4$, and $V_s' = 11.4$ m/s.

Now that V_L, V_s, and V_s' are all determined, Equation 6.2 can be used, which yields $V_L' = 94.6$ m/s. The high value of the pickup velocity in air obtained here is due to a much higher settling velocity (42 times higher) in air than in water. The high density of the solids to be transported (specific gravity equals 2.4) also contributes to the high pickup velocity in air.

6.7.2 DENSITY AND PICKUP VELOCITY VARIATION ALONG PIPELINE

From Chapter 3, it is known that when any gas flows through a pipe of constant diameter, the pressure of the gas decreases along the pipe, which causes a decrease in the density of the gas and a corresponding increase in the gas velocity along the pipe. If the temperature is isothermal, as is often the case for pipelines, the density decrease and the velocity increase will be directly proportional to the decrease in the absolute pressure along the pipe. This relationship holds with or without solids transported through the pipe. For example, if a pneumatic pipeline that conveys solids requires a pressure drop of 5 atmospheric pressure (from 6 atmosphere to 1 atmosphere in terms of absolute pressure), under constant temperature the gas density will decrease by a factor of 6, and the gas speed will increase by a factor of 6, provided that the pipe diameter remains constant. Such high-speed conveying (at 6 times the original velocity) not only causes unnecessary energy loss (headloss), but also abrasion of the pipeline and possible damage to the product (solids) transported. To solve this problem, one may design the pipeline to have stepped-up diameters, with larger diameters for the downstream part of the pipe. The increase in diameter should be based on what is needed to provide the minimum velocity to suspend the solids, namely, the saltation velocity or pickup velocity. The following example illustrates how this pickup velocity can be determined for the increased pipe diameter along the pipe.

Example 6.3 A pneumatic pipeline starts with a 6-inch-diameter pipe, and the upstream pressure (at the pump outlet) is 6 atmosphere absolute. The outlet pressure is 1 atmosphere absolute. This makes the density of the air near the outlet to be the same as that for atmospheric air, and the density of the air upstream to be 6 times that of downstream. Suppose that the pickup velocity of the solids upstream (at 6 times the atmospheric pressure) is 20 m/s. (a) What should be the pickup velocity of the air near the outlet? (b) What should be the pipe diameter downstream near the outlet in order to match the pickup velocity there?

[Solution] From discussion in the previous example, it is seen that the pickup velocity, the same as the limit-deposit velocity of slurry pipeline, is expected to be proportional to the square root of $D(S-1)$, where D is the pipe diameter, and S is the density ratio ρ_s/ρ, namely,

$$V_L' \propto \sqrt{D(S-1)} \propto \sqrt{DS} \propto \sqrt{D(\rho_s/\rho)} \propto \sqrt{D/\rho} \qquad \text{(a)}$$

The sign $\propto$ in Equation a is the proportionality sign. The various steps in Equation a are justified for pneumatic transport of solids because $S \gg 1$, and ρ_s is constant (the solid density does not change with pressure change). Using Equation a for both the upstream point, 1, and the downstream point, 2, yields

$$\frac{V_1}{V_2} = \left(\frac{D_1}{D_2} \frac{\rho_2}{\rho_1}\right)^{1/2} = \left(\frac{D_1}{D_2} \times \frac{1}{6}\right)^{1/2} \qquad \text{(b)}$$

From the continuity equation $\rho_1 V_1 A_1 = \rho_2 V_2 A_2$ given in Chapter 3, we have

$$\frac{V_1}{V_2} = \left(\frac{D_2}{D_1}\right)^2 \times \left(\frac{1}{6}\right) \qquad \text{(c)}$$

Solving Equations b and c simultaneously yields $D_2/D_1 = 1.431$, and $V_2/V_1 = 2.93$. This means $D_2 = 1.431 D_1 = 8.59$ inches, and $V_2 = 2.93 V_1 = 58.6$ m/s. Because pipe diameter does not come in 8.59 inches, either an 8-inch pipe or a 10-inch pipe should be used downstream. The best for this pipeline is to use pipes of 3 sizes: 6-inch pipe for the upstream one third of the pipeline, 8-inch pipe for the middle one third of the pipeline, and 10-inch pipe for the downstream one third of the pipeline.

6.7.3 LOADING RATIO

The loading ratio, N_{lr}, is defined as the mass flow rate of the solids transported in a pipe, divided by the mass flow rate of the gas used for the transport, namely,

$$N_{lr} = \frac{\rho_s Q_s}{\rho Q} \qquad (6.5)$$

where Q and Q_s are the volumetric flow rate of the fluid and the solids, respectively. But, by definition, the weight concentration of the solids in the pipe, C_w, is the weight flow rate of the solids, divided by the weight flow rate of the mixture in the pipe, namely,

$$C_w = \frac{\rho_s g Q_s}{\rho_s g Q_s + \rho g Q} = \frac{\rho_s Q_s}{\rho_s Q_s + \rho Q} \qquad (6.6)$$

Substituting Equation 6.5 into Equation 6.6 yields

$$C_w = \frac{N_{lr}}{1 + N_{lr}} \quad \text{or} \quad N_{lr} = \frac{C_w}{1 - C_w} \qquad (6.7)$$

Substituting Equation 5.6 from Chapter 5 into the above equation yields

$$C_v = \frac{N_{lr}}{S + N_{lr}} \quad \text{or} \quad N_{lr} = \frac{SC_v}{1 - C_v} \qquad (6.8)$$

Example 6.4 A pneumatic pipeline transports fine coal particles from the coal pulverizer of a power plant to a boiler of the same plant for combustion to generate electricity. The coal particles through the pipe have a volume concentration rate of 10%, and the specific gravity of the coal particles is 1.4. If the air in the pipe has a density of 1 kg/m³, what is the S value and the loading factor N_{lf} for this case?

[Solution] Because the coal has a specific gravity of 1.4, the density of the coal is $\rho_s = 1.4 \times 998 = 1397$ kg/m³. The density of the air is $\rho = 1.0$ kg/m³. Therefore, $S = 1397/1.0 = 1397$. From Equation 6.8, $N_{lr} = (1397 \times 0.1)/(1 - 0.1) = 155$ approximately. This shows that for pneumotransport, even a relatively small volume concentration, such as 10%, results in a very large loading ratio. This is, of course, due to the small density of air as compared to that of solids.

Note that even after the pickup velocity of a given case is determined, one cannot determine what the loading ratio is because there is no direct or unique relationship between the pickup velocity and the loading ratio. In fact, for any given pickup velocity, one can design and operate a pipeline at any loading ratio. However, if the loading ratio is too small, the pipe will be transporting little solids, and hence will not be economical. On the other hand, if the loading ratio used is too large, either the pipe may get clogged by the solids, or too much friction will be generated between the solids and the pipe, causing excessive headloss. Only through experience learned from similar systems and from tests can one determine what the optimum loading ratio is for a given case. Table 6.1 lists the range of loading ratio commonly used in pneumotransport. Note than the loading ratio for negative pressure systems is limited to about 30. This is due to the rather limited pressure drop (less than one atmosphere) that can be derived from such systems for transporting solids.

6.7.4 Pressure Drop along Pipe in Dilute-Phase Transport

The approach to determine the total pressure drop along a pneumatic pipeline is similar to that for single-phase flow, in that one must evaluate separately the **pipe loss** along the entire length of the pipe, and then add to it the **local losses** due to local effects such as pipe entrance, bends, valves, etc. They are separately discussed below.

6.7.4.1 Pipe Loss (Loss in Straight Uniform Pipe)

Consider the general case of a long straight pipe of uniform diameter, with the pipe inclined at an angle α to the horizontal plane (see Figure 6.8). A solid–fluid mixture is moving through the pipe going up the slope. Under steady and fully developed flow conditions, the pressure gradient in the pipe for the mixture is

$$i_m = \frac{\Delta p_m}{L} = (i_m)_s + (i_m)_d \tag{6.9}$$

where the first and second terms on the right side of the above equation are the pressure gradient due to static and dynamic effects, respectively.

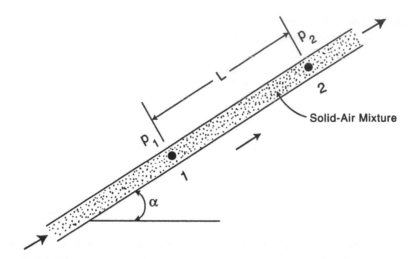

FIGURE 6.8 Analysis of pneumatic conveying of solids through an inclined pipe.

The static pressure gradient, caused by gravity and elevation change of the two ends of the pipe (due to the pipe slope), includes two parts. The part on the solids is $(1 - \varepsilon)(\sin \alpha)\rho_s g$, and the part on the fluid is $\varepsilon \rho g(\sin \alpha)$, in which ε is the **void ratio**, which is the space in the pipe filled by the fluid instead of the solids, divided by the total volume of space in the pipe. Combining the two parts yields:

$$(i_m)_s = [S(1-\varepsilon)+\varepsilon]\rho g(\sin \alpha) \qquad (6.10)$$

Likewise, the dynamic pressure gradient, needed to overcome the frictional force on the mixture by the pipe due to motion of the mixture, can also be broken down into two parts. The part due to the fluid is $f\rho V^2/(2D)$, and the part due to the solids in the fluid is $f_s \rho_s (1 - \varepsilon) V_p^2 /(2D)$. Note that f is the friction factor that can be found from the Moody diagram, and f_s is the friction factor for the solids, which can be found from experience or test data. The quantity V_p is the mean velocity of the particles moving through the pipe. Combining the two parts yields

$$(i_m)_d = \frac{f_s \rho_s (1-\varepsilon) V_p^2}{2D} + \frac{f\rho V^2}{2D} \qquad (6.11)$$

Substituting Equations 6.10 and 6.11 into Equation 6.9 yields the total pressure gradient i_m.

In using the above approach to find the pressure gradient i_m of any horizontal pipe, the static pressure term becomes zero since α is zero. When the pipe is vertical with the flow going upward, $\alpha = 90°$, and $\sin \alpha = 1$. Therefore, for vertical pipe the right-hand side of Equation 6.10 reduces to $\rho g[S(1 - \varepsilon) + \varepsilon]$. Furthermore, with sufficient solids in the conveying gas, the second term on the right side of Equation 6.11, which represents the pressure gradient due to the gas, is much smaller than the first term, which represents the pressure gradient due to solids. Thus, the second term can be neglected for pneumatic conveying. Various empirical formulas have been proposed in the literature for determining the values of f_s and V_p for solids–air mixtures in horizontal pipe; they are discussed in Reference 4.

6.7.4.2 Local Losses

Local losses due to fittings, such as pipeline inlet, bends, pipe expansions, diverters, open valves, etc., may not be important for long and straight pneumatic pipelines, but they can be the dominating factor in the flow behavior when the pipeline is short. Therefore, these losses must be properly taken into account in the design of short pneumatic pipelines. Unfortunately, insufficient information exists in the literature to enable the designer to have a good estimate of the local losses of various types of fittings under various conditions such as loading ratio and conveying velocity. The designer must either rely on a pilot plant test, or use the design a similar system that has been tested before. Books on pneumatic conveying, such as References 3 and 4, contain useful information that can help the designer. Only two specific local losses, namely, the inlet loss and the bend loss, will be discussed here.

6.7.4.2.1 Inlet Loss Due to Particle Acceleration

At the inlet of any pneumatic conveying system, both the solids to be transported and the fluid (normally air) must be fed into the pipeline, either through two separate inlets (Type A feeder, Figure 6.6) or through a common inlet (Type B feeder). Either way, the solids enter the pipe laterally at zero horizontal velocity, and they must accelerate in the pipe to attain a final velocity of the mixture flow. The energy expended by the flow to cause the particles to accelerate and reach the final mixture velocity causes a headloss or pressure drop that is in addition to the pipe loss discussed previously. This inlet loss due to particle acceleration can be quite substantial for a relatively short pipeline. A number of empirical relations have been proposed in the literature to determine this inlet loss; they are listed and discussed in References 3 and 4. In what follows, a theoretical analysis is provided to determine this pressure loss due to particle acceleration. Future researchers should check the correctness of the following theory through experiments.

Consider the inlet region of a pneumatic pipeline as shown in Figure 6.9. Sections 1, 2, and 3 are, respectively, a section of the pipe upstream of the particle injection, a section downstream in the pipe where the acceleration of the particles has ceased and the particles have reached the mixture velocity, and the feeder outlet. The solids and fluid in the pipe between these three sections constitute the control volume to be analyzed. Assume steady operation with constant inflow of air through Section 1, constant inflow of solids through Section 2, and constant outflow of mixture through Section 3. From fluid mechanics, the power of the airflow at Section 1 is

$$P_1 = p_1 Q + \frac{\rho V^2 Q}{2} \tag{6.12}$$

where P_1, p_1, Q, ρ, and V are, respectively, the power, pressure, discharge, density, and velocity of the fluid (air) at Section 1. Likewise, the power at Section 2 is

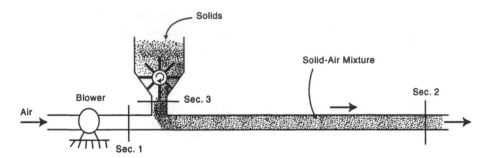

FIGURE 6.9 Analysis of pressure drop near pipe intake in pneumatic conveying due to solids acceleration.

$$P_2 = p_2 Q_m + \frac{\rho_m V_m^2 Q_m}{2} \tag{6.13}$$

where the subscript m denotes mixture. Note that potential energy is not included in the two foregoing equations because the pipeline near the intake is horizontal — no change of potential energy from Section 1 to 2.

From the above results, the power loss of the flow between Sections 1 and 2 is

$$\Delta P = P_1 - P_2 = p_1 Q - p_2 Q_m + \frac{\rho V^2 Q}{2} - \frac{\rho_m V_m^2 Q_m}{2} \tag{6.14}$$

On the other hand, the power gained by the solids through acceleration in the pipe is

$$P_s = \frac{\rho_s V_m^2 Q_s}{2} \tag{6.15}$$

where the subscript s denotes solids. It is assumed in the above equation that the solids will accelerate to the mixture velocity, V_m, when they reach Section 2.

Knowing that the power lost in the pipe, ΔP, must equal to the power gained by the solids through acceleration, P_s, namely, $\Delta P = P_s$, the last two equations can be combined to yield

$$P_2 = P_1 \left(\frac{Q}{Q_m} \right) + \frac{\rho V^2}{2} \left(\frac{Q}{Q_m} \right) - \frac{\rho_s V_m^2}{2} \left(\frac{Q_s}{Q_m} \right) - \frac{\rho_m V_m^2}{2} \tag{6.16}$$

To solve the above equation, one also needs two other equations derived from continuity consideration. They are

$$Q + Q_s = Q_m \tag{6.17}$$

and

$$\rho Q + \rho_s Q_s = \rho_m Q_m \tag{6.18}$$

Note that Equation 6.17 is justified only when the pressure loss from Section 1 to Section 2 is not severe, namely, $(p_1 - p_2)$ is much less than p_1 absolute. Otherwise, the compressibility of the air in this region, between Sections 1 and 2, must be taken into account.

Example 6.5 A positive-pressure pneumatic conveying system has an intake similar to that shown in Figure 6.9. The pipe diameter is 4 inches, and the inflow air velocity is 20 m/s. The pipeline is at a constant temperature of 20°C, which is the same as the ambient temperature. Solids of a density of 798 kg/s are fed into the pipe at a rate of 15 kg/s. Air is pumped into the pipe at an absolute pressure of 300 kPa. The outlet of the pipe is atmospheric, at 101 kPa. Determine: (a) the mass flow rate of air through the pipe; (b) the mass flow rate of the mixture through the pipe; (c) the volumetric flow rates of the air, solids, and mixture — Q, Q_s, and Q_m, respectively; (d) the mixture velocity V_m and the mixture density ρ_m; (e) the loading ratio; and (f) the pressure drop due to particle acceleration.

[Solution] (a) The engineering gas constant R for air, in SI units, is 287 J/kg K. The air temperature is $T = (20 + 273) = 293$ K. Therefore, from the equation of state of ideal gas, the air density in the pipe at the location of Section 1 in Figure 6.9 is $\rho = p/RT = 300 \times 10^3/(287 \times 293) = 3.57$ kg/m³. The pipe diameter is $D = 4$ inches $= 0.1016$ m, and the pipe cross-sectional area is $A = 0.00811$ m². Therefore, the mass flow rate of air through the pipe is $\rho VA = 0.579$ kg/s. (b) The mass flow rate of solids through the pipe is given as 15 kg/s. Therefore, from Equation 6.18, the mass flow rate of the mixture is $\rho_m Q_m = 0.579 + 15 = 15.6$ kg/s. (c) The volumetric flow rate is the mass flow rate divided by density. Therefore, the volumetric flow rates for the air and the solids are, respectively, $Q = 0.579/3.57 = 0.162$ m³/s, and $Q_s = 15/798 = 0.0188$ m³/s. Then, from Equation 6.17, $Q_m = 0.162 + 0.0188 = 0.181$ m³/s. (d) The mixture velocity is $V_m = Q_m/A = 0.181/0.00811 = 22.3$ m/s. The mixture density is $\rho_m = \rho_m Q_m/Q_m = 15.6/0.181 = 86.1$ kg/m³. (e) The loading ratio is $N_{lr} = \rho_s Q_s/\rho Q = 15/0.579 = 25.9 \approx 26$. (f) Now that all the values of the quantities on the right side of Equation 6.16 have been determined, they can be substituted into Equation 6.16, which yields $p_2 = 227$ kPa. From this calculation, we see that the drop of pressure due to particle acceleration in this case is $\Delta p = 300 - 227 = 73$ kPa, which is quite substantial.

6.7.4.2.2 Loss along Bends

Particles going around bends impinge on the outer wall of the bend. This causes particle deceleration in the bend, which must be followed by acceleration after passing through the bend, in a way similar to the acceleration near the inlet. Because in a working bend solids do not come to a stop in the bend (otherwise the pipe would be clogged and the bend would not be working), the headloss and pressure drop along the bend will be substantially less than that caused by particle acceleration near the inlet of the same pipeline. However, there is no analytical method to predict such headloss or pressure drop, and experimental data exist only for certain solids under limited conditions. Therefore, accurate prediction of headloss (or pressure drop) for bends must be tested in the laboratory under similar conditions of that encountered in the field. The loss or pressure drop depends on the bend angle, bend radius, particle loading ratio, operating velocity, and particle characteristics. Vertical bends also behave differently from horizontal bends. For vertical bends, downward-flow bends produce larger headloss than upward-flow bends. Generally, larger bend angle, smaller bend radius, and larger loading ratio all contribute to greater pressure drop. Optimum bend radius is often found in the neighborhood of 10 diameters of the pipe.

6.7.5 VERTICAL CONVEYING

Most horizontal pneumatic pipelines also have short sections of vertical pipe, in which the flow may either be upward or downward. Because the pickup velocity in a given horizontal pipe is normally significantly higher than the velocity required to operate the same pipe in the vertical position, once a mainly horizontal system is adequately designed for the horizontal part, the system will normally be adequate for the vertical part as well. This is true provided that the bends from horizontal to vertical and vice versa are well designed, and do not cause blockage. For pipes that are primarily vertical and upward, one must determine what the minimum velocity is to move the solid upward — the **choking velocity**. The system must be operated at a velocity significantly greater than the choking velocity.

If all the solid particles in a vertical flow were identical in size, shape, and density, and if the velocity profile of the upward flow were constant across the vertical pipe, then the choking velocity would be simply the settling velocity (terminal velocity) of the particles in the same fluid. However, due to the fact that the particles in a pneumatic conveying system vary considerably in size and shape, and sometimes in density as well, the settling velocities of different particles are different. Furthermore, the boundary layer across the pipe causes the velocity of the air to vary from a maximum at the pipe centerline to zero velocity at the wall. Due to these complexities, the upward flow of particles in pneumatic conveying behaves in a complex manner. Particles tend to rise upward in the center portion of the pipe where the velocity is high, whereas they tend to fall in the region near wall. This causes circulation of fluid and mixing of particles in the pipe. Under such a complex situation, prediction of the choking velocity from analysis is difficult, and many empirical formulas have been proposed in the literature. Readers interested in this subject should consult Reference 4.

6.7.6 DENSE-PHASE FLOW

Dense-phase flow of pneumatic conveying has been classified in various ways by different investigators. A way to classify dense phase is by the loading ratio: the

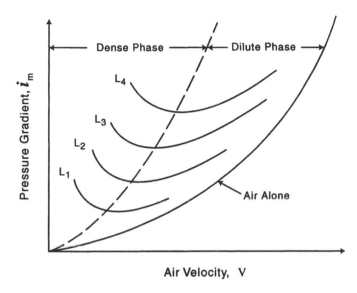

FIGURE 6.10 State diagram of a typical pneumatic conveying pipeline. L_1 through L_4 represent different mass flow rates.

conveying is dense phase when the loading ratio is greater than 30 [3]. Another way is by the volume concentration of the flow, C_w, such as the flow is considered dense phase when C_w is greater than 5%. All such classifications are somewhat arbitrary, and they should not be considered as clear cut lines to distinguish dense phase from the dilute phase. A more scientific way to distinguish the dense phase from the dilute phase is to use the *state diagram*, which is the graph of pressure gradient plotted against fluid velocity (see Figure 6.10). Note the similarity between Figure 6.10 and Figure 5.2 for hydrotransport. From Figure 6.10, for each solid mass flow rate, the pressure gradient first decreases with increasing air velocity. After the pressure gradient has reached a minimum, further increase in velocity causes the pressure gradient to increase. The dashed line connecting the minimum of pressure gradients for different mass flow rates of the solids marks the demarcation between the dilute phase and the dense phase, with the dense phase being on the left and the dilute phase being on the right of the line.

Note that dense-phase pneumatic conveying corresponds to the moving-bed regime of hydrotransport of solids (see Figure 5.2). In this phase, particles move in a variety of ways as they do in hydrotransport. They may move as dunes, or slugs. Calculation of such motion and the resultant pressure drop in the pipe are highly empirical. Readers interested in such calculations may want to consult References 3 through 5.

PROBLEMS

6.1 Calculate the settling velocity of sand particles in air for two different sand grain sizes: 0.1 mm and 1 mm. The specific gravity of the sand is 2.4, and the particles fall in the air at standard atmospheric conditions. Assume that the particles are spherical, and the sizes cited are the diameters.

6.2 Work the previous problem except that the sand particles are falling in water instead of air. Again assume standard atmospheric temperature and pressure. Compare the result with that of the last problem obtained in air.

6.3 Work the same problem as the one given in Example 6.2, except for the fact that the sand particles used in the pneumotransport are those given in the two previous problems, having sizes of 0.1 mm and 1 mm. Note the strong effect of particle size on pickup velocity. Do you expect the same trend when the particle size is much smaller than 0.1 mm? Why?

6.4 Use a computer to generate a table on the loading ratio, N_{lr}, as a function of the solid weight concentration, C_w. Vary the values of C_w from 0 to 0.50, at constant intervals of 0.02.

6.5 Solid particles are to be transported by a negative-pressure pneumatic pipeline, using air at 20°C and absolute pressure of 101 kPa at the inlet. If the specific gravity of the solids is 0.8, what is the loading ratio at the inlet? What would be the loading ratio near the outlet where the air pressure falls to 40 kPa absolute?

6.6 Work the same problem as given in Example 6.5, except for the fact that the solid load is decreased by half to 7.5 kg/s. Find the pressure drop due to particle acceleration near the inlet, and compare the result with that of Example 6.5. Determine whether the pressure drop is linearly proportional to the solid load.

REFERENCES

1. Kraus, M.N., *Pneumatic Conveying of Bulk Materials*, 2nd ed., McGraw-Hill, New York, 1980.
2. Williams, O.A., *Pneumatic and Hydraulic Conveying of Solids*, Marcel Dekker, New York, 1983.
3. Marcus, R.D. et al., *Pneumatic Conveying of Solids*, Chapman & Hall, London, 1990.
4. Shamlow, P.A., *Handling of Bulk Solids: Theory and Practice*, Butterworths, London, 1988.
5. ASCE, Sedimentation Engineering, Task Committee Report, American Society of Civil Engineers, New York, 1975.

7 Capsule Pipelines

7.1 INTRODUCTION AND HISTORY

A **capsule pipeline** transports material or cargo in capsules propelled by fluid flowing through a pipeline. The cargo may either be contained in capsules (e.g., wheat enclosed inside sealed cylindrical containers), or may itself be the capsules (e.g., coal compressed into the shape of a cylinder or capsule). There are two general types of capsule pipelines: **hydraulic capsule pipelines (HCP),** which use liquid (usually water) to suspend and propel the capsules, and **pneumatic capsule pipelines (PCP),** which use gas (usually air) to propel the capsules.

According to a 1976 study by Zandi [1], pneumatic capsule pipeline (PCP) technology was invented by the Danish engineer Medhurst in 1810. It was used extensively in Europe and the U.S. in the 19th and early 20th centuries for transporting mail, telegrams, and other lightweight cargoes. Until about 1950, five major cities in the U.S. — New York, Boston, Philadelphia, Chicago, and St. Louis — used PCP (then called **pneumatic tubes**) for transporting mail and parcels between the central post office and the branch offices in each of these cities. The tubes used were small (less than 12-inch diameter), and the systems were crude by modern standards, yet they were used successfully for decades. The growing popularity of automobiles in the mid-20th century greatly affected the continued use of PCP. The old PCP systems were not replaced by new PCPs but rather by trucks. The U.S. Library of Congress also used pneumatic tubes for transporting books between buildings until the 1960s, when that system was replaced by a conveyor belt system that transported books much more slowly than the pneumatic tube system. This can hardly be called progress. Reasons given for the shift from the old pneumatic tube to the new conveyor belt were that the old tube system was too cumbersome to use and caused occasional damage to the books. The advent of computer control systems has greatly enhanced PCP, making the current systems fully automatic, convenient, and reliable (free from cargo damage).

The PCP systems currently (2003) used in the U.S. are primarily used for transporting only lightweight cargoes over short distances (e.g., within a building or between buildings). Such systems use nonwheeled capsules in pipes of diameter up to approximately 12 inches (30 cm). They are widely used at drive-up tellers (for transporting cash and documents between customers in their cars and tellers in the bank), in hospitals (for transporting blood samples, medicine, and supplies between buildings or between rooms within a building), in factories (for transporting tools), at airline terminals (for transporting cash and documents between ticket

175

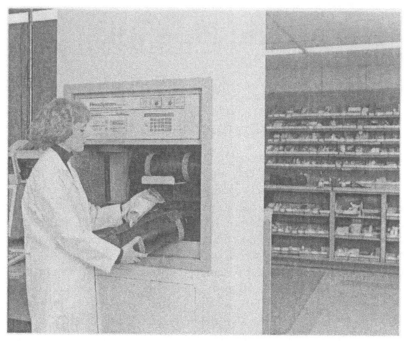

FIGURE 7.1 PEVCO's computerized PCP (pneumatic tube) system used at the Johns Hopkins Bayview Medical Center in Baltimore, Maryland, U.S.A. (Courtesy of PEVCO.)

counters and accounting offices), and in office buildings (for transporting documents). An example of such a system used at a major hospital is shown in Figure 7.1. Similar systems are also used in other parts of the world (see an example in Figure 7.2).

To transport heavy cargo by PCPs, the capsules must have wheels. The capsules are wheeled vehicles moving through pipe propelled by airflow. Both Japan and the former Soviet Union have built and used such systems [2,3]. An example is the 3.2-km pipeline of 1-m diameter built by Sumitomo Metal Industries, Ltd. in Japan for transporting limestone from a mine to a cement plant [2]. The system transports 2 million tons of limestone per year. It has been used very successfully since its construction in 1983. The pipeline is highly automated, and it has an availability record of 98%. A picture of the inlet loading station is shown in Figure 1.6 (see Chapter 1); a close-up view of the loader and the capsules loaded with limestone are shown in Figure 7.3. Similar systems were developed in the U.S. and the former Soviet Union. While the Soviet systems (Lilo 1 and 2) were used successfully for years for transporting rocks, the U.S. system (Tube Express) was not accepted commercially even though the technology was sufficiently developed for use.

In contrast to the long history of PCP, the hydraulic capsule pipeline (HCP) is a new technology first considered by the British Authority for military use during World War II. A military consultant, Geoffrey Pyke, who was considered a genius with many inventions and patents, proposed to the commander of the Allied forces fighting Japan in the Pacific to use HCP to transport ammunitions, supplies, and

FIGURE 7.2 A computerized non-wheeled PCP (pneumatic tube) system used for transporting tools at the BMW factory in Dingolfing, Bavaria, Germany. (Courtesy of Hoertig Rohrpost Air Tube Transport System.)

even troops in Burma [4]. However, his recommendations were not implemented because the technology was not yet ready. It was judged that by the time the technology was sufficiently developed and ready for use the war would be over. The concept was reinvented independently in Canada in 1958, and explored by researchers at the Alberta Research Council between 1958 and 1978 [5]. Since 1970, interest in HCP has spread to the U.S., Japan, the Netherlands, South Africa, Australia, and other nations. In 1991, the U.S. National Science Foundation (NSF) founded a Capsule Pipeline Research Center (CPRC) at the University of Missouri-Columbia (UMC), which resulted in 10 years of extensive research and development in HCP and PCP — substantially improving the knowledge and know-how of capsule pipeline technology.

A derivative of HCP is the **coal log pipeline (CLP)**, invented jointly by Professors Henry Liu and Thomas R. Marrero at UMC [6]. The capsules in a CLP consist of coal particles compressed together into large cylindrical objects — the **coal logs** (see Figure 7.4). Coal logs can be injected into and transported hydraulically through underground pipelines from large coal mines to major power plants. With wear-resistant coal logs, no containers are needed to enclose and separate the coal logs from the water in the pipe. This alleviates the need for using capsule shells for containing the coal, and the need for returning the empty containers through a separate pipeline. Consequently, the economics of CLP appears more

FIGURE 7.3 Loading of limestone into capsules for transport by a 1-m-diameter PCP system in Kuzuu, Japan. The pipeline supplies all the limestone needed for a large cement factory. (Courtesy of Sumitomo Metal Industries, Ltd.)

promising than that of other types of HCP and PCP that require the use of containers to enclose cargo. In 2002, a major facility (pilot plant) for testing CLP was built at the University of Missouri-Columbia to test an entire CLP system — including coal log fabrication, injection of logs into the pipeline, pumping coal logs through the pipeline, and automatic control of the entire system (see Figure 7.5).

FIGURE 7.4 Coal logs (left) compacted at the University of Missouri-Columbia by using a specially designed, double-action hydraulic press (right). The 5.4-inch-diameter logs were tested in a 6-inch-diameter HCP.

FIGURE 7.5 A CLP pilot plant facility constructed and demonstrated at the University of Missouri-Columbia in 2002. The photograph on the right shows members of the American Pipeliners Union Local 798 donating services to complete construction of the facility. (Photograph courtesy of American Pipeliners Union Local 798.)

Both HCP and PCP hold promise for the 21st century for the following reasons:

1. Recent advancements in pipeline technologies and computer technologies have made it possible to develop highly efficient and fully automated systems of HCP and PCP. The costs of transporting cargo in HCP and PCP continue to drop as the technology improves.
2. Increased congestion on highways and city streets has made HCP and PCP appealing. Because HCP and PCP are buried underground, they help alleviate congestion caused by trucks on highways and streets.
3. The public is increasingly concerned about accidents and air pollution caused by trucks. HCP and PCP are extremely safe and nonpolluting. Air pollution in cities will be reduced because while trucks are fueled by diesel, HCP and PCP are powered by electricity.
4. Recent concerns about transportation security have made the capsule pipeline system appealing. Unlike planes and trucks, which can be hijacked and used by terrorists as weapons of mass destruction, capsule pipelines that are fixed underground cannot be hijacked and used as weapons. They are also far more difficult to be damaged or destroyed by terrorists than surface structures are. Furthermore, use of PCP in cities reduces the number of trucks clogging streets and hence facilitates evacuation following any terrorist attack.

7.2 PNEUMATIC CAPSULE PIPELINE (PCP)

7.2.1 SYSTEM DESCRIPTION

Large PCP systems that transport heavy cargo, such as the systems developed and used by the Sumitomo Metal Industries, Ltd. in Japan, use wheeled vehicles — **capsules** — to carry cargo through the pipe. The wheels are unguided by tracks and

FIGURE 7.6 PCP capsules in circular (left) and rectangular (right) conduits.

roll directly on the pipe wall. To facilitate handling, several capsules are often tied together to form a capsule train in the pipe. A blower blows air, which in turn drives the capsule train through the pipeline. The train unloads its cargo at the end of the pipe and then returns to the pipeline inlet, often through a separate return pipeline. It is also possible to use the same pipeline for transporting both the loaded capsules and the returning empty capsules, by reversing the direction of the flow in the PCP. However, this reduces the throughput capacity of the pipe by 50%.

The pipe or conduit used for PCP can be of either circular cross section (see Figure 7.6, left), or rectangular cross section (see Figure 7.6, right). For the type that uses circular pipe, the wheel assemblies of each capsule must be mounted at the centerline of the two ends of the capsule, in the form of gimbals. This enables the wheel assemblies to rotate freely in the pipe while the capsule body remains stable. Otherwise (with wheels mounted on capsule bottom), the capsules would spin and spill their cargo. In contrast, for capsules in rectangular or square pipes (Figure 7.6b), the capsules can use bottom wheels without becoming unstable, as in the case of ordinary vehicles. This simplifies the wheel design of capsules in rectangular or squared PCPs. Bottom wheels are also subjected to less wear and tear, enabling the capsules to travel at high speeds — just like automobiles or trucks — without overheating and damaging the tires. However, rectangular conduits cannot withstand high internal pressure and therefore cannot be used for high-pressure systems. Whether to use a circular or rectangular pipe for PCP is a design decision that must be based on careful consideration and comparison of all the pros and cons of the two systems for any given application.

7.2.2 ANALYSIS

7.2.2.1 Capsule Pressure Drop and Drag

The pressure drop across any capsule is

$$\Delta p_c = C_D \frac{\rho(V - V_C)^2}{2} \tag{7.1}$$

where C_D is the drag coefficient of the capsule; ρ is the air density in the pipe; V is the bulk velocity of the fluid (air); and V_c is the capsule velocity or speed. The drag

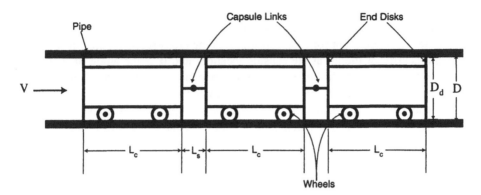

FIGURE 7.7 Basic configuration of a PCP capsule train with each capsule containing two end disks.

coefficient C_D depends to a large extent on the size of the end disks and the degree of seal provided by the end disks (see Figure 7.7). Accurate determination of C_D requires testing of prototype capsules in a pipe. The following equation by Kosugi [7] can predict C_D within 20% error:

$$C_D = \frac{2k_d^4}{\left(1-k_d^2\right)^2} \tag{7.2}$$

where k_d is the disk diameter ratio, which is the diameter of the disk, D_d, divided by the pipe inner diameter, D. Note that Equation 7.2 gives the drag coefficient of a single capsule having a single end disk. For a capsule with two end disks (one in the front and one in the back as is usually the case), the factor 2 in the numerator of Equation 7.2 must be doubled. Furthermore, for a train consisting of N capsules with each capsule having two disks, the value of C_D for the entire train must be the value found from Equation 7.2 multiplied by $2N$.

From Equation 7.1, the drag force F_D developed by the flow on the capsule is

$$F_D = A\Delta p_c = AC_D \frac{\rho(V-V_c)^2}{2} \tag{7.3}$$

Note that once the drag force on each capsule is determined, the power consumed by each capsule moving through the pipe is simply $P = V_c F_D$.

7.2.2.2 Steady-State Capsule Velocity

When a capsule (or capsule train) is moving at a constant (steady) velocity through a PCP that is sloped upward, application of Newton's second law yields

$$V_c = V - V_d \quad \text{where} \quad V_d = \sqrt{\frac{2W_c(\eta\cos\alpha + \sin\alpha)}{C_D A\rho}} \tag{7.4}$$

In Equation 7.4, V_d is the differential velocity between V and V_c; η is the contact friction coefficient between the capsule and the pipe; W_c is the weight of each capsule; C_D is the drag coefficient for each capsule; A is the cross-sectional area of the pipe; ρ is the density of the air at the capsule location; and α is the angle of incline of the pipe with respect to the horizontal plane. Once the velocity of the air in the pipe, V, is known, Equation 7.4 can be used to calculate the capsule velocity, V_c.

For downward inclined PCP, when the slope angle α is smaller than $\tan^{-1}\eta$, the capsule velocity V_c is smaller than the air velocity V, and the following relationships hold:

$$V_c = V - V_d \quad \text{where} \quad V_d = \sqrt{\frac{2W_c(\eta\cos\alpha - \sin\alpha)}{C_D A\rho}} \tag{7.5}$$

If, on the other hand, when the downward slope angle α is greater than $\tan^{-1}\eta$, the capsule speed will exceed the air speed in the pipe, and the equation becomes

$$V_c = V + V_d \quad \text{where} \quad V_d = \sqrt{\frac{2W_c(\sin\alpha - \eta\cos\alpha)}{C_D A\rho}} \tag{7.6}$$

For a horizontal pipe, α is zero and the term $(\sin\alpha + \eta\cos\alpha)$ in Equation 7.4 is simply η. Consequently, the equation becomes

$$V_c = V - V_d \quad \text{where} \quad V_d = \sqrt{\frac{2\eta W_c}{C_D A\rho}} \tag{7.7}$$

Note that regardless whether it is an upward or downward slope, the angle α in Equations 7.5 to 7.7 is always considered positive.

Example 7.1 A capsule with two end disks is moving through a horizontal PCP of 1-m diameter. The diameter of the end disks is 0.98 m. The air velocity in the pipe at the location of the capsule is 10 m/s. The capsule weight is 1.5 tonnes (metric tons), the rolling friction coefficient of the capsule wheels is 0.01, and the air density at the capsule location is 1.3 kg/m³. Determine the capsule velocity, the pressure drop across the capsule, the drag on the capsule, and the power consumed by the capsule.

[Solution] $\eta = 0.01$, $W_c = 1.5$ tonnes = 1500 kg = 1500 x 9.81 N = 14,715 N, $D_d = 0.98$ m, $D = 1.00$ m, $k_d = D_d/D = 0.98$, $A = \pi D^2/4 = 0.7854$ m², $\rho = 1.3$ kg/m³, and $V = 10$ m/s. From Equation 7.2, for a two-disk capsule $C_D = 2353$. Then, Equation 7.7 yields $V_d = 0.350$ m/s, and $V_c = V - V_d = 9.65$ m/s. Equation 7.1 yields $\Delta p_c = 187$ Pa, and Equation 7.3 yields $F_D = 147$ N, which is equivalent to 15.0 kg. Finally, the power consumed by the capsule is $P_{capsule} = V_c F_D = 9.65 \times 147 = 1419$ W = 1.419 kW.

7.2.2.3 Pressure Variation along PCP

The pressure drop in any capsule-free regions of the pipe can be calculated from the compressible isothermal flow in pipe, Equation 3.27, restated as follows:

$$\frac{fL}{D} = \frac{1}{kM_1^2}\left[1 - \frac{p_2^2}{p_1^2}\right] + 2\ln\frac{p_2}{p_1} \tag{7.8}$$

where f is the Darcy-Weisbach friction factor of the pipe flow; L is the length of the pipe between points 1 and 2, which are an upstream point and a downstream point, respectively; D is the pipe diameter; k is the adiabatic exponent, which is equal to 1.4 for air; M_1 is the Mach number of the upstream point; and p_1 and p_2 are the pressures at points 1 and 2, respectively.

Assuming that the pressure and density changes across the capsule are adiabatic,

$$\frac{p_2}{p_1} = \left(\frac{\rho_2}{\rho_1}\right)^k \tag{7.9}$$

Once the pressure drop across each capsule (or capsule train) and the pressure drop across each capsule-free part of the pipeline are determined from the above equations, the total pressure drop along the pipeline can be determined from

$$\Delta p_T = \Sigma\Delta p_c + \Sigma\Delta p_F \tag{7.10}$$

where Δp_T is the total pressure drop along a PCP; $\Sigma\Delta p_c$ is the sum of all the pressure drops across capsules in the line; and $\Sigma\Delta p_F$ is the sum of all the pressure drops along the remaining length of the pipe occupied only by fluid (air).

The airflow going through the blower may be considered as isentropic ideal gas — the same as discussed in Chapter 3 for compressors. Therefore, Equation 7.9 also holds for the air going through the blower used in a PCP system.

To illustrate the use of the foregoing equations to solve any PCP problem, consider a single capsule (or a capsule train) moving through a PCP of total length L and diameter D, as shown in Figure 7.8. Points 1 through 6 are the pipeline inlet, blower inlet, blower outlet, capsule rear (tail), capsule front (face), and pipeline outlet, respectively, the pressure variation along the pipe is sketched above the pipeline in the figure. Application of the foregoing equations to various parts of this pipeline yields

$$\frac{fL_{12}}{D} = \frac{1}{kM_1^2}\left[1 - \frac{p_2^2}{p_1^2}\right] + 2\ln\frac{p_2}{p_1} \tag{7.11}$$

$$p_{23} = p_3 - p_2 = p_{blower} \tag{7.12}$$

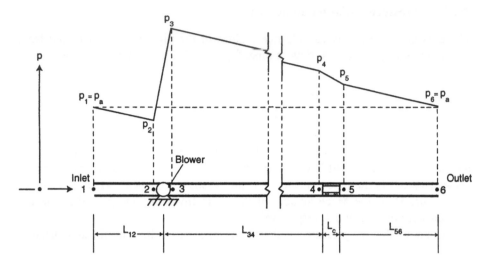

FIGURE 7.8 Analysis of a PCP system with a single capsule or capsule train.

$$\frac{fL_{34}}{D} = \frac{1}{kM_3^2}\left[1 - \frac{p_4^2}{p_3^2}\right] + 2\ln\frac{p_4}{p_3} \tag{7.13}$$

$$\Delta p_c = p_4 - p_5 = C_D \frac{\rho_4(V_4 - V_c)^2}{2} \tag{7.14}$$

$$\frac{fL_{56}}{D} = \frac{1}{kM_5^2}\left[1 - \frac{p_6^2}{p_5^2}\right] + 2\ln\frac{p_6}{p_5} \tag{7.15}$$

$$V_c = V_4 - \sqrt{\frac{2\eta W_c}{C_D A \rho_4}} \tag{7.16}$$

$$\frac{p_5}{p_4} = \left(\frac{\rho_5}{\rho_4}\right)^k \tag{7.17}$$

$$\frac{p_3}{p_2} = \left(\frac{\rho_3}{\rho_2}\right)^k \tag{7.18}$$

$$\rho_2 V_2 = \rho_3 V_3 = \rho_4 V_4 = \rho_5 V_5 = \rho_a V_1 = \text{constant} \tag{7.19}$$

The subscripts 1, 2, 3, etc. used in Equations 7.11 to 7.19 represent values taken at points 1, 2, 3, etc. along the pipe. The above is a set of 12 equations because

Equation 7.19 contains four equations. In these equations, $p_1 = p_6 = p_a$, where p_a is the ambient or atmospheric pressure at the two ends of the pipe, which is known; ρ_a is the density of air at atmospheric pressure p_a and at temperature T, and so it can be determined from the equation of state of ideal gas. $M_1 = V_1 / \sqrt{kRT}$ where V_1 is the design value (say, $V_1 = 10$ m/s) and T is also given (say, $T = 18°C$); hence M_1 is known. The blower is selected so that p_{blower} is known. The drag coefficient C_D is also known from Equation 7.2. Therefore, the unknown quantities in the above set of 12 equations are p_2, p_3, p_4, p_5, M_3, M_5, ρ_2, ρ_3, ρ_4, ρ_5, V_4, and V_c. Since there are 12 independent equations with 12 unknowns, the equations can be solved numerically with a computer. Detailed treatment of various cases of compressible PCP flow can be found in York [8].

7.2.2.4 Power of PCP

From Equation 3.68, the power output of the blower, P_{blower}, can be calculated from

$$P_{blower} = \dot{m}w \tag{7.20}$$

where

$$\dot{m} = \rho_a V_1 A \quad \text{and} \quad w = \frac{k}{(k-1)} \frac{1716 p_2}{GR\rho_2} \left[\left(\frac{p_3}{p_2} \right)^{\frac{k-1}{k}} - 1 \right] \tag{7.21}$$

In the two above equations, P_{blower} is the output power of the blower; $\dot{m}$ is the mass flow rate through the pipe; w is the work per unit mass of the air; G is the gas gravity, which equals 1.0 for air; and R is the engineering gas constant. Dividing P_{blower} by the efficiency of the blower, η_b, yields the input power to the blower. Note that Equation 7.21 is written in ft-lb units. To use the equation in SI units, the constant 1716 must be changed to 287.

Example 7.2 A PCP of 1-m diameter is powered by a blower. At the pipeline inlet, the air density is 1.2 kg/m^3, and the air velocity is 12 m/s. The absolute pressures at the blower inlet and outlet are 98 kPa and 120 kPa, respectively. The air density at the blower inlet is 1.18 kg/m^3. Determine the mass flow rate of air through the pipe, the compression work done by the blower for unit mass of the air through the blower, and the power output of the blower.

[Solution] For this problem, $A = \pi D^2/4 = 0.7854$ m^2, $V_1 = 12$ m/s, $\rho_a = 1.2$ kg/m^3. Therefore, from Equation 7.21 the mass flow rate is $\dot{m} = 11.31$ kg/s. The work done by compression on unit mass of the airflow going through the blower can be computed from the second part of Equation 7.21. Because $p_2 = 98$ kPa, $p_3 = 120$ kPa, $\rho_2 = 1.18$ kg/m^3, $k = 1.4$, $G = 1.0$, $R = 287$ J/kg-K, the equation yields $w = 17,320$ J/kg. Therefore, from Equation 7.20, the output power of the blower is $P_{power} = 195,840$ W $= 195.8$ kW.

FIGURE 7.9 A capsule containing grain tested at Capsule Pipeline Research Center, University of Missouri-Columbia.

7.3 HYDRAULIC CAPSULE PIPELINE (HCP)

7.3.1 SYSTEM DESCRIPTION

In contrast to PCP, which uses wheeled vehicles as capsules, the capsules in HCP (including CLP) are plain cylinders without wheels. Due to the high density of the fluid (water) used in HCP, it is possible to convey nonwheeled heavy capsules through pipe hydraulically without large contact friction. As will be shown in the next section, the high density of the fluid (liquid) in HCP generates a large buoyancy force and a large lift force to suspend the capsules. Also, while in PCP the capsules only occupy a small length of the pipeline (i.e., have low linefill rate), in HCP over 80% of the length of the pipeline is occupied by capsules. Figure 7.9 shows a capsule tested in an 8-inch HCP at the University of Missouri-Columbia.

The capsules in HCP move through the pipe at a velocity of 6 to 10 fps, which is much slower than PCP capsules, which normally move at 30 to 50 fps. Due to its low speed, HCP is not suitable for transporting cargoes that must reach their destinations promptly, such as mail or fresh fruits. PCP is much more suitable for transporting such cargoes. At the speed of 50 fps (34 mph), capsules can reach their destination in about the same time that a truck can in many cases, due to the need for the trucker to stop for rest, and running into traffic inside cities. Even higher speed PCP systems can be developed in the future for interstate freight transport using large rectangular conduits. Capsules in such PCPs may move at the same speed of trucks, except nonstop until the destination is reached. On the other hand, because HCP is much slower than PCP, from fluid mechanics it uses much less energy than PCP does to transport the same cargo over the same distance. Also, HCP has higher linefill rates, which often result in a higher freight throughput than that of PCP of the same diameter. Thus, the unit cost of freight transport using HCP, in dollars per ton of cargo transported, is usually lower than that of PCP. This economic incentive favors HCP over PCP, unless speedy delivery is required, or when there are problems in using HCP, such as when the cargo size is larger than can be contained in a small-diameter HCP.

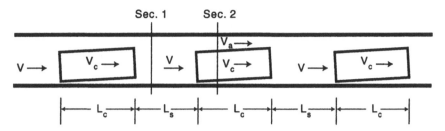

FIGURE 7.10 Analysis of steady capsule flow in HCP.

7.3.2 ANALYSIS

7.3.2.1 Basic Capsule Flow Relationships

Some basic relationships of HCP flow can be derived with the use of Figure 7.10. By choosing the fluid and part of the capsule between Sections 1 and 2 in Figure 7.10 as the control volume, application of the continuity equation for incompressible flow yields

$$VA = V_c A_c + V_a (A - A_c) \quad \text{or} \quad V_a = \frac{V - k^2 V_c}{1 - k^2} \qquad (7.22)$$

In the above equation, A, A_c, and $(A - A_c)$ are the cross-sectional areas of the pipe, the capsule, and the capsule–pipe annulus, respectively; V, V_c, and V_a are the mean velocities across the pipe, the capsule, and the annulus, respectively; and k is the **diameter ratio**, which is the capsule diameter D_c divided by the pipe inner diameter D. Equation 7.22 is a useful relationship often used in the analysis of capsule flow. For instance, by using Equation 7.22, one can prove and conclude the following:

- When V_c is smaller than V (i.e., $V_c < V$), then $V_a > V$, or $V_a > V > V_c$.
- When V_c is larger than V (i.e., $V_c > V$), then $V_a < V$, or $V_c > V > V_a$.
- When V_c is equal to V (i.e., $V_c = V$), then $V_a = V$, or $V_a = V = V_c$.

Another useful relationship that can be derived by using Figure 7.10 is the following. Consider the steady-state operation of an HCP with the pipeline filled with capsules except for an average spacing L_s between neighboring capsules. Suppose that the capsules move at a constant velocity V_c in the pipe. The total length of the pipeline occupied by the capsules divided by the total length of the pipeline is called the **linefill rate** or simply **linefill**, designated by λ. From Figure 7.10,

$$\lambda = \frac{L_c}{L_c + L_s} \qquad (7.23)$$

If the average period for capsules to pass any point in the pipe is T_c, and the average frequency of capsule passages at any point is N_c, from Figure 7.10,

$$T_c = \frac{L_c + L_s}{V_c} = \frac{L_c}{\lambda V_c} \quad \text{and} \quad N_c = \frac{1}{T_c} = \frac{V_c}{L_c + L_s} = \frac{\lambda V_c}{L_c} \tag{7.24}$$

Because capsule flow in pipe is a two-phase flow with intermittent passages of capsules at any location in pipe, it is important to distinguish the various discharges and mean velocities of the flow. First, the cross-sectional mean velocity of the fluid between capsules, V, shown in Figure 7.10, is often referred to in the literature as the **bulk velocity**, and the discharge $Q = VA$ as the **bulk discharge**. Likewise, the cross-sectional mean velocity of the fluid in the capsule–pipe annulus (i.e., the **annulus velocity**) is V_a, and the discharge of the flow through the annulus (i.e., the **annulus discharge**) is $Q_a = V_a A_a = V_a(A - A_c)$. Moreover, the **capsule velocity** is V_c, and the **capsule discharge** is $Q_c = V_c A_c$. From Equation 7.22, $Q = Q_a + Q_c$.

The foregoing velocities and discharges at any given point in pipe vary with time in the form of a square wave as shown in Figure 7.11. Based on the wave shape, the time-averaged liquid discharge, $\bar{Q}$, can be found as follows:

$$\bar{Q} = Q\left(\frac{L_s}{L_c + L_s}\right) + Q_a\left(\frac{L_c}{L_c + L_s}\right) = Q(1 - \lambda) + Q_a \lambda \tag{7.25}$$

Likewise, the time-averaged capsule discharge, $\bar{Q}_c$, can be found from

$$\bar{Q}_c = Q_c\left(\frac{L_c}{L_c + L_s}\right) = Q_c \lambda \tag{7.26}$$

Equations 7.25 and 7.26 can be used to determine the throughputs of liquid and cargo by capsule pipelines, equally valid for HCP and PCP. From Equations 7.25 and 7.26, it can be proved that the total throughput of an HCP, including both capsules and the liquid, is simply $Q = VA$.

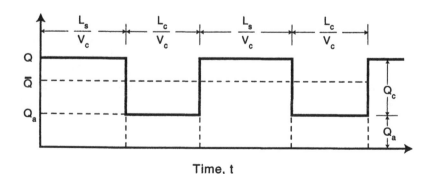

FIGURE 7.11 Liquid discharge variation with time in HCP.

Example 7.3 An HCP of 11.5-inch inner diameter transports capsules of 10-inch diameter and 20-inch length. The fluid is water and the average spacing between capsules in the pipe is 4 inches. The capsules are coal logs of an average specific gravity of 1.32. The bulk velocity of the water is 7.2 ft/s, and the capsule velocity is 8.0 ft/s. Determine: (a) the annulus velocity; (b) the average linefill rate; (c) the period and frequency of capsule passage at any point in the pipe; (d) the bulk discharge, the annulus discharge, and the capsule discharge; (e) the throughput of capsules and the throughput of water, in both cfs and MT/yr (million tonnes per year); and (f) the number of capsules to be handled per day (24 h).

[Solution] (a) For this problem, $V = 7.2$ fps, $V_c = 8.0$ fps, $D = 11.5$ inches = 0.9583 ft, $D_c = 10$ inches = 0.8333, $k = D_c/D = 0.870$. Therefore, from Equation 7.22, $V_a = 4.73$ fps. (b) $L_c = 20$ inches = 1.667 ft, and $L_s = 4$ inches = 0.3333 ft. Hence, from Equation 7.23 the linefill rate is $\lambda = 0.8333$. (c) From Equation 7.24, $T_c = 0.250$ s, and $N_c = 4.0$ capsules per second. (d) The pipe cross-sectional area is $A = \pi D^2/4 = 0.721$ ft², the capsule cross-sectional area is $A_c = \pi D_c^2/4 = 0.545$ ft², and the annulus area is $A_a = A - A_c = 0.176$ ft². Therefore, the bulk discharge $Q = VA = 5.19$ cfs, the annulus discharge is $Q_a = V_a A_a = 0.833$ cfs, and the capsule discharge is $Q_c = V_c A_c = 4.36$ cfs. (e) From Equation 7.26, the capsule throughput is $\overline{Q}_c = 3.633$ cfs. Assuming continuous operation (around the clock and 365 days a year), this throughput is equivalent to 114.6 Mcf/yr (million cubic feet per year). Because the average specific gravity of the capsules is 1.32, each cubic foot of capsule weighs $1.32 \times 62.4 = 82.4$ lb. Therefore, the coal throughput in terms of weight is $114.6 \times 82.4 = 9443$ Mlb/yr (million pounds per year) which is equal to 4.72 MT/yr (million short tons per year). From Equation 7.25, the water throughput is $\overline{Q} = 1.559$ cfs. This is equivalent to a weight throughput of 97.3 lb/s = 3068 Mlb/yr = 1.53 MT/yr. (f) Because $N_c = 4$ capsules/s, the number of capsules handled per day is $N_c = 4 \times 3600 \times 24 = 345,600$ capsules/day.

7.3.2.2 Four Regimes of HCP Flow

The hydraulics of HCP (including CLP) can be explained by dividing the flow into four distinctly different regimes [9]. In regime 1 (see Figure 7.12a), the bulk fluid velocity is so small that a denser-than-fluid capsule will remain stationary on the pipe floor, prevented to move by the contact friction between the capsule and the pipe floor. As the fluid velocity increases and exceeds a certain value V_i, called the **incipient velocity**, the capsule will start to slide along the pipe floor, and the flow enters regime 2 (see Figure 7.12b). In regime 2, the fluid velocity is higher than the capsule velocity ($V > V_c$). The total drag force on the capsule (i.e., pressure drag plus the skin drag due to shear) is balanced by the contact friction force encountered by the sliding capsule. As the fluid velocity continues to increase, the hydrodynamic lift on the capsule increases, which reduces the buoyant weight of the capsule and diminishes the contact friction between the capsule and the pipe floor. As the fluid velocity exceeds a certain value, the contact friction on the capsule reduces to the

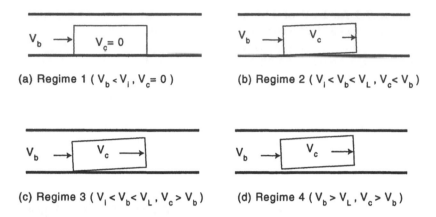

(a) Regime 1 ($V_b < V_i$, $V_c = 0$)   (b) Regime 2 ($V_i < V_b < V_L$, $V_c < V_b$)

(c) Regime 3 ($V_i < V_b < V_L$, $V_c > V_b$)   (d) Regime 4 ($V_b > V_L$, $V_c > V_b$)

FIGURE 7.12 Four regimes of HCP flow.

extent that it is less than the drag on the capsule. Now the capsule must accelerate and exceeds the bulk fluid velocity (i.e., $V_c > V$), and the shear on capsule is in the opposite direction of the flow. This is regime 3 in Figure 7.12c. The demarcation between regimes 2 and 3 is $V = V_c$; the fluid velocity at this stage is called **critical velocity**, V_o. As the fluid velocity continues to increase in regime 3, the capsule velocity increases, and ($V_c - V$) also increases. The point is reached when the hydrodynamic lift on the capsule is so large that the capsule becomes waterborne — lifted off. The fluid velocity at this point is called the **lift-off velocity**, V_L. At the lift-off velocity, the capsule front rises as in the case of an airplane taking off on the runway when the plane has reached the take-off velocity. Experimental evidence shows that at V_L, the capsule velocity is about 15% higher than the fluid velocity, namely, ($V_c - V_L)/V_L \approx 0.15$. Regime 4, depicted in Figure 7.12d, is reached when the bulk fluid velocity V is greater than V_L. In regime 4, the capsule is completely waterborne. Due to the lack of lateral restraints and the turbulent wake trailing the capsule, the capsule in regime 4 is unstable. It flutters (rocks and rolls) and impacts the wall. Operation of HCP in regime 4 should be avoided to minimize abrasion and headloss. This is especially true for CLP since the coal logs are relatively fragile and must be protected from excessive wear and damage.

The phenomenon of capsule lift-off was first reported and studied by R.T. Little in Canada in 1968 [5]. Since then, many investigators have observed capsule lift-off in laboratory tests. The phenomenon that at high liquid velocities the capsule velocity exceeds the liquid velocity has also been reported by many investigators.

7.3.2.3 Incipient Velocity

The incipient velocity V_i of a capsule in HCP including CLP can be calculated approximately from the following equation [10]:

$$V_i = \sqrt{\frac{2g(S-1)L_c(\sin\alpha + \eta\cos\alpha)}{C_D}} \tag{7.27}$$

where

$$C_D = \left[1 + K_c + f_a \left(\frac{L_c}{D - D_c} - 3 \right) \right] \Big/ \beta^2 \qquad (7.28)$$

In the above two equations, g is gravitational acceleration; S is the density ratio, which is the capsule density ρ_c divided by the fluid density ρ (S becomes the specific gravity of the capsule when the fluid is water); L_c is the capsule length; α is the pipe incline angle, considered positive for upward slope and negative for downward slope; β is the clearance ratio $(A - A_c)/A$; η is the contact friction between the capsule and the pipe, which is usually in the range 0.45 to 0.55; f_a is the Darcy-Weisbach friction factor for the annular flow through the capsule–pipe clearance; and D_c is the capsule diameter.

The quantity K_c is to be determined from

$$K_c = \left(\frac{1}{C_c} - 1 \right)^2 \qquad (7.29)$$

where C_c is the contraction coefficient for the flow near the capsule entrance (i.e., upstream end of the capsule). For a sharp-edged entrance, values of C_c and K_c are listed as a function of β in Table 7.1. The value of f_a can be found either from $f_a = 64/\mathfrak{R}$ if the flow in the annulus is laminar, or from the Moody diagram if the flow in the annulus is turbulent. The Reynolds number of the annular flow is

$$\mathfrak{R}_a = \frac{V_a(D - D_c)}{v} = \frac{VD}{(1+k)v} \qquad (7.30)$$

where v is the kinematic viscosity of the fluid (water). The constant k is the diameter ratio, which is the capsule diameter D_c divided by the pipe diameter D. When dealing with HCP, k denotes the diameter ratio rather than the adiabatic exponent used before

---

TABLE 7.1
Contraction Coefficient, C_c , and Headloss Coefficient, K_c , for Capsules of Sharp-Edged (90°-Corner) Entrance (Upstream End)

Clearance Ratio $\beta = (A - A_c)/A$	0.1	0.2	0.3	0.4	0.5	0.6	0.7	0.8	0.9	1.0
Blockage Ratio $b = A_c/A$	0.9	0.8	0.7	0.6	0.5	0.4	0.3	0.2	0.1	0
Contraction Coefficient										
C_c		0.624	0.632	0.643	0.659	0.681	0.712	0.755	0.813	0.892	1.0
K_c		0.363	0.339	0.308	0.268	0.219	0.164	0.105	0.053	0.015	0

for gas flows. The switch of notation here is warranted in order to conform to conventional use of k in different fields, and to keep notations simple.

For a horizontal pipe, $\alpha = 0°$, and Equation 7.27 reduces to

$$V_i = \sqrt{\frac{2g(S-1)L_c\eta}{C_D}} \tag{7.31}$$

Example 7.4 A researcher conducted a set of laboratory tests to determine the accuracy of Equation 7.27. In one of the tests, he used a horizontal Plexiglas pipe of 54-mm I.D. The capsule was a plastic cylinder of 1.169 specific gravity, 47.7-mm diameter, and 62-mm length. The contact friction coefficient between the capsule and the pipe was 0.421. By running water at 20°C through the pipe containing the capsule, he found that the minimum velocity of the fluid (water) to cause the capsule to move was 0.047 m/s. How much does this measured incipient velocity differ from the one predicted from Equation 7.31?

[**Solution**] For this problem, $D = 54.0$ mm $= 0.054$ m, $D_c = 47.7$ mm $= 0.0477$ m, $k = D_c/D = 0.883$, $\beta = (A - A_c)/A = (D^2 - D_c^2)/D^2 = 0.220$, $\eta = 0.439$, $g = 9.81$ m/s², $S = 1.169$, $L_c = 62$ mm $= 0.062$ m. From Table 7.1, $C_c = 0.634$ and $K_c = 0.333$. Substituting the foregoing values into Equation 7.31 yields

$$V_i = \frac{0.0660}{\sqrt{1.333 + 6.841f}} \tag{a}$$

Since the f value depends on the value of V_i, Equation a can only be solved by iteration. Assuming arbitrarily that $f_a = 0.015$, Equation a yields $V_i = 0.0551$ m/s. The kinematic viscosity of water at 20°C is $\nu = 1.007 \times 10^{-6}$ m²/s. From Equation 7.30, $\Re_a = 1569$. Since this is laminar flow, $f_a = 64/\Re_a = 0.0408$. Substituting this f_a value into Equation a yields $V_i = 0.0520$ m/s. Thus, the new Reynolds number is $\Re = 1480$. This yields $f_a = 64/\Re_a = 0.0432$, and $V_i = 0.0517$ m/s. After two more iterations, the final values of $f_a = 0.0435$ and $V_i = 0.0517$ m/s are reached. Comparing this V_i with the measured value of 0.047 m/s, the discrepancy is $(0.0517 - 0.047)/0.047 = 0.10$ or 10%. Note that Equation 7.28 is not precise; a more complicated equation is available, which yields better results [11].

7.3.2.4 Lift-Off Velocity

Optimum operation of HCP (especially CLP) requires that the velocity V to be slightly below the lift-off velocity. At such a velocity, the contact friction between the capsule and the pipe is minimum, and the capsule is moving through the pipe in a stable nose-up position. Both the headloss (pressure drop across a capsule) and the abrasion of capsules in pipe are a minimum. The lift-off velocity for horizontal HCPs can be predicted from Liu's equation, as follows [12]:

$$V_L = 7.2\sqrt{|S-1|gak(1-k^2)D} \tag{7.32}$$

where a is the aspect ratio, which is the capsule length, L_c, divided by the capsule diameter. The absolute sign enclosing $(S-1)$ is used so that the same equation can be used for both denser-than-fluid and lighter-than-fluid capsules. Tests of CLP showed that coal log abrasion is a minimum when V is in the range of 85 to 95% of V_L, as predicted from the above equation.

Example 7.5 A cylindrical capsule of 22-inch diameter and 40-inch length is used to carry grain for transport by a horizontal HCP of 23.5-inch I.D. The fluid is water, and the capsule has a specific weight of 65 lb/ft³. Determine the lift-off velocity of this capsule.

[Solution] For this problem, $S = 65/62.4 = 1.0417$, $D_c = 22$ in $= 1.833$ ft, $D = 23.5$ inches $= 1.958$ ft, $L_c = 40$ inches $= 3.333$ ft, $k = D_c/D = 0.9362$, $a = L_c/D_c = 1.818$, and $g = 32.2$ ft/s². From Equation 7.32, $V_L = 5.36$ ft/s.

7.3.2.5 Critical Velocity

Based on a 1999 study by Gao [13], the critical velocity V_o that marks the boundary between regimes 2 and 3 can be predicted from

$$V_o = \sqrt{\frac{2g(S-1)L_c(\eta\cos\alpha + \sin\alpha)\beta}{kaf_a}} \tag{7.33}$$

where f_a can be determined in the same manner as discussed before for calculating the incipient velocity, except that the Reynolds number in this case is

$$\Re_a = \frac{V_a(D-D_c)}{\nu} = \frac{D(V_o - k^2V_c)}{\nu(1+k)} = \frac{DV_o(1-k)}{\nu} \tag{7.34}$$

Example 7.6 Calculate the critical velocity of the capsule treated in the previous example. Assume that the contact friction coefficient between the capsule and the pipe is 0.45, and the water temperature is 70°F.

[Solution] In this case, $g = 32.2$ ft/s², $S = 1.0417$, $L_c = 3.333$ ft, $k = D_c/D = 0.9362$, $\beta = (A - A_c)/A = 1 - k^2 = 0.1236$, $\eta = 0.45$, $\alpha = 0°$, $\sin\alpha = 0$, $\cos\alpha = 1$, and $a = L_c/D_c = 1.818$. Substituting these values into Equation 7.33 yields

$$V_o = \frac{0.5408}{\sqrt{f_a}} \tag{a}$$

At 70°F, the kinematic viscosity of the water is 1.059×10^{-5} ft²/s. Thus, the Reynolds number is, from Equation 7.34,

$$\mathfrak{R}_a = 11,796V_o \tag{b}$$

The friction factor f_a can be determined through iteration as follows: Assume that $f_a = 0.02$. From Equation a, $V_o = 3.82$ fps. Then, from Equation b, $\mathfrak{R}_a = 4.51 \times 10^4$. Assume that the roughness of the capsule is the same as the roughness of the pipe, namely, $e = 0.00015$ ft. Then, the relative roughness is $e/D_c = 0.00015/1.833 = 0.000082$. From the Moody diagram, $f_a = 0.0216$, and from Equations a and b, $V_o = 3.68$ ft/s, and $\mathfrak{R}_a = 4.34 \times 10^4$. From the Moody diagram, the new value of f_a found is 0.0218. Then, from Equation a, $V_o = 3.66$ ft/s. Comparing this with the lift-off velocity of 5.36 ft/s, $V_o/V_L = 3.66/5.36 = 0.683$, namely, V_o is approximately 68% of V_L.

7.3.2.6 Capsule Velocity

The capsule velocity V_c is always smaller than the fluid velocity V when V is smaller than V_o, and the opposite holds when V is greater than V_o, namely,

$$V_c < V \text{ when } V < V_o, \quad V_c > V \text{ when } V > V_o,$$
$$\text{and } V_c = V \text{ when } V = V_o \tag{7.35}$$

Based on Equation 7.35, and based on the experimental evidence reported by many investigators that the value of V_c is always linearly proportional to V, it can be deduced that if we plot the values of V_c as a function of V for any given capsule in a given pipe, the result will be a straight line as shown in Figure 7.13. Written mathematically,

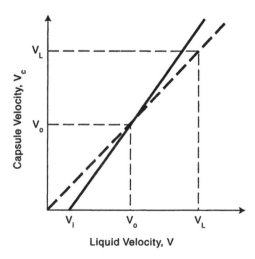

FIGURE 7.13 Variation of capsule velocity with liquid velocity in HCP.

$$V_c = C_1 V - C_2 \quad \text{where} \quad C_1 = \frac{V_o}{V_o - V_i}, \text{ and } \quad C_2 = \frac{V_o V_i}{V_o - V_i} \tag{7.36}$$

7.3.2.7 Steady Flow Analysis of HCP System

The steady flow of an entire HCP system can be analyzed by assuming that both the fluid and the capsules are incompressible. Assume that water and capsules are pumped from reservoir 1 (intake reservoir) to reservoir 2 (outlet reservoir), over a distance L and an elevation increase $z_2 - z_1 = h_s$, which is the static head. From the one-dimensional energy equation,

$$h_p = h_s + f \frac{L}{D} \frac{V^2}{2g} + N_c \frac{\Delta p_{cn}}{\rho g} \tag{7.37}$$

where h_p is the pump head; N_c is the number of capsules in the pipe; ρ is the water density; and Δp_{cn} is the net capsule pressure drop, which is the actual pressure drop across a single capsule, Δp_c, minus the water pressure drop across the capsule distance without the capsule present, Δp_w, namely,

$$\Delta p_{cn} = \Delta p_c - \Delta p_w, \text{ where } \quad \Delta p_w = f \frac{L_c}{D} \frac{\rho V^2}{2} \tag{7.38}$$

Note that in Equation 7.37, all the heads are expressed in terms of water column, irrespective of the number of capsules present in the pipe. If more than one pump is needed in series, or if booster pumps are scattered along the pipe, the head h_p in Equation 7.37 becomes the sum of the heads of all the pumps used in the pipeline.

Finally, the power output of the pump is

$$P_o = h_p Q \gamma \tag{7.39}$$

where γ is the specific weight of the water, and Q is the total throughput including both capsules and the water.

7.3.2.8 Pressure Gradient in HCP

Accurate prediction of the net capsule pressure gradient along an HCP, the quantity Δp_{cn} in Equation 7.37, is possible — see, for instance, Reference 9. However, due to the complex nature of the computation, details cannot be discussed here. Those interested in such predictions (calculations) should consult References 9, 13, and 14.

7.3.3 CAPSULE INJECTION AND EJECTION

7.3.3.1 Injection

There are various methods of injecting capsules into the pipeline [15]. Only the most practical and generally applicable system, the **multi-lock type**, is discussed here.

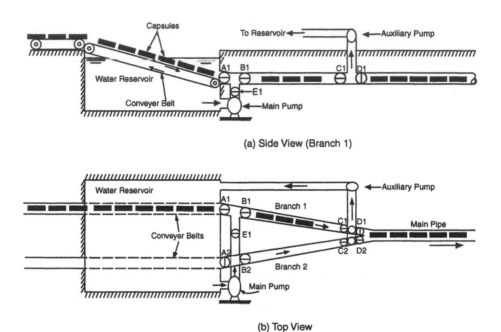

(a) Side View (Branch 1)

(b) Top View

FIGURE 7.14 Injection system of HCP and CLP (multi-lock type).

This system uses a set of parallel launching tubes (locks) to receive capsules from conveyor belts, and to launch capsules into a common pipeline — the main pipeline. Usually, each injection system contains four to six locks. However, for simplicity, Figure 7.14 is shown with only two locks. As shown in Figure 7.14, the locks are either horizontal or nearly horizontal lines, with their downstream ends connected to the main pipeline through a set of Y-joints or bifurcation. The upstream end of each lock is connected to either a common water reservoir or separate reservoirs, whichever is more practical for a given case. Capsules are first loaded on a set of conveyor belts, each of which is connected to the inlet of a lock, to bring capsules into the lock. Connection between the conveyors and the locks requires that each conveyor be tilted at a slope of about 30° angle, with the end part of the conveyor in the reservoir underwater. A low-head pump (auxiliary pump) has its suction side connected to the downstream ends of the locks, and its discharge side connected to the reservoir(s). By opening the valve connected to a given lock, the auxiliary pump draws capsules from the corresponding conveyor belt into the lock. A high-head pump (main pump) has its discharge side connected to the upstream ends (entrance regions) of the locks, and its suction side connected to the reservoir(s). By opening the discharge valve connected to any given lock, this pump drives the capsules out of the lock and into the main pipeline downstream. During normal mode of operation, both pumps are on continuously, but valves are frequently switched. By alternately opening and closing valves, capsules can be drawn into the locks and then driven into the main pipeline one train at a time. Each train of capsules entering the main pipeline consists of the capsules drawn into the lock at an earlier time. There will be some spacing between any two neighboring trains in the pipe, but there will be

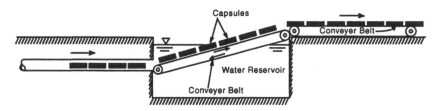

FIGURE 7.15 Outlet of HCP and CLP.

little spacing between individual capsules in a train. Having multiple parallel locks reduces the speed needed for the feeding capsules by conveyors. For instance, if the capsule speed in the main pipeline is 8 fps, by having four parallel locks the conveyor speed will be reduced four times, or be 2 fps. This makes it much more manageable to load capsules onto conveyors at the inlet station. A special advantage of the injection system discussed above is that the capsules never go through the pumps. As such, any ordinary water pumps, centrifugal or positive-displacement types, can be used and located at the pipeline inlet. However, careful design of the system, including proper sizing of the diameter and the length of the locks to avoid cavitation, proper design of the Y diverters to avoid excessive abrasion, and proper design of the automatic control system to open and close valves alternatively, is a must for trouble-free operation.

7.3.3.2 Ejection

Ejection of capsules at any pipeline outlet station can be done in a reverse manner as injection, except that no pumps are needed and only one conveyor is required, as shown in Figure 7.15 for a CLP system. Note that the foregoing discussions of injection and ejection systems are applicable to all types of HCP including CLP. The only restriction is that the capsule specific gravity must be greater than 1.0 so that the capsules will stay on the conveyor by gravity. A different design of the conveyors in the reservoirs is required if the capsule specific gravity is less than 1.0, such as by using an upside-down conveyor belt for the part where the capsules are underwater.

7.3.4 CAPSULE PUMPS

Many different types of capsule pumps have been studied and reported by different researchers. A discussion of various types is presented in Reference 15. Only the two most practical types are discussed briefly in the following subsections.

7.3.4.1 Pump Bypass

This is a system invented in the 1960s in Canada at the Alberta Research Council, and later (in the 1990s) improved by researchers at Capsule Pipeline Research Center, University of Missouri-Columbia, for use in association with coal log pipeline (CLP). The basic system includes two long parallel pipes (locks), having a length sufficiently long to hold an entire train of capsules in each lock. The two locks are connected to a booster pump and a set of eight valves as shown in Figure 7.16. By alternately

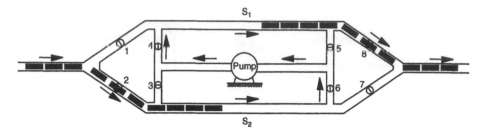

FIGURE 7.16 Pump bypass of HCP and CLP.

opening and closing two sets of valves (the four odd-numbered valves and the four even-numbered valves), capsule trains bypass the booster pump without affecting the pump's ability to put energy into the liquid (water), which in turn carries the capsules through the booster station. The design of the pump-bypass is complicated by the unsteady flow and water hammer generated by rapid switching of valves. This requires careful and sophisticated analysis and optimization by using the method of characteristics modified to incorporate capsules in the flow. It also involves the use and analysis of surge tanks or air chambers to minimize water hammer. Those interested in such design and analysis should consult literature such as References 16 and 17.

7.3.4.2 Electromagnetic Capsule Pumps

An electromagnetic capsule pump (EMC pump) consists of a coil (or a set of coils) wrapped around the pipe through which capsules and fluid pass. The coil(s) may be connected to an AC or DC source to energize the coil(s), and to create a traveling magnetic field inside the pipe. By using capsules that have metallic wall, the traveling magnetic field interacts with the capsule, causing a thrust on the capsule in the direction of flow. This thrust causes the capsules to move through the pump, and the moving capsules in turn generate thrusts on the fluid, causing the fluid in the pipe to move in the direction of capsule motion. By using an EMC pump having a bore slightly smaller than the inner diameter of the pipe connected to the pump, a small clearance is created between the capsules and the pump wall, which in turn creates large electromagnetic thrusts on capsules, and a large pressure rise across each capsule in the pump. The capsules passing through the pump behave like one-way pistons, forcing the fluid to move through the entire pipeline. By using a small clearance in the EMC pump, high pump head can be generated by the pump. This concept of pumping capsules electromagnetically was co-invented by the writer [18,19], and subsequently studied extensively at the Capsule Pipeline Research Center, University of Missouri-Columbia. Figure 7.17 illustrates the concept.

EMC pumps can be of several types [18]. The first is the DC type, which uses a set of solenoids arranged in series and connected to a common DC source. By switching from one solenoid to another sequentially and synchronizing them with the capsule motion, magnetic thrusts develop on a capsule whose wall is made of a ferromagnetic material such as steel, forcing the capsule through the pump and generating a pump head. Two AC types have been studied for EMC pumps. The first uses the principle of linear induction motor (LIM). The LIM is a coil or a set of coils connected to an

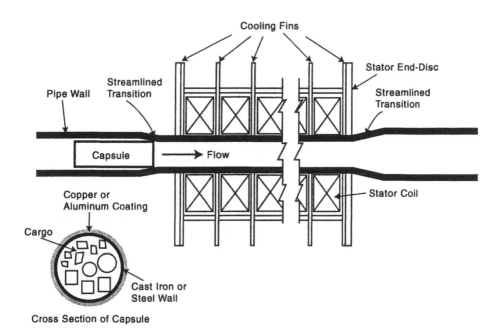

Cross Section of Capsule

FIGURE 7.17 LIM capsule pump configuration.

AC source, often using 3-phase AC of high voltage (240 to 480 V) in order to attain best efficiency. The AC current in the coil(s) creates a traveling magnetic field in the flow direction, and an electrical eddy current in the capsule wall that interacts with the traveling magnetic field to create a thrust on the capsule wall in the flow direction. For best efficiency, the LIM pump used should have small clearance between the capsule and the LIM, and the capsules should have a two-layer wall: an inner wall made of a ferromagnetic material such as steel, and an outer wall that consists of a thin layer of a good conductor such as aluminum. Furthermore, during normal operation the linear motor should have small slip of the order of 3 to 5%. Slip is defined here as $S = (V_s - V_c)/V_s$, where V_s is the synchronous speed of the electromagnetic field created by the LIM. Another AC type EMC pump uses the principle of linear synchronous motor (LSM). For this system to work, the capsule speed going through the pump must match the synchronous speed of the motor, with the slip being zero or nil. A comparison of the pros and cons of the LIM and LSM systems is provided next.

1. The EMC pump based on LIM is simpler in construction than that based on LSM. For instance, LSM requires each capsule to carry either a set of permanent magnets on the capsule surface or a set of electromagnets powered by an on-board battery or generator. Either way, it is complicated, costly, and subject to wear and frequent maintenance. In contrast, the capsules of LIM need no permanent magnets or electromagnets.
2. Because LSM requires synchronism and no slip, either the speed of the capsules entering the pump must be adjusted to match the synchronous speed of the motor, or the latter must be adjusted to match the former.

Both are expensive to do. The usual solution is to adjust the synchronous speed of the LSM, which requires sophisticated and expensive electronics. No such speed adjustment is needed for LIM, which can tolerate a large variation of speed.

3. LSM is not self-starting. Following a power failure and stoppage of capsules, it may be difficult for a LSM-based EMC pump to restart. In contrast, no such problem exists with LIM since it creates maximal thrusts on the capsules at capsule standstill.

4. LSM systems are more efficient than LIM systems, and they can tolerate larger clearance between the capsule and the pump — the so-called **air gap** by electrical engineers.

Based on the above comparison, researchers at the University of Missouri-Columbia opted to study and develop LIM-based EMC pumps for both HCP and PCP [20–22]. In contrast, another group of researchers in the U.S. has developed a LSM-based pump for PCP to transport phosphates [23].

7.4 COAL LOG PIPELINE (CLP)

Coal log pipeline (CLP) is a special type of HCP that does not use containers (capsules) to enclose the cargo (coal in this case). Instead, the coal is compressed into a capsule (cylinder) shape (see Figure 7.4), using high compaction pressure on the order of 10,000 to 20,000 psi. Because no containers are used and there is no need for a return pipeline to bring empty capsules back to the pipeline inlet, the cost of CLP is expected to be much lower than that for an equivalent HCP system that uses containers and requires a second pipeline for returning empty capsules.

Extensive research and development was conducted on CLP at the Capsule Pipeline Research Center (CPRC), University of Missouri-Columbia, during the period 1991 to 1999. The research and development encompassed a variety of subjects concerning CLP — including hydrodynamics, compaction of coal logs, pumping of coal logs (the pump-bypass system), injection and ejection of coal logs, coal log abrasion in pipeline, use of polymers and fiber for drag reduction in CLP, automatic control of CLP systems, handling of coal logs at power plants and treatment of CLP effluent, development of a dielectric sensor for detecting coal logs and capsules in pipelines, economics of CLP, and legal issues including eminent domain rights, water rights, rights to cross railroads, etc. The results of this research are reported in more than 100 publications in technical journals, conference proceedings, M.S. theses, Ph.D. dissertations, and CPRC technical reports. Only a few key findings are discussed here. A brief review of the state of development of CLP prior to 1997 is given in Reference 24.

Researchers at CPRC found that high-pressure compaction (10,000 to 20,000 psi) is a necessity for producing strong, water-resistant, and wear-resistant coal logs. In addition to high pressure, the coal may need a small amount (1 to 3%) of binder, or to be heated to about 80°C, or both, before strong logs can be produced. About 30 factors affect the strength of compacted coal logs, such as pressure, binder, temperature, coal particle size distribution, coal type, mold shape and materials, piston (plunger) shape and material, clearance between the piston and mold wall, compaction time including pressure holding time, ejection force and speed, application of backpressure during ejection from mold, lubrication of mold, cleanness of mold interior, etc. It was found that depending on the coal type and the compaction process used, the quality of the coal logs produced varies greatly. The worst coal logs produced disintegrated in pipes within a few minutes. The best logs, produced from a subbituminous coal mined in Powder River Basin in Wyoming, lasted more than one day when circulated through a steel pipe over a distance of 200 mi with less than 5% weight loss. It was also found that coal logs last three times longer in a plastic pipe than in a steel pipe. This mean that in future commercial use of CLP, one should consider using either a plastic pipe such as polyethylene (PE) pipe, or a steel pipe lined with a layer of plastic, hard elastomer, or resins, in order to minimize coal log abrasion and damage during hydrotransport via pipelines. In addition to making strong logs, other factors that minimize coal log abrasion and damage in pipeline include: the aspect ratio, a, of the coal logs should be in the range of 1.6 to 2.0; the diameter ratio k should be in the range of 0.85 to 0.93; the flow velocity V should be between 85 and 95% of the lift-off velocity, V_L; the temperature of water in the pipe should be relatively low (less than 70°F); the pipe joints should be smooth and without misalignment; the pipe interior should be smooth and lined with a plastic or another relatively soft material, etc. Conditions that favor the use of CLP include relatively short pipes (less than 100 mi of transportation distance); new coal mines in remote areas without existing railroads to transport the coal; coal mines on high mountains where the pipeline will have a relatively large downward slope — this can reduce pressure in the pipe, lengthen the distance between booster pump stations, and saves pumping energy; and coal transportation in areas where it is undesirable to use trucks or trains to transport coal, for safety and/or environmental reasons.

7.5 CONCLUSION

According to a 1998 task committee report of the American Society of Civil Engineers (ASCE) [25], capsule pipeline is a rapidly advancing new technology with strong implications for the future of many nations. As highways and railroads become increasingly congested and overused, capsule pipelines can provide a safe and environmentally desirable means of moving freight underground. Anticipated future use of HCP when the technology is fully developed includes, in addition to coal transport by CLP, the transportation of grain and other agricultural products over long distances (e.g., hundreds of miles from the Grain Belt in the U.S. Midwest to major cities for domestic consumption, and to seaports for export). This can be done by using metallic capsules driven by LIMs, and using pipelines of relatively large diameter — say, 3 ft. Other potential uses of HCP include transporting construction materials (such as sand and gravel) municipal solid wastes over intermediate distances (20 to 200 miles) using smaller pipes — say 1-ft diameter. As to PCP, Japan has demonstrated that large size PCPs (of 1-m diameter or 1 m × 1 m cross section) that are a few kilometer in length and driven by blowers can compete economically with traditional modes of freight transport in certain cases — for transporting limestone to cement plants, and for transporting excavated materials (soil and rocks) and construction materials in the construction of long tunnels [26]. In 2001, Japan has also used PCP to construct vertical mine shafts for deep burial of hazardous solid wastes [27]. With the development of the LIM capsule pumps, a new opportunity is created to extend PCPs to long distances for intercity freight transport. Such future use has enormous potential and strong implications for the world. Basic information about HCP and PCP can be found in References 28 through 30.

PROBLEMS

7.1 Compare HCP with PCP in terms of fluid types, fluid density, capsule shape, need for wheels, capsule speed, linefill, energy intensiveness, and respective niches of potential future markets.

7.2 Discuss why contemporary PCP systems have low linefill, in the neighborhood of only 3%. How can this linefill be improved in order to improve the system efficiency for transporting freight?

7.3 For the PCP analyzed in Example 7.1, if the end disk diameter is reduced to 0.96 m, what will be the capsule velocity, capsule pressure drop, and the drag on the capsule? Compare the results with those given in the example to determine whether such a small change in end disk diameter has a strong effect on the system performance. Discuss the implications.

7.4 In a future PCP system to be used for interstate freight transport, the pipeline used is a rectangular conduit of 1.83 m × 1.83-m cross section. Each capsule used in the system has eight wheels mounted on the bottom of the capsule, and the rolling friction coefficient of the wheels is 0.01. Each capsule is 7.0 m long, and it weighs 18 tons when loaded with cargo. Each capsule has two end disks, each having a cross section of 1.77 m × 1.77 m. The air velocity in the conduit is 22 m/s, and the air density at the capsule location is 1.5 kg/m³. Determine: (a) the drag coefficient of each capsule, (b) the capsule velocity in the pipe under steady-state condition, (c) the drag on each capsule, (d) the pressure drop across each capsule, and (e) the power consumed by each capsule.

7.5 A large HCP system is to be designed for transporting grain from Kansas City to New Orleans, over a distance of 650 mi. Commercial steel pipe with a relatively smooth interior is used, and the fluid is water at 60°F. The pipe inner diameter is 35 inches, and each capsule has an outer diameter of 33 inches, and a length of 8 ft. When filled with grain, each capsule has a specific gravity of 1.05. The contact friction coefficient between the capsules and the pipe is 0.45. Due to the rounding of the capsule edges, the contraction coefficient for this case is 0.71, which is significantly larger than that given in Table 7.1 for sharp-edged capsules. The pipeline is to be run at a water velocity equal to 90% of the lift-off velocity, and at a linefill of 90%. Determine: (a) the incipient velocity, (b) the lift-off velocity, (c) the operation velocity, (d) the critical velocity, (e) the capsule velocity under steady condition, (f) the number of capsules that must be loaded at the pipeline inlet and injected into the pipeline per second in steady operation, (g) the total number of capsules in the pipe at any given time, (h) the grain throughput in MT/yr assuming that 60% of the capsule weight is that of the grain, (i) the net capsule pressure drop if it is equivalent to 30% of the pressure drop by water without capsule, over a distance of one capsule length, (j) the total pump head needed for this pipeline considering the fact that the elevation of Kansas City is about 750 ft higher than New Orleans, and (k) the total power to be consumed in MW (mega watts) for this project assuming that the average efficiency of the pump is 80%.

REFERENCES

1. Zandi, I., Transport of Solid Commodities via Freight Pipeline, Vol. 2, Freight Pipeline Technology, U.S. Department of Transportation, Report No. DOT-TST-76T-36, Washington, D.C., 1976.
2. Kosugi, S., A Capsule Pipeline System for Limestone Transportation, Proceedings of the International Conference on Bulk Materials Handling and Transportation, Institution of Engineers, Australia, July 1992, pp. 13–17.
3. Jvarsheishvilli, A.G., Pneumo-capsule pipelines in U.S.S.R., *Journal of Pipelines*, 1(1), 109–110, 1981.
4. Lampe, D., *Pyke: The Unknown Genius*, Evans Brothers, London, 1959.
5. Brown, R.A.S., Capsule pipeline research at the Alberta Research Council, 1958–1978, *Journal of Pipelines*, 6, 75, 1987.
6. Liu, H. and Marrero, T.R., Coal log pipeline system and method of operations, U.S. Patent No. 4946317, 1990.
7. Kosugi, S., A basic study on the design of the pneumatic capsule pipeline system, Ph.D. dissertation, Osaka University, Osaka, Japan, 1985.
8. York, K., Predicting the performance of a PCP system using a linear induction motor for capsule propulsion, M.S. thesis, Department of Civil Engineering, University of Missouri-Columbia, 1999.
9. Liu, H., Hydraulic behavior of coal log flow in pipe, in *Freight Pipelines*, Round, G.F., Ed., Elsevier Science, 1993, pp. 215–230.
10. Liu, H. and Richards, J.L., Hydraulics of stationary capsule in pipe, *Journal of Hydraulic Engineering*, 120(1), 22–40, 1994.
11. Gao, X. and Liu, H., Predicting incipient velocity of capsules in pipe, *Journal of Hydraulic Engineering*, 126(6), 470–473, 2000.
12. Liu, H., A theory on capsule lift-off in pipeline, *Journal of Pipelines*, 2(1), 22–33, 1982.
13. Gao, X., Hydrodynamics of HCP with slopes and bends, Ph.D. dissertation, Department of Civil Engineering, University of Missouri-Columbia, 1999.
14. Richards, J.L., Behavior of coal log trains in hydraulic transport through pipe, M.S. thesis, Department of Civil Engineering, University of Missouri-Columbia, 1992.
15. Liu, H., Hydraulic capsule pipeline, *Journal of Pipelines*, 1(1), 11–23, 1981.
16. El-Bayya, M.M., Unsteady flow of capsules in a hydraulic pipeline: theory and experiment, Ph.D. dissertation, Department of Civil Engineering, University of Missouri-Columbia, 1994.

17. Wu, J.P., Dynamic modeling of an HCP system and its control, Ph.D. dissertation, Department of Civil Engineering, University of Missouri-Columbia, 1994.
18. Liu, H. and Rathke, J.E., Electromagnetic Capsule Pumps, paper presented at International Symposium on Freight Pipelines, Washington, D.C., 1976.
19. Liu, H., Gibson, D.L., Cheng, H.S., and Rathke, J.E., Pipeline transportation system, U.S. Patent No. 4437799, 1984.
20. Assadollahbaik, M., Linear induction motor for pumping capsules in pipes, Ph.D. dissertation, Department of Civil Engineering, University of Missouri-Columbia, 1984.
21. Plodpradista, W., Study of tubular linear induction motor for pneumatic capsule pipeline system, Ph.D. dissertation, Department of Electrical Engineering, University of Missouri-Columbia, 2002.
22. Liu, H., O'Connell, R., Plodpradista, W., and York, K., Use of Linear Induction Motors for Pumping Capsules in Pneumatic Capsule Pipeline, Proceedings of the 1st International Symposium on Underground Freight Transport by Capsule Pipelines and other Tube/Tunnel Systems, Columbia, Missouri, 1999, pp. 84–94.
23. Montgomery, D.B., Fairfax, S., Beals, D., Smith, B., and Whitley, J., Electromagnetic Pipeline Transport System for the Phosphate Industry, Proceedings of the 1st International Symposium on Underground Freight Transport by Capsule Pipelines and other Tube/Tunnel Systems, Columbia, Missouri, 1999, pp. 74–83.
24. Liu, H. and Marrero, T.R., Coal log pipeline technology: an overview, *Powder Technology*, 94, 217–222, 1997.
25. ASCE Task Committee on Freight Pipelines, Freight pipelines: current status and anticipated future use, *Journal of Transportation Engineering*, 124(4), 300–310, 1998.
26. Kosugi, S., Pneumatic Capsule Pipelines in Japan and Future Developments, Proceedings of the 1st International Symposium on Underground Freight Transport by Capsule Pipelines and Other Tube/Tunnel Systems, Columbia, Missouri, 1999, pp. 61–73.
27. ASCE, Pneumatic capsule pipeline removes soil vertically, *Civil Engineering*, 72(3), 22, 2002.
28. Freight Pipelines, in *Encyclopedia Britannica*, Encyclopedia Britannica, Chicago, 1993, pp. 861–864.
29. Pipeline, in *McGraw-Hill Yearbook of Science and Technology*, S.P. Parker, Ed., McGraw-Hill, New York, 1994, pp. 305–307.
30. Pipelines, in *Kirk-Othmer Encyclopedia of Chemical Technology*, John Wiley & Sons, New York, 1997, pp. 102–110.

Part II

Engineering Considerations

8 Pipes, Fittings, Valves, and Pressure Regulators

8.1 TYPES OF PIPE

Different applications or operating conditions call for the use of different types of pipes for pipeline projects. An engineer cannot make a wise selection of the best pipe needed for a given project without some understanding of the characteristics of the various types of pipes commercially available. A brief discussion of each type is provided here. Two broad classifications are used: metallic and nonmetallic pipes.

8.1.1 METALLIC PIPES

Most metallic pipes are stronger and harder to break, but they are more conductive to heat and electricity and less corrosive-resistant than nonmetallic pipes. Commercially available metallic pipes are discussed in the following sections.

8.1.1.1 Ordinary Steel Pipe

Ordinary steel pipes are made of wrought (carbon) steel. They may be either **seamless** or **seamed (welded)**. The seamed steel pipes are made of steel sheets or steel plates rolled or press formed into circular shape, with the edge (seam) of each pipe closed by welding. Four types of welding are shown in Figure 8.1: (a) butt weld, (b) lap weld, (c) electric arc weld (single-welded joint), and (d) electric arc weld (double welded joint). The first two are furnace welded, mostly for small pipes (up to 4 inches); the last two do not require a furnace but do require weld filler metal, and they are mostly for large pipes. Two alternatives to electric arc weld with weld filler metal are electric resistance welding and electric induction welding, which do not require weld filler metal. They are all referred to as **fusion-welded pipes**, in contrast to **furnace-welded pipes**. Figure 8.2 shows how a seamless pipe is made by piercing a cold rod through a hot rod (billet) of steel at a temperature of the order of 2000°F. The two rollers in the figure grip and turn the billet, causing the billet to rotate and advance over the piercer point, forming a hole through the length of the billet. Figure 8.3 shows how a spiral steel pipe is made. Two types of welding are used for spiral or helical seam steel pipes: butt weld and lap weld. Depending on the type of weld used, spiral weld pipes are used for both high pressure and low pressure services.

(a) Butt-Welded Pipe

(b) Lap-Welded Pipe

(c) Electric Arc-Welded Pipe (Single-Welded Joint)

(d) Electric Arc-Welded Pipe (Double-Welded Joint)

FIGURE 8.1 Different types of welding for seamed steel pipe.

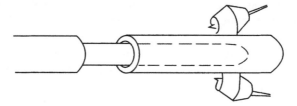

FIGURE 8.2 Formation of seamless pipe.

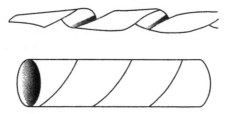

FIGURE 8.3 Spiral-welded pipe.

Steel pipes are structurally strong and ductile; they do not fracture easily. However, unless they are coated or lined* with an inert material, or protected by other means they can be corroded badly. Steel pipes come in a great variety of sizes and strengths. They are widely used for transporting natural gas, petroleum products, air, and water. When steel pipes are used for water, preventive measures must be taken against not only external corrosion but also internal corrosion. Chapter 11 discusses various types of corrosion and measures to control them.

Steel pipes use various grades of steel with yield strength in the range of 30,000 to 70,000 psi. The design stress is normally based on a certain percentage of the pipe yield strength or tensile strength and it varies with the requirements of the applicable design code. For internal pressure below 150 psi, the wall thickness of steel pipe required is so small that design is normally controlled by the external pressure and the desired minimum deflection. For internal pressure greater than 150 psi, design is often controlled by the hoop tension generated by internal pressure. See Chapter 13 for design details.

8.1.1.2 Corrugated Steel Pipe

Corrugated steel pipes are thin-wall, large-diameter pipes made of galvanized steel sheets having either helical or annular corrugations. Due to its low cost, the corrugated steel pipe is used extensively in sewer and drainage systems where both the internal pressure (water pressure) and the external pressure (soil pressure) are low, and where leakage will not cause serious problems. Some corrugated steel pipes come with perforations to allow drainage into the pipe. The water in a large diameter corrugated pipe is often gravity flow (open channel flow) rather than pressure flow (pipe flow). Corrugated pipes come in a great range of sizes, and have a large variety of fittings. Some corrugated pipes are fabricated to have a pecan-shaped (arched) instead of circular cross section.

8.1.1.3 Cast-Iron Pipe

There are two types of cast-iron pipe: the ordinary or **gray cast-iron** pipe, and the **ductile-iron** pipe. The ordinary cast-iron pipe is made of iron containing 3 to 4% of carbon in the form of graphite flakes. The pipe is cast either by using a stationary mold (horizontal or vertical) or a centrifugal mold. The mold can be either a metal mold cooled by water, or a sand-lined mold. The centrifugal mold usually produces better results than the stationary mold. There are two strength designations for cast-iron pipes: 18/40 and 21/45. In the first designation, the number 18 means that the minimum bursting tensile strength is 18,000 psi, and the number 40 means that the minimum **modules of rupture** (i.e., the tensile stress that causes failure due to bending) is 40,000 psi. The meaning of the 21/45 designation is similar.

The gray cast-iron pipe has relatively strong corrosion-resistance ability and long life. It is used in a number of applications, such as for water supply or wastewater. The pipe is often lined and/or coated with cement or another inert

---

* While **coating** pertains to covering the pipe exterior, **lining** pertains to coating the pipe interior. Lining and coating are discussed in more detail in Chapter 11.

nonmetallic material when a corrosive fluid and/or corrosive environment is encountered. The main disadvantage of the cast-iron pipe is that it is not ductile — the pipe fractures under excessive loads or impact.

8.1.1.4 Ductile-Iron Pipe

The ductile-iron pipe is made of iron containing approximately 3.5% of carbon in spheroidal or nodular form, and a magnesium alloy. It is rather ductile, and does not rupture easily. It combines the advantages of the gray cast-iron pipe, including corrosion-resistance and long life, with the advantage of steel pipe — ductility. Like the gray cast-iron pipe, the ductile-iron pipe is often lined and/or coated with cement mortar. The strength designation of ductile iron pipe is 60-42-10. The number 60 means a minimum tensile strength of 60,000 psi; the number 42 means a minimum yield strength of 42,000 psi; and the number 10 means 10% minimum elongation. The ductile-iron pipe has largely replaced the gray cast-iron pipe in recent years. It is used extensively in sanitary engineering works for water supply and wastewater.

8.1.1.5 Stainless Steel Pipe

The most extensively used stainless steel pipes are those in the 300 series such as SS304 or SS316. They are made of steel that contains chrome-nickel alloys, and they are corrosion resistant. However, due to its high price, stainless steel pipe is used only in special applications such as when the fluid or environment is rather corrosive, or when no rusting of pipe can be tolerated such as in pharmaceutical or food industries.

8.1.1.6 Aluminum Pipe

They are corrosion resistant and are used in certain food plants and chemical plants. Different types of aluminum are used for making pipes. For instance, aluminum 1100 is low in strength but easy to weld. High-number aluminums, such as those in the 3000 or 6000 series, are stronger mechanically and more corrosion-resistant, although they are harder to weld. Aluminum pipes and tubings are usually formed by a drawing or extrusion process; they are seamless.

8.1.1.7 Copper Pipe

Copper is corrosion-resistant but expensive. It is used only for small pipes such as those used in plumbing. Copper pipes can be formed by cold drawing. Note that many of the stainless steel, aluminum and copper pipes are actually tubings rather than pipes. See Section 8.1.3 for a discussion of the difference between tubing and pipe.

8.1.1.8 Other Metal Pipes

Many other metal pipes are available. They are made from different alloys and are used for different purposes, such as corrosion resistance to a particular fluid, high-temperature resistance, low-temperature resistance, etc. Some examples are listed in Table 8.1.

TABLE 8.1
Some Metal Pipes for Special Applications

Admiralty metal	Resistant to corrosive water, including saltwater
Aluminum 1100	Most weldable type of aluminum, low strength, and resistant to formaldehyde, ammonia, phenol, and hydrogen sulfide; used in food plants
Aluminum 3003	Superior mechanical properties; contains manganese; used in chemical plants
Aluminum 6061 or 6063	Contains silicon and magnesium; highly corrosion resistant
Brass and silicon	Resistant to corrosive waters, including brines, sugar water and organic acids
Copper	Resistant to corrosive water
Cast iron	Corrosion resistant; used for water and gas distribution, and sewage systems
Ductile iron	Similar corrosion-resistant properties as cast iron, but is more ductile
Monel	Nickel-based alloy; high strength; corrosion resistant to alkaline solutions and air free acids; used only when copper contamination is not a problem
Nickel	Highly corrosion resistant
Stainless steel (type 304)	Corrosion resistant; commonly used in processing food and medicine; extra low carbon (ELC) grade provides good weldability
Stainless steel (types 316, 321, or 347)	Also available in ELC grades; good corrosion resistance at elevated temperatures and pressures; available in ELC grade for good weldability
Tantalum	Resistant to nitric and other acids
Titanium	Highly resistant to corrosion in oxidizing media; resistant to sulfuric acid and perchlorites; resistant to abrasion and cavitation

Source: Data from Nayyar, M.L., *Piping Handbook*, 6th ed., McGraw-Hill, New York, 1992.

8.1.2 NONMETALLIC PIPES

Although nonmetallic pipes may not be as strong as metallic pipes structurally, they may be lighter in weight, more economical, or may have certain other advantages such as being more corrosion resistant. A brief discussion of various nonmetallic pipes is provided next.

8.1.2.1 Concrete Pipe

Concrete pipes can be divided into the **nonpressure** (or **low-pressure**) and the **pressure** (or **high-pressure**) types. The nonpressure type is normally made of plain concrete. It is used in applications such as sewers or culverts, which do not operate under high or even moderate internal pressure. Because concrete has high compressive strength, the plain concrete pipes can withstand high external pressure, such as that imposed on it by the earth and traffic above it. However, without reinforcement the concrete is easy to crack or break when under tension.

Concrete pipes can be made to withstand moderate to high internal pressure by placing reinforcement in the concrete — either by using prestressed concrete or ordinary reinforced concrete. They are normally called **concrete pressure pipes**, and are used extensively in water and wastewater works.

Concrete pressure pipes can be divided into several subgroups. The first is **prestressed concrete cylinder pipe (PSCCP)**, which is fabricated by casting a thick layer of concrete lining inside a thin-wall steel cylinder. After the concrete is set, the pipe is helically wrapped with high strength hard-drawn wire, with the wire spacing accurately determined to produce a predetermined residual compression in the concrete core. Then, a layer of cement–mortar is coated outside to cover the wire. Figure 8.4a shows the details of the pipe. Another design of PSCCP is to have the steel cylinder embedded in the concrete instead of outside the concrete, and having the prestressed wire outside the concrete instead of in contact with the cylinder. This is shown in Figure 8.4b. Note that PSCCP has been designed for operating pressures greater than 400 psi and earth covers in excess of 100 ft.

Another type of concrete pressure pipe is **reinforced concrete cylinder pipe (RCCP)**. For this pipe, mild steel reinforcement bar cages are cast into the wall of

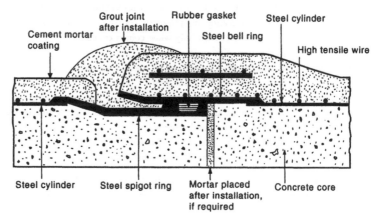

(a) Lined Cylinder

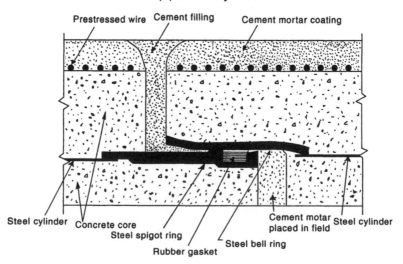

(b) Embedded Cylinder

FIGURE 8.4 Two types of prestressed concrete cylinder pipe (PSCCP).

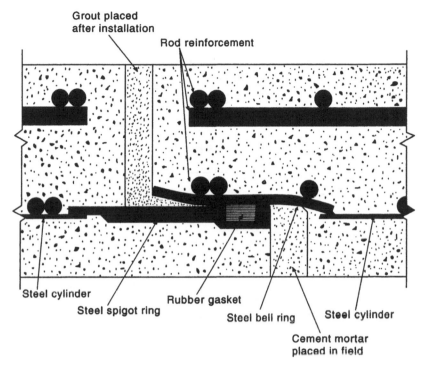

FIGURE 8.5 Reinforced concrete cylinder pipe (RCCP).

the pipe instead of prestressing with wire (see Figure 8.5). This pipe should not be used for internal pressures greater than 250 psi, nor in trenches exceeding 20 ft of earth cover.

Another type is **reinforced concrete noncylinder pipe (RCNCP)**. Because it does not contain a watertight steel cylinder, it is used for pressures below 60 psi. Figure 8.6 shows this pipe. Its use has been declining since the introduction of PSCCP.

Yet another type is the **pretensioned concrete cylinder pipe (PTCCP)** (see Figure 8.7). In this pipe, a thick layer of cement mortar lines the inside of a steel cylinder. After the cement is hardened, the cylinder is wrapped with a smooth, hot-rolled steel bar, using moderate tension in the bar. Then, the cylinder and the bar wrapping are covered with a cement slurry and a dense mortar coating that is rich in cement. This pipe is somewhat similar to the PSCCP, and it can be designed for internal pressures as high as 400 psi.

A summary of the types of concrete pipes, their pressure ratings, and typical applications can be found in Table 8.2.

Concrete pipes have the advantages of being structurally strong, corrosion-resistant and economical. They can also be cast on site if necessary or desirable. Their disadvantages include being heavy (and hence being costly to transport and difficult to move), hard to cut (and hence must be used in standard lengths), come in shorter sections than metal or plastic pipes (and hence require more joints), and more limited in the kinds of joints that can be used. More and more pressure concrete pipes are used for longer and longer distances and higher and higher pressure. They should

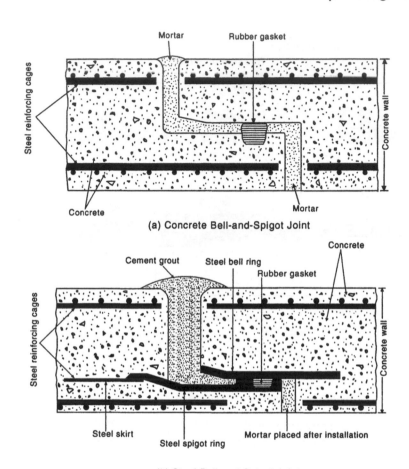

(a) Concrete Bell-and-Spigot Joint

(b) Steel Bell-and-Spigot Joint

FIGURE 8.6 Two types of reinforced concrete noncylinder pipe (RCNCP).

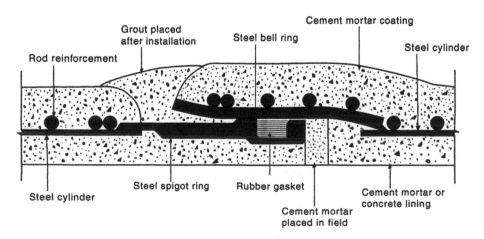

FIGURE 8.7 Pretensioned concrete cylinder pipe (PTCCP).

TABLE 8.2
Types of Concrete Pipes and Applications

Type	Maximum Pressure Allowed (psi)	Typical Applications
Plain concrete pipe (PCP)	Practically 0	Gravity flow or nonpressure flow, as for certain sewers and culverts
Reinforced concrete noncylinder pipe (RCNCP)	60	Sewers, storm drain, irrigation pipe, etc.
Reinforced concrete cylinder pipe (RCCP)	250	Sewers, water mains, etc.
Prestressed concrete cylinder pipe (PSCCP)	400	High-pressure water and sewer lines
Pretensioned concrete cylinder pipe (PTCCP)	400	Same as for PSCCP

be considered for possible use in any water or wastewater project that requires a large (say, greater than 2-ft-diameter) pipe.

8.1.2.2 Plastic Pipe

Three types of plastic pipes used commonly are **PVC (polyvinyl chloride)**, **PE (polyethylene)**, and **PP (polypropylene)**. They are used for water, wastewater, natural gas, and certain other fluids that do not dissolve or chemically interact with the plastic material.

Plastic pipes are low cost, lightweight, easy to cut and join, and corrosion resistant. Due to these advantages, plastic pipes are being used increasingly. However, users also must know that plastic pipes are usually not as strong as metal pipes, deform easily, expand five times as much as steel pipes as a result of temperature changes, soften or burn at high temperature, and become brittle in very cold weather. Besides, not enough data have been accumulated about plastic pipes to know their long-term performance.

PVC pipe is currently very popular for sewers of relatively small diameter — up to about 24 inches (610 mm). For larger diameter sewers, concrete pipes are more popular for they cost less. PVC pipe is also used for water supply and indoor plumbing. It is relatively hard (as compared to PE pipes), and hence is highly abrasion (wear) resistant — more so than concrete pipes. It is corrosion resistant to a number of chemicals, and it will not deteriorate under the attack of bacteria and other microorganisms, macroorganisms, or fungi. This property makes PVC pipe especially suited for use as sewers. Because PVC pipe has a hard, inert, and smooth interior, soluble solids such as calcium carbonate cannot accumulate and attach to the interior surface of PVC pipes when the water or sewage in the pipe is flowing. Consequently, PVC pipes are free from the tuberculation problem that often exists in water works using steel or other metal pipes. A weakness of PVC and other plastic pipes such as PE is that they deteriorate when exposed to ultraviolet light. Therefore, they should not be exposed to direct sunlight when used or stored over extended time. They are most suited for underground pipelines. PVC pipes come in segments no longer than 40 ft (12.2 m) each — most often 20 ft (11.1 m) or less. They can be joined together either by using solvent-cement, or gaskets. The former is for flat ends of pipes such as those due to cuts, whereas the latter is for full lengths of pipes with bell and spigot ends. The gasketed joints allow for thermal expansion. PE pipe is very flexible, and can be coiled for compactness in storage and transportation. Hundreds of feet of PE pipes, up to 6-inch (152 mm) diameter, can be rolled around a single spool. This greatly reduces the number of joints, and facilitates transportation and laying of pipe. The pipe can be easily laid on surface or underground without having joints over a long distance — several hundred feet or meters. Its flexibility allows the pipe to follow the curvature of topography easily. Joining is usually done by thermal fusing — heating and melting the pipe ends while holding them together. PE pipe is widely used for natural gas distribution lines up to 10-inch (254 mm) diameter. It is also used for water lines and other applications. According to their density (or specific gravity), PE pipes are classified as low density polyethylene (LDPE) when specific gravity is in the range of 0.910 to 0.925, medium density polyethylene (MDPE) in the specific gravity range of 0.926 to 0.940, and high density polyethylene (HDPE) in the range 0.941 to 0.965. All PE pipes are highly corrosion resistant and abrasion resistant. A main weakness of PE is that the material deteriorates under exposure to sunlight.

8.1.2.3 Clay (Ceramic) Pipe

Vitrified clay pipes ("clay pipes" for short) are used for sewers and drain tiles. Their advantages include low cost and high corrosion resistance. The disadvantages include fragility (break easily during handling) and low resistance against internal pressure and bending. In recent years, high-quality ceramic materials have been developed to make pipes, parts of valves, etc. The high compressive strength of such ceramic materials has enabled the development of clay pipes for pipe-jacking operations.

8.1.2.4 Wood and Bamboo Pipes

Wood pipes are mostly made of wood staves. They were used extensively prior to the 20th century for water transmission lines, for conveying salt water for extraction of salt or bromide, and for piping wood pulp. Their use since the invention of steel pipes has steadily declined. They are largely a remnant of the past. Bamboo pipes were used extensively in the Orient, especially in China, for conveying water. Their use in modern times has greatly declined. As with the wood pipes, the bamboo pipes have become a remnant of the past. However, due to their low cost and environmental friendliness, they are still being used by some farmers in certain remote areas of China for irrigation on small farms.

8.1.2.5 Graphite and Carbon Pipes

They are very brittle, and are used only in special applications involving very high temperature.

8.1.2.6 Asbestos Cement Pipe

For many years, asbestos cement pipes (i.e., pipes made of cement mixed with asbestos fiber) were widely used for sewers and water lines. They had a number of advantages and hence were widely used in the U.S. and around the world. However, they are no longer used in the U.S. because inhaling asbestos has proved to cause lung cancer. Current regulations in the U.S. do not allow use of most asbestos products including pipes. Still, there are many existing pipes in use in the U.S. that are asbestos cement pipes, or pipes lined with asbestos cement. They pose little threat to public health because the only proven danger from asbestos is through inhalation.

8.1.2.7 Rubber and Elastomer Piping

Natural rubber and synthetic rubbers (elastomers) are used for making flexible hoses. They are needed for small-scale, special applications, such as gardening and hydraulic machineries.

8.1.2.8 Glass Pipe

The disadvantages of glass pipes are that they are extremely brittle, difficult to tap or handle, and expensive. The advantages are that they are inert to corrosive fluids, are clean, and can withstand relatively high temperature (up to about 200°C). They are used mainly in food processing plants and pharmaceutical plants for their cleanliness, and in chemical plants for their ability to handle corrosive fluids such as strong acids.

8.1.3 TUBING

Tubings are similar to pipes except that they are normally smaller in diameter, and have thinner walls than those of pipes. Although there is an overlap of the size of tubings with pipes, their designations are different (e.g., a 4-inch tubing does not have the same inner or outer diameter of a 4-inch pipe). Due to this difference, fittings and valves for pipes cannot be used for tubings without some modification of either the tubing or fitting (or valve connection).

As in the case of pipes, tubings are made of various materials such as steel, stainless steel, aluminum, copper, and plastic. They are manufactured for special purposes such as for use in drilling and maintenance of oil and gas wells, and food processing or pharmaceutical plants.

8.2 PIPE DESIGNATION

A common designation of pipe size in the U.S. is the **nominal pipe size (NPS)**. In this system, the nominal diameter of pipes is given in inches. The weight (or thickness) of the pipe is given in **schedule number** (for steel, stainless steel pipes, and PVC), or **class number** (for cast-iron or ductile-iron pipes).

Table 8.3 gives the designations for steel pipe. As can be seen from the table, for diameters equal to or greater than 14 inches, the nominal diameter becomes the same as the outer diameter (O.D.) of the pipe, given in inches. On the other hand, for diameters smaller than 14 inches, the nominal diameter is somewhat different from either the O.D. or the I.D. of the pipes. Table 8.3 shows that the weight or thickness of a steel pipe can be specified in terms of either a schedule number (from 10 to 160) or another system having three designations: **STD (standard)**, **XS (extra strong)**, and **XXS (double extra strong)**. An exception to the foregoing designation system for steel pipes is light-gauge piping such as the spiral-welded pipe. For sizes from $1/2$ inch to 12 inches, the wall thickness of spiral-welded pipes is the same as schedule 10S (S stands for stainless steel), and for sizes from 14 to 24 inches, it is the same as schedule 10.

The designations for stainless steel pipes and aluminum pipes are similar to that of steel pipe except that a suffix S is affixed to the schedule number, as for instance the designation schedule 40S. Table 8.4 lists the designations for stainless steel pipes.

Note that the American Petroleum Institute (API) does not use schedule number for steel pipes. Instead, it directly lists the O.D., I.D., and the thickness of the pipe, in two extensive tables — one for welded pipe, and one for threaded pipe.

Cast-iron pipes and ductile-iron pipes are usually designated in **class numbers**. The American Water Works Association (AWWA) has established standards for such designations.

Although ANSI has no standard for nonferrous metal pipes such as brass, copper, and nickel pipes, manufacturers use the ANSI standard for stainless steel for these pipes. Finally, the nominal size for tubing is the same as the O.D. of the tubing. For the wall thickness of tubing, the BWG (Birmingham Wire Gage) is often used.

Example 8.1 For an NPS 8 Schedule 20 steel pipe, determine the following: (a) the O.D. (outside diameter) of the pipe, (b) the wall thickness of the pipe, (c) the I.D. (inside diameter) of the pipe, and (d) the weight of the pipe for a 20-ft length of the pipe.

[Solution] (a) From Table 8.3, the O.D. of the NPS 8 Schedule 20 steel pipe is 8.625 inches. (b) From the same table, the wall thickness of the pipe is δ = 0.250 inch. (c) Therefore, the I.D. of the pipe should be I.D. = O.D. $- 2\delta$ = $8.625 - 2 \times 0.250 = 8.125$ inches. (d) From Table 8.3, the unit weight of the NPS 8 Schedule 20 steel pipe is 22.36 lb/ft. Therefore, for a 20-ft segment, the weight of the pipe is W = $20 \times 22.36 = 447$ lb approximately.

8.3 CONNECTIONS (JOINTS)

Pipes can be joined (connected) in several ways, as follows:

1. **Bonded (welded) joints** — They include welding for steel pipe, brazing or soldering for brass, copper and lead pipes, and fusing of plastic pipes. Figure 8.8 shows different ways of welding steel pipes. Joining plastic pipes (PVC and PE) were discussed in Section 8.1.2.2.

TABLE 8.3
Dimensions and Weights of Welded Wrought Steel Pipe

Nominal Pipe Size	Outside Diameter	STD Wall	STD Wt.	XS Wall	XS Wt.	XXS Wall	XXS Wt.	10 Wall	10 Wt.	20 Wall	20 Wt.	30 Wall	30 Wt.
$1/8$	0.405	0.068	0.24	0.095	0.31								
$1/4$	0.540	0.088	0.42	0.119	0.54								
$3/8$	0.675	0.091	0.57	0.126	0.74								
$1/2$	0.840	0.109	0.85	0.147	1.09	0.294	1.71						
$1/4$	1.050	0.113	1.13	0.154	1.47	0.308	2.44						
1	1.315	0.133	1.68	0.179	2.17	0.358	3.66						
$1\frac{1}{4}$	1.660	0.140	2.27	0.191	3.00	0.382	5.21						
$1\frac{1}{2}$	1.900	0.145	2.72	0.200	3.63	0.400	6.41						
2	2.375	0.154	3.65	0.218	5.02	0.436	9.03						
$2\frac{1}{2}$	2.875	0.203	5.79	0.276	7.66	0.552	13.70						
3	3.500	0.216	7.58	0.300	10.25	0.600	18.58						
$3\frac{1}{2}$	4.000	0.226	9.11	0.318	12.51								
4	4.500	0.237	10.79	0.337	14.98	0.674	27.54						
5	5.563	0.258	14.62	0.375	20.78	0.750	38.55						
6	6.625	0.280	18.97	0.432	28.57	0.864	53.16						
8	8.625	0.322	28.55	0.500	43.39	0.875	72.42			0.250	22.36	0.277	24.70
10	10.750	0.365	40.48	0.500	54.74	1.000	104.13			0.250	28.04	0.307	34.24
12	12.750	0.375	49.56	0.500	65.42	1.000	125.49			0.250	33.38	0.330	43.77
14	14.000	0.375	54.57	0.500	72.09			0.250	36.71	0.312	45.61	0.375	54.57
16	16.000	0.375	62.58	0.500	82.77			0.250	42.05	0.312	52.27	0.375	62.58
18	18.000	0.375	70.59	0.500	93.45			0.250	47.39	0.312	58.94	0.438	82.15

TABLE 8.3 *(Continued)*
Dimensions and Weights of Welded Wrought Steel Pipe

Nominal Pipe Size	Outside Diameter	Pipe Thickness						Schedule Number						
		STD		XS		XXS		10		20		30	
		Wall	Wt.	Wall	Wt.	Wall	Wt.	Wall	Wt.	Wall	Wt.	Wall	Wt.
20	20.000	0.375	78.60	0.500	104.13			0.250	52.73	0.375	78.60	0.500	104.13
22	22.000	0.375	86.61	0.500	114.81			0.250	58.07	0.375	86.81	0.500	114.81
24	24.000	0.375	94.62	0.500	125.49			0.250	63.41	0.375	94.62	0.562	140.68
26	26.000	0.375	102.63	0.500	136.17			0.312	85.60	0.500	136.17		
28	28.000	0.375	110.64	0.500	146.85			0.312	92.26	0.500	146.85	0.625	182.73
30	30.000	0.375	118.65	0.500	157.53			0.312	98.93	0.500	157.53	0.625	196.08
32	32.000	0.375	126.66	0.500	168.21			0.312	105.59	0.500	168.21	0.625	209.43
34	34.000	0.375	134.67	0.500	178.89			0.312	112.25	0.500	178.89	0.625	222.78
36	36.000	0.375	142.68	0.500	189.57			0.312	118.92	0.500	189.57	0.625	236.13
38	38.000	0.375	150.69	0.500	200.25								
40	40.000	0.375	158.70	0.500	210.93								
42	42.000	0.375	166.71	0.500	221.61								
44	44.000	0.375	174.72	0.500	232.29								
46	46.000	0.375	182.73	0.500	242.97								
48	48.000	0.375	190.74	0.500	253.65								

TABLE 8.3 (Continued)
Dimensions and Weights of Welded Wrought Steel Pipe

Pipe Thickness — Schedule Number

Nominal Pipe Size	40 Wall	40 Wt.	60 Wall	60 Wt.	80 Wall	80 Wt.	100 Wall	100 Wt.	120 Wall	120 Wt.	140 Wall	140 Wt.	160 Wall	160 Wt.
$\frac{1}{8}$	0.068	0.24			0.095	0.31								
$\frac{1}{4}$	0.088	0.42			0.119	0.54								
$\frac{3}{8}$	0.091	0.57			0.126	0.74								
$\frac{1}{2}$	0.109	0.85			0.147	1.09							0.188	1.31
$\frac{3}{4}$	0.113	1.13			0.154	1.47							0.219	1.94
1	0.133	1.68			0.179	2.17							0.250	2.84
$1\frac{1}{4}$	0.140	2.27			0.191	3.00							0.250	3.76
$1\frac{1}{2}$	0.145	2.72			0.200	3.63							0.281	4.86
2	0.154	3.65			0.218	5.02							0.344	7.46
$2\frac{1}{2}$	0.203	5.79			0.276	7.66							0.375	10.01
3	0.216	7.58			0.300	10.25							0.438	14.31
$3\frac{1}{2}$	0.226	9.11			0.318	12.51								
4	0.237	10.79			0.337	14.98			0.438	18.98			0.531	22.52
5	0.258	14.62			0.375	20.78			0.500	27.04			0.625	32.96
6	0.280	18.97			0.432	28.57			0.562	36.39			0.719	45.35
8	0.322	28.55	0.406	35.64	0.500	43.39	0.534	50.95	0.719	60.71	0.812	67.76	0.906	74.69
10	0.365	40.48	0.500	54.75	0.594	64.43	0.719	77.03	0.844	89.29	1.000	104.13	1.125	115.65
12	0.406	53.53	0.562	73.15	0.688	88.63	0.844	107.32	1.000	125.49	1.125	139.67	1.312	160.27
14	0.438	63.44	0.594	85.05	0.750	106.13	0.938	130.85	1.094	150.79	1.250	170.21	1.406	189.11
16	0.500	82.77	0.656	107.50	0.844	136.61	1.031	164.82	1.219	192.43	1.438	223.64	1.594	245.25
18	0.562	104.67	0.750	138.17	0.938	170.92	1.156	207.96	1.375	244.14	1.562	274.22	1.781	308.50

TABLE 8.3 (*Continued*)
Dimensions and Weights of Welded Wrought Steel Pipe

Nominal Pipe Size	Pipe Thickness						Schedule Number									
	40		60		80		100		120		140		160	
	Wall	Wt.	Wall	Wt.	Wall	Wt.	Wall	Wt.	Wall	Wt.	Wall	Wt.	Wall	Wt.
20	0.594	123.11	0.812	166.40	1.031	208.87	1.281	256.10	1.500	296.37	1.750	341.09	1.969	379.14
22			0.875	197.41	1.125	250.81	1.375	302.88	1.625	353.61	1.875	403.00	2.125	451.06
24	0.688	171.29	0.969	238.35	1.219	296.58	1.531	367.39	1.812	429.39	2.062	483.12	2.344	542.13
26														
28														
30														
32	0.688	230.08												
34	0.688	244.77												
36	0.750	282.35												

Note: The pipe size, diameter, and wall thickness are in inches; the weight is in lb/ft.

Source: Reprinted from ASME B 36.10-1979 by permission of the American Society of Mechanical Engineers. All rights reserved.

TABLE 8.4
Dimensions and Weights of Stainless Steel Pipe

		Nominal Pipe Schedule							
		55		105		405		805	
Pipe Size	Outside Diameter	Wall	Wt.	Wall	Wt.	Wall	Wt.	Wall	Wt.
$1/8$	0.405			0.049	0.19	0.068	0.24	0.095	0.31
$1/4$	0.540			0.065	0.33	0.088	0.42	0.119	0.54
$3/8$	0.675			0.065	0.42	0.091	0.57	0.126	0.74
$1/2$	0.840	0.065	0.54	0.083	0.67	0.109	0.85	0.147	1.09
$1/4$	1.050	0.065	0.69	0.083	0.86	0.113	1.13	0.154	1.47
1	1.315	0.065	0.87	0.109	1.40	0.133	1.68	0.179	2.17
$1^1/4$	1.660	0.065	1.11	0.109	1.81	0.140	2.27	0.191	3.00
$1^1/2$	1.900	0.065	1.28	0.109	2.09	0.145	2.72	0.200	3.63
2	2.375	0.065	1.61	0.109	2.64	0.154	3.65	0.218	5.02
$2^1/2$	2.875	0.083	2.48	0.120	3.53	0.203	5.79	0.276	7.66
3	3.500	0.083	3.03	0.120	4.33	0.216	7.58	0.300	10.25
$3^1/2$	4.000	0.083	3.48	0.120	4.97	0.226	9.11	0.318	12.51
4	4.500	0.083	3.92	0.120	5.61	0.237	10.79	0.337	14.98
5	5.563	0.109	6.36	0.134	7.77	0.258	14.62	0.375	20.78
6	6.625	0.109	7.60	0.134	9.29	0.280	18.97	0.432	28.57
8	8.625	0.109	9.93	0.148	13.40	0.322	28.55	0.500	43.39
10	10.750	0.134	15.23	0.165	18.70	0.365	40.48	0.500	54.74
12	12.750	0.156	22.22	0.180	24.20	0.375	49.56	0.500	65.42
14	14.000	0.156	23.04	0.188	27.71				
16	16.000	0.165	27.88	0.188	31.72				
18	18.000	0.165	31.40	0.188	35.73				
20	20.000	0.188	39.74	0.218	46.02				
24	24.000	0.218	55.32	0.250	63.35				
30	30.000	0.250	79.36	0.312	98.83				

Note: The pipe size, diameter, and wall thickness are all in inches; the weight is in lb/ft.

Source: Reprinted from ASME B36.19-1976 by permission of the American Society of Mechanical Engineers. All rights reserved.

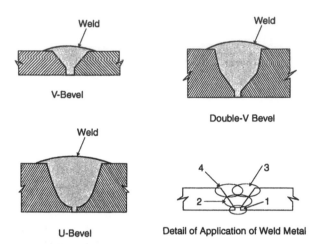

Weld
V-Bevel

Weld
Double-V Bevel

Weld
U-Bevel

Detail of Application of Weld Metal

FIGURE 8.8 Welding types and processes.

2. **Threaded joints** — Used for connecting threaded pipe sections together, or connecting a threaded pipe to a threaded coupling or fitting.

3. **Flanges** — Flanges are the most common way to provide a strong joint without permanently joining the pipe sections together as done in welding. They are used extensively for steel pipes at both the inlet and the outlet of pumps, valves, flowmeters, and other fittings. Use of flanges allows such devices be easily installed in the pipeline and easily disconnected from the pipe.

4. **Mechanical joints** — Various mechanical connectors exist for ease in assembly/disassembly.

5. **Bell-and-spigot joints** — See Figure 8.7 for those used in pressure concrete pipes. Other pipes, including glass pipes and plastic pipes, also use bell-and-spigot joints.

6. **Push-on joints** — Joints that can be connected together simply by pushing two pipe sections (segments) against each other. Likewise, the sections can be disconnected simply by pulling them apart.

8.4 FITTINGS

Screwed pipe fittings are to be used with threaded pipes. Welding fittings and socket-welding fittings, on the other hand, are used with nonthreaded pipes. The purposes of different fittings are described briefly in Table 8.5. Detailed discussion of fittings and valves, with detailed drawings for each, can be found in piping handbooks such as References 1 and 2. Readers interested in such details should consult those books.

8.5 VALVES

Various types of valves are used for various purposes and in different situations. The headloss coefficients of various types of valves are given in Table 2.2 of Chapter 2.

TABLE 8.5
Types of Fittings and Their Purposes

Fitting Type	Purpose
Bushing	To connect a threaded small pipe to a larger one
Cap	To seal the end of a pipe
Coupling	To connect two threaded pipes of the same size together
Half coupling	One end threaded, and the other plain. The plain end can be welded to, for example, a tank to form a pipe entrance
Cross	To connect a pipe to three others
Elbow	To change flow direction
Nipples	To tap a pipe (small tap)
Plug	To seal the end of a threaded pipe
Reducer	To change (reduce or enlarge) pipe diameter
Saddle	To tap a pipe
Sleeve	To connect two pipes together
Tee	To connect a 90° branch
Union	To connect two threaded pipes of the same size together without having to turn the pipes — just turn the union
Y	To connect two pipes to one pipe in the shape of a Y

Detailed configurations of different types of valves are given in piping handbooks such as references 1 and 2. A brief discussion of each type follows.

1. **Gate valve** — A gate valve is closed and opened by turning the handle connected to it, which raises or lowers a stem (shaft) connected to the gate. It takes many turns to completely open or close a gate valve. The headloss of the valve is small when the gate is fully open. The gate in the valve may be a wedge or a disk (for nonslurry), or a knife (for slurry). The gate valve used in oil or natural gas pipelines has a conduit with a full round bore for smooth passage of pigs or scrapers. They are called **conduit gate valves** or **full-bore gate valves**.

2. **Globe valve** — Outside is globe shaped. Flow changes direction as it goes through the valve. Consequently, large headloss is generated even when the valve is fully open. This valve gives better control of flow than gate valves — good for flow throttling.

3. **Angle valve** — Same as globe valve except that the flow direction is changed by 90° as the flow leaves the valve. It is used only in locations of 90° bends and is good for flow throttling.

4. **Ball valve** — The gate is a large bead (i.e., a large sphere having a central piercing). The gate is turned from a completely closed to a fully open position in 90°. When fully open, it causes little blockage to the flow and hence has little headloss. It is used mainly for on-off operations. Figure 8.9 shows a ball valve from a main manufacturer of this type of valve.

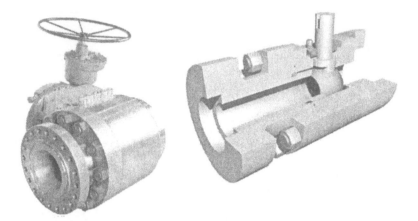

FIGURE 8.9 A typical ball valve that has full bore. (Courtesy of Valvtechnologies, Inc.)

FIGURE 8.10 A typical butterfly valve. (Courtesy of Tyco Valves & Controls.)

5. **Plug valve** — Similar to ball valve except that instead of a pierced ball, a pierced plug is used. Used mainly in small lines (tubings) where it is usually referred to as a **cock valve**. The valve can be either lubricated or nonlubricated. It can have multiple ports, such as a 3-way valve.

6. **Butterfly valve** — Uses a center-pivoted disk gate (see Figure 8.10). This is the most economical type for use in large pipes as in penstocks. Must be closed slowly or else the valve can be damaged easily. Not for pipeline transportation of solids due to wear by solids.

7. **Diaphragm valve** — A diaphragm separates the valve from the fluid. Consequently, the valve can be used with corrosive fluids or abrasive slurry.

8. **Pinch valve** — Pinches a flexible tube to restrict or control the flow; is suitable for small tubes only.

9. **Check valve** — Flow cannot reverse through a check valve; the valve produces unidirectional flow. Three types of check valves are **swing check valves** (horizontal or vertical lift types), **tilt disk check valve** (does not slam), and **ball check valve.**

10. **Foot valve** — A special vertical-lift type check valve embedded in the vertical end of a pipe connected to a reservoir below. The foot valve is used there to prevent the pump from losing priming when the flow is stopped (see discussion in Chapter 9).

8.6 PRESSURE RELIEF VALVES AND PRESSURE REGULATING VALVES

Both **pressure relief valve** and **pressure regulating valves** are for the same purpose, which is to keep the pressure in a pipe within a certain limit, so that the pipe and equipment connected to the pipe will not be damaged by unexpected high pressure generated in the pipeline, such as due to water hammer (pressure surges), or an accidental closure of a valve downstream while the pump is running. The two types of valves differ in their configuration, sophistication, and ability to control pressure in prescribed ranges.

The pressure relief valve, also called the **safety relief valve,** or simply the **safety valve,** is a small valve mounted on the wall of a pipeline (see Figure 8.11, left). The pressure relief valve is usually a spring-loaded valve with the spring preset to withstand certain pressure. When the pressure in the pipeline is within the preset value of the valve, the valve is closed, and it has no influence on the flow in the pipe. However, when the pressure in the pipe rises to a dangerous level, which is the preset pressure of the safety valve, the spring is pushed back by this pressure, and the valve is opened. The open valve draws a portion of the flow away from the pipe, which in turn causes the pressure in the pipe to fall, and reduce the dangerous pressure surge. Just because a safety valve of a given pressure rating is used in a pipe does not mean the pressure in the pipe will never exceed the preset value. If the safety valve is too small, opening the valve will have little effect on the pipeline pressure. One must make sure that the safety valve is of an adequate size so that it can reduce the pressure surge in the pipeline sufficiently. Safety valves are good protection against slowly rising pressure in the pipe, but ineffective in protecting rapid rise of pressure — large surges. Because safety valves alone cannot guarantee the safety of a piping system, other devices, such as pressure regulating valves, come to play.

Pressure regulating valve (PRV) is an automated special valve mounted in a pipeline to regulate the pressure (i.e., maintain the pressure within a predetermined range) in the pipe downstream of the regulator. Unlike the safety valve, which is a small valve mounted on the pipe wall and can affect only a small portion of the flow through the pipe, the PRV is mounted in the main pipe (see Figure 8.11, right). Because the entire flow in the pipe passes through the PRV, operation of the PRV has a strong impact on the pipe flow and the pressure in the pipe. For instance, by throttling the flow through the PRV, the pressure downstream of the valve will be reduced. The operational principle of PRVs can be stated briefly as follows. The

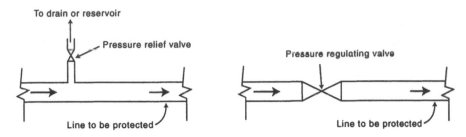

FIGURE 8.11 Mounting positions of pressure relief valve (left) and pressure regulating valve (right).

main valve (control valve) of a PRV is partially closed under normal flow conditions, so that it can either be closed more, or opened more, on demand. The valve is controlled by the PRV outlet (or downstream) pressure. When the outlet pressure exceeds a preset safe value of PRV, the valve will close more until the outlet pressure has dropped to below the preset value. On the other hand, when the downstream pressure is lower than the preset value, the valve will open more to allow more flow through the pipe, causing the downstream pressure to rise. This feedback mechanism keeps the downstream pressure approximately constant. For most commercial PRV systems, the energy to open and close the control valve comes from the flow itself. The outlet pressure is utilized to drive the valve; no outside energy is required. In some larger systems, an electrical motor is used to drive the valve, controlled by high-tech electronics and sensors.

There are two general types of PRV: **direct-operated** and **pilot-operated**. The direct-operated type has the valve closure member (controller) in contact with the fluid pressure of the PRV outlet. The fluid pushes the controller, which in turn closes or opens the valve, to the extent needed to maintain a constant pressure on the outlet side. This type is simple, less expensive but less accurate than the pilot-operated type. The pilot-operated type has two valves in a single housing: the pilot valve is in contact with the outlet pressure, and the control valve contains a closure member. The pilot valve senses the variation of outlet pressure and magnifies the closure member travel. Consequently, the pressure in the line can be regulated quickly.

Selection of PRV is complicated. It depends on the maximum and minimum flow rates through the pipe, the relative steadiness of the flow, the location of the PRV in the pipe — whether near the beginning or the end of a pipe branch, the maximum pressure expected at the PRV inlet, the outlet pressure that the PRV must maintain, the maximum pressure drop generated by the PRV, the accuracy of the pressure to be maintained or regulated, the speed of response to change in line pressure, and the allowed **fall-off**, which is the difference between the design pressure (i.e., the pressure to be maintained) and the actual outlet pressure.

PRVs are used in many places. An important application is the reduction of water and natural gas pressure in distribution lines before they enter a home or other building, so that the water and gas will be at a safe pressure within the building.

PROBLEMS

8.1 Find the outside diameter, wall thickness, inside diameter, and the weight of an NPS 20 schedule 80 steel pipe having a length of 40 ft. Convert the answers into SI units.

8.2 An NPS 10 steel pipe is designated as extra strong. What is its inside diameter and weight for a 10-ft section?

8.3 A stainless steel pipe is designated as NPS 5, 10S. What are its outside diameter, wall thickness, inside diameter, and weight per linear foot of length?

REFERENCES

1. Nayyar, M.L., *Piping Handbook*, 6th ed., McGraw-Hill, New York, 1992.
2. *Valves, Piping & Pipelines Handbook*, The Trade & Technical Press Limited, Crown House, London, 1986.

9 Pumps and Turbines

9.1 ENERGY CONVERSIONS BY PUMPS AND TURBINES

In a broad sense, the word *pump* refers to any machine in a pipeline that forces the fluid, be it a liquid or gas, to move through the pipe.* The basic function of a pump is to convert mechanical energy or power to fluid energy or power. As can be seen from the energy equation of incompressible flow (Equation 2.19), the pump head h_p generated by a pump causes an increase in the total energy of the flow, namely,

$$h_P = \left(h_2 - h_1\right) + h_L \tag{9.1}$$

where h is the total head (i.e., the total energy per unit weight of the fluid passing through the pump) represented by

$$h = \frac{V^2}{2g} + \frac{p}{\gamma} + z \tag{9.2}$$

Subscripts 1 and 2 in Equation 9.1 represents two points, one upstream and the other downstream of the pump, respectively. The quantity h_L is the headloss along the pipe between points 1 and 2. If 1 is taken in the pipe immediately upstream of the pump while 2 is taken immediately downstream of the pump, h_L is zero and Equation 9.1 reduces to $h_p = h_2 - h_1$, where h_1 and h_2 are then referred to as the **suction head** and the **discharge head** of the pump, respectively. Thus, the pump head is simply the discharge head minus the suction head, namely, the difference between the two heads. It is the energy imparted by the pump to the flow in the pipe for unit weight of the fluid flowing through the pipe. The power output of the pump is

$$P_o = \gamma Q h_P \tag{9.3}$$

where Q is the discharge ($Q = VA$).

---

* In a narrow sense, the word **pump** often refers to liquid pumps, whereas air or gas pumps are referred to as **compressors** (for high pressure), or **blowers** (for medium pressure), or **fans** (for low pressure).

In contrast, a turbine is a machine that does the opposite of a pump in terms of energy or power conversion. It converts fluid energy or power into mechanical energy or power. From Equation 2.17,

$$h_T = (h_1 - h_2) - h_L \qquad (9.4)$$

where h is the total energy defined by Equation 9.2, and subscripts 1 and 2 refer to points upstream and downstream of the turbine, respectively. Again, when 1 and 2 are points immediately upstream and downstream of a turbine, respectively, h_L becomes zero and $h_T = h_1 - h_2$, where h_1 and h_2 become, respectively, the inlet head and discharge head of the turbine.

Most commonly, a pump is mechanically connected to an electrical motor through a shaft. The motor drives the pump, which in turn drives the fluid. In doing so, electrical energy or power is converted to mechanical energy or power, which in turn is converted to fluid energy or power. Alternatively, a pump may be driven by something other than an electric motor, such as a gasoline or diesel engine which converts the energy or power derived from burning fuel into mechanical energy or power to drive the pump. Likewise, a turbine is often connected mechanically through a shaft to a generator. The fluid forces the turbine to rotate, which in turn forces the generator to rotate and generate electricity. In doing so, fluid energy or power is converted to mechanical energy or power, which in turn is converted to electrical energy or power. Alternatively, the mechanical energy or power generated by a turbine may be used directly to do useful work such as propelling an aircraft (in the case of a jet engine), or rotating a mill (for grinding grain).

Example 9.1 A 4-inch-diameter steel pipe is used to convey water from a reservoir having a water level at 800 ft elevation to another reservoir having water level at 1000 ft elevation. The total length of the pipe is 3000 ft, and both ends of the pipe are submerged under 5 ft of water. The pump is properly located near the pipe inlet so that no cavitation occurs. Determine the pump head required to produce a flow through the pipe at 6 fps. Also determine the power output of the pump.

[Solution] Selecting point 1 to be at the water surface of the inlet reservoir, and point 2 to be at the water surface of the outlet reservoir, $V_1 = V_2 = 0$, $p_1 = p_2 = 0$, and $z_2 - z_1 = 1000 - 800 = 200$ ft. Therefore, from Equation 9.1,

$$h_p = 200 + h_L \qquad (a)$$

Next, assume that the local loss is negligible in this case, and h_L can be calculated by using the Darcy-Weisbach formula given in Chapter 2, namely,

$$h_L = f \frac{L}{D} \frac{V^2}{2g} \qquad (b)$$

Since $V = 6$ fps, $L = 3000$ ft, $D = 4$ inches $= 0.3333$ ft, and $g = 32.2$ ft/s^2, Equation b yields $h_L = 5031 f$. Substituting this result into Equation a yields

$$h_p = 200 + 5031 f \qquad\qquad (c)$$

Assume that the water temperature is 70°F. The kinematic velocity, v, of the water is 1.059×10^{-5} ft^2/s, and the Reynolds number is $\Re = VD/v = 1.89 \times 10^5$. The relative roughness of the pipe is $e/D = 0.00015/0.3333 = 0.00045$. Therefore, from the Moody diagram, $f = 0.0188$. Using this value of f, Equation c yields $h_p = 295$ ft. Next, the cross-sectional area of the pipe is $A = 0.0873$ ft^2, and the discharge is $Q = VA = 0.524$ cfs. Using Equation 9.3, the power output of the pump is $P_o = 62.4 \times 0.524 \times 295 = 9646$ ft-lb/s $= 17.5$ hp. This shows that the pump should operate at a head of 295 ft and have a power output of 17.5 horsepower.

9.2 TYPES OF PUMPS AND TURBINES

Pumps and turbines can be categorized in different ways. Based on the kind of fluid being pumped, pumps can be classified as water pumps, oil pumps, air pumps, etc. Likewise, hydraulic turbines are those that derive power (hydropower) from water, gas turbines are those that derive power from gas or steam, etc. Based on the level of pressure that they generate, pumps are often classified as high-pressure (or high-head), intermediate-pressure (intermediate-head), or low-pressure (or low-head). Likewise, hydraulic turbines are often classified as high-head, intermediate-head and low-head, depending on the water head (pressure) that drives the turbines. In fluid mechanics, pumps and turbines are both referred to as turbomachines. They come in several general types or designs, which are discussed in the following sections.

9.2.1 CENTRIFUGAL PUMPS

9.2.1.1 Main Components

The three main components of a typical centrifugal pump include (1) the **impeller**, which is the rotating part of the pump that contains curved vanes and generates the centrifugal force needed to increase the fluid pressure at the pump outlet; (2) the **casing** that encloses the impeller and contains the liquid, having an axial inlet connected to the suction pipe and a tangential outlet connected to the discharge pipe; and (3) the rotating shaft connected to the impeller that transmits the torque and power from the driver (prime mover) to the pump. Associated with these three main components are parts including the volute, bearings, packing, stuffing box, mechanical seal, shaft sleeve, air vent, and tubes connected to the housing for various purposes, such as supplying the liquid needed to lubricate the packing. Note that the **volute** is the spiraling pipe with expanding cross section that forms the outlet of the casing. Bearings are needed to minimize the friction between the shaft and the stationary parts that surround and support the shaft. **Packing** is the ring-shaped element around the shaft used for sealing that is preventing leakage through the small space around the rotating shaft. The **stuffing box** is the box that contains the packing. The **mechanical seal** is an alternative to packing, providing an air tight seal of the shaft. The **shaft sleeve** is a sleeve attached to the shaft, for the purpose of protecting the shaft from damages caused by abrasion, corrosion, etc. An air vent is often mounted on the top of the pump casing to vent any air trapped in the pump, especially during pump start-up. Finally, tubes are often connected to pumps, to supply water needed for cooling or for lubricating the packing.

It should be kept in mind that centrifugal pumps are the most widely used pumps, and that they come in various forms. For instance, a large centrifugal pump may have double suction, with suction inlets on both sides of the pump casing. This arrangement not only doubles the total cross-sectional area of the inlet and hence facilitates the flow, but it also balances the large thrust generated on the opposite sides of the impeller, resulting in negligible net thrust on the impeller and the casing, and the net axial shear force between the impeller and the shaft. An alternative to the volute-type centrifugal pump is the diffuser pump, which uses a set of stationary vanes surrounding the impeller having an expanding cross section (diffuser) instead of a volute. The diffuser pump used to be called a **turbine pump** because the stationary vanes make the pump look like a Francis turbine (see Section 9.8). However, in recent years manufacturers of multistage vertical centrifugal pumps have been calling their products *turbine pumps*. For this reason, the term turbine pump now has a very different connotation.

9.2.1.2 Fluid Mechanics of Centrifugal Pumps

The most common type of pumps is the centrifugal pump whose name is derived from the centrifugal force generated in the pump, which in turn generates the pump head (pressure). The cross section of a typical centrifugal pump is shown in Figure 9.1a. Fluid enters the pump at its center axially (perpendicular to the paper in Figure 9.1), and exits the pump tangentially. The rotating vanes (impellers) cause the fluid to rotate in the pump and generate a centrifugal force that causes the pressure to rise at the pump outlet. The impeller vanes are shaped and designed in such a manner (with a backward curvature as shown in Figure 9.1b) that the fluid enters the vanes (at their base) radially and leaves the vanes (at their tips) tangentially. This generates a maximum torque and power according to the following formula that can be derived from the angular momentum equation and the continuity equation given in basic fluid mechanics texts:

$$T_p = \rho Q r_o V_t \tag{9.5}$$

$$P_{ip} = \omega_p T_p = \omega_p \rho Q r_o V_t = \rho Q U V_t \tag{9.6}$$

$$P_{op} = \rho g Q H \tag{9.7}$$

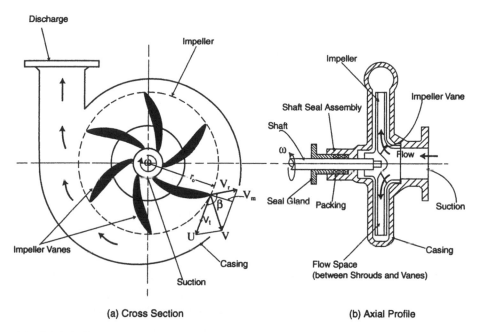

(a) Cross Section

(b) Axial Profile

FIGURE 9.1 Centrifugal pump.

$$\eta_p = \frac{P_{op}}{P_{ip}} = \frac{\rho g Q H}{\omega_p T_p} \tag{9.8}$$

$$H = h_2 - h_1 = \frac{P_{op}}{\rho g Q} = \frac{\omega_p r_o V_t \eta_p}{g} = \frac{U V_t \eta_p}{g} \tag{9.9}$$

In the above equations, T_p is the torque generated on the pump's impeller shaft; P_{ip} is the power delivered by the impeller (i.e., the pump's input power or the *brake horsepower*); P_{op} is the output power of the pump; η_p is the pump efficiency; ρ is the liquid density; Q is the volumetric flow rate (discharge) through the pump; r_o is the radius of the impeller measured from the center axis of the impeller to the tip of the vanes; ω_p is the angular velocity of the pump in radians per second ($\omega_p = 2\pi N_p$ where N_p is the angular speed of the pump in revolutions per second or rps); V_t is the tangential (circumferential) component of the fluid velocity at the radial distance r_o; U is the circumferential velocity of the vane tip, $\omega_p r_o$; and H is the pump head which is the same as h_p. For simplicity and in conformity with nomenclature used commonly in pump literature, henceforth in this chapter when dealing with pumps, h_p will be written as H. Also, ω_p, T_p, P_{ip}, P_{op}, and η_p will be written simply as ω, T, P, P_o, and η, respectively. The subscript p will be used when the parameters of pumps are compared to those of the electrical motors that drive the pumps in a later section.

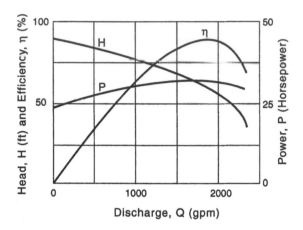

FIGURE 9.2 Characteristic curves of a centrifugal pump.

Equations 9.5 through 9.9 give the torque, input power, output power, efficiency and head of an ideal centrifugal pump with a purely radial flow entering the impeller. In reality, the flow entering the impeller has not only a radial component but also a tangential component, which decreases the values of T, P, and H. This is true especially when the pump is operating at a discharge Q quite different from the discharge used in the design of the pump — the operating discharge. This explains why the efficiency of a centrifugal pump varies with the discharge Q, having maximum efficiency at the design discharge which yields an almost radial flow at the vane inlet (base). Figure 9.2 shows the characteristic curves of a typical centrifugal pump, giving the variations of H, P, and η with Q.

9.2.1.3 Euler's Pump Performance Curves

From a vector diagram showing the relationship among the tangential velocity component V_t, the meridional (radial) velocity component V_m, the velocity of the vane tip U ($U = \omega_p r_o$), and the relative velocity of the liquid to the moving vane V_r, it can be proved that [1]

$$V_t = U + V_m \cot \beta \tag{9.10}$$

where β is the angle between V_r and U.

Substituting Equation 9.10 into Equation 9.5 and using the condition $Q = 2\pi r_o b V_m$ (where b is the thickness of the pump) yields

$$T = B_1 Q + B_2 Q^2 \tag{9.11}$$

where

$$B_1 = \rho \omega r_o^2 \quad \text{and} \quad B_2 = \frac{\rho \cot \beta}{2\pi b} \tag{9.12}$$

Likewise, substituting Equation 9.10 into Equation 9.9 yields

$$H = \eta \left[\frac{U^2}{g} + \frac{U \cot \beta}{2\pi r_o b g} Q \right] \tag{9.13}$$

For a given centrifugal pump running at a constant speed (angular velocity) ω or circumferential velocity U, all the quantities on the right side of Equation 9.13 except Q are constant. This means Equation 9.13 is of the following form:

$$H = C_1 + C_2 Q \tag{9.14}$$

where both C_1 and C_2 are constant.

For backward-curved impeller vanes, $\beta > 90°$, cot β is negative, and C_2 is negative. This means that H decreases linearly with Q. For radial vanes, $\beta = 90°$, cot β is zero, C_2 is zero, and H is constant (i.e., independent of Q). For forward-curved vanes, $\beta < 90°$, cot β is positive, C_2 is positive. This means H increases with Q linearly. The three cases are illustrated in Figure 9.3, which is often referred to as the Euler performance curves for centrifugal pumps. Although the Euler performance curves do not describe accurately the relationship between H and Q for real (actual) centrifugal pumps, they show correctly the general trends such as when β is less than 90° (backward vanes), H generally increases with Q and so on.

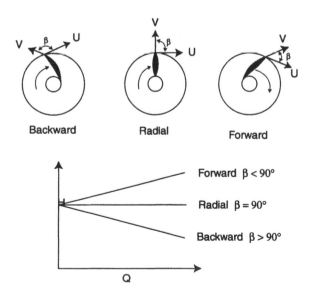

FIGURE 9.3 Euler performance curves of centrifugal pump.

For actual centrifugal pumps the relationship between H and Q is nonlinear. For instance, with backward vanes, the $H \sim Q$ curve can be fitted approximately by the following equation:

$$H = C_1 - C_3 Q^n \text{ (pump curve)} \tag{9.15}$$

where both C_1 and C_2 are positive constants, and the power (exponent) n is positive and greater than 1.0. The values of C_1, C_2, and n must be determined from curve fitting. Note that centrifugal pumps are generally used with backward-curved vanes not only because they yield better efficiency but also because they are more stable in operation (see next subsection).

Example 9.2 A centrifugal pump is designed with the following physical properties: $r_o = 1.5$ ft, $b = 0.5$ ft, and $\beta = 120°$. The pump is to be used for water, and is to operate at a speed of 875 rpm. Determine from theory the pump characteristic curves for torque T, pump head H, and input power P.

[Solution] Based on the given conditions, $\rho = 1.94$ slug/ft³, $\omega = 2\pi N = 2 \times 3.1416 \times 875/60 = 91.6$ rad/s, $r_o = 1.5$ ft, $b = 0.5$ ft, $g = 32.2$ ft/s², $U = \omega r_o = 137.4$ fps, $\beta = 120°$, and $\cot \beta = -0.577$. Substituting these values into Equation 9.12 yields $B_1 = 400$ and $B_2 = -0.356$. Then, from Equations 9.11, 9.13, and 9.6, we have, respectively,

$$T = 400Q - 0.356Q^2 \tag{a}$$

$$H = 586 - 0.522Q \tag{b}$$

$$P = 36,640Q - 32.6Q^2 \tag{c}$$

It can be seen from the foregoing calculation that the three characteristic curves of any centrifugal pump — for pump torque T, pump head H, and brake horsepower P — can be determined approximately from theory. In addition, if the efficiency of the pump as a function of Q is known from experiments, then the output power P_o at any discharge can also be calculated.

9.2.1.4 Pump and System Curves

9.2.1.4.1 For a Single Pump

In the selection of a pump for a given pipeline system, one can plot the pump curve $(H \sim Q)$ versus the system (pipeline) curve (also $H \sim Q$) as shown in Figure 9.4. The intersecting point of the two curves gives the operating point of the pump used in a given system (pipeline). This constitutes the graphical way to solve the problem. While the $H \sim Q$ curve for the pump is given by the pump manufacturer through testing of the pump, the $H \sim Q$ curve of the system is obtained from Equations 9.1

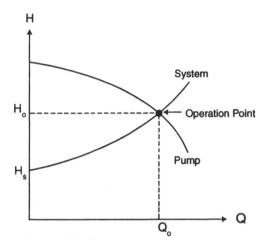

FIGURE 9.4 $H \sim Q$ curve for pipeline system and centrifugal pump.

and 9.2. If the flow is incompressible and the pipe diameter is constant, $V_1 = V_2$. For an entire pipeline system, with each end of the pipe connected to a reservoir, points 1 and 2 are taken at the free surfaces of the reservoirs, which yields $p_1 = p_2 = 0$ (atmospheric). Consequently, Equations 9.1 and 9.2 yield

$$H = (z_2 - z_1) + h_L - H_s + \left(f \frac{L}{D} + \sum K \right) \frac{V^2}{2g} \qquad (9.16)$$

Substituting $V = Q/A$ into Equation 9.16 yields

$$H = H_s + C_4 Q^2 \quad \text{(system curve)} \qquad (9.17)$$

in which $H_s = (z_2 - z_1)$ is the static head of the system that the pump must overcome, and $C_4 Q^2$ is the dynamic head of the system that the pump must also supply. The constant C_4 is given by

$$C_4 = \frac{\left(f \frac{L}{D} + \sum K \right)}{2gA^2} \qquad (9.18)$$

Note that C_4 is known if pipe length L, pipe inner diameter D, Darcy-Weisbach friction coefficient f, total local headloss coefficients due to fittings $\sum K$, gravitational acceleration g, and pipe cross-sectional area A are all known. An alternative to solving any pipeline-with-pump problem graphically is to solve simultaneously Equations 9.15 and 9.17 (with the values of C_4 found from Equation 9.18). This can be done easily using a computer.

Example 9.3 Suppose that the pump selected for the system described in Example 9.1 has an $H \sim Q$ curve that can be approximated by the equation $H = 400 - 360Q^2$. Determine whether this pump is adequate for supplying the required discharge, 0.524 cfs (at 6 fps), through the system.

[**Solution**] From Equations a and b of Example 9.1, the system curve for the pipeline is

$$H = 200 + 18,340\, fQ^2 \tag{a}$$

Strictly speaking, f is a function of the Reynolds number, which in turn depends on Q (through V). However, in the range of practical interest to this problem (Q between 0.1 and 1.0 cfs), the value of f is approximately constant and equal to 0.0188, as found from the Moody diagram in Example 9.1. Therefore, Equation a reduces to

$$H = 200 + 345Q^2 \text{ (system curve)} \tag{b}$$

In contrast, the pump curve is

$$H = 400 - 360Q^2 \text{ (pump curve)} \tag{c}$$

Solving Equations b and c simultaneously yields $Q = 0.532$ cfs, and $H = 298$ ft. Since this discharge is about 15% higher than the desired discharge of 0.524 cfs, it is considered to be adequate for the project. Note that a discharge slightly (10 to 20%) higher than the design discharge is good because it is on the safe side to provide adequate flow rate. Lower discharges can be achieved easily either by throttling the flow with a valve, or using a variable speed pump.

9.2.1.4.2 For Booster Pumps

For long-distance pipelines, it is not possible to place all the pumps near the pipeline inlet because that would cause enormous pressure in the pipe and the pumps, requiring unreasonably thick pipe wall and unreasonably strong pump casing. To avoid this problem, the pumps are normally spaced along the pipeline at more or less equal distance (spacing) apart. Pumps spaced along the pipeline, other than the one (or ones) at the inlet, are called **booster pumps**. With booster pumps, the term h_p (or H) in Equation 9.1 must represent the combined head of all the pumps in the system, and the pressure rise across each pump can be analyzed as follows.

Suppose that a long pipeline has a pump at the inlet and two booster pumps downstream. The first booster pump is located at a distance L_1 from the pipeline inlet, the second booster is at a distance L_2 from the first booster, and the pipeline outlet is at a distance L_3 further downstream. The total length of the pipe is $L = L_1 + L_2 + L_3$. If the three pumps have different $H \sim Q$ curves, how can the discharge Q through the pipeline be determined? Writing the one-dimensional energy equation for the entire pipeline system from point 1 (the inlet) to point 2 (the outlet), Equation 9.17 yields

$$H_1 + H_2 + H_3 = h_s + C_4 Q^2 \tag{9.19}$$

From the pump manufacturer, the characteristic curves of all the three pumps are found. They are

$$H_1 = a_1 - b_1 Q^{n1} \tag{9.20}$$

$$H_2 = a_2 - b_2 Q^{n2} \tag{9.21}$$

$$H_3 = a_3 - b_3 Q^{n3} \tag{9.22}$$

for pumps 1, 2, and 3, respectively.

Note that the quantities h_s, C_4, a_1, a_2, a_3, b_1, b_2, b_3, n_1, n_2, and n_3 are all known for a given system. As the foregoing four equations have four unknowns (H_1, H_2, H_3, and Q), they can be solved simultaneously with a computer to determine the values of H_1, H_2, H_3, and Q for the system. The procedure involves iteration because the value of friction factor f must be assumed initially. This shows how to determine the flow through a long pipeline that has booster stations. If each booster station uses more than one pump in combination, then the values of H_1, H_2, and H_3 should be those for the combined pumps at each station, obtained in a manner discussed later in Section 9.2.1.5.

Example 9.4 A 12-inch-diameter steel pipe is used to pump gasoline over a distance of 120 mi against a static head of 300 ft. Three pumps are placed at various locations along the pipe as listed in the table below. Find the discharge Q through the pipeline and the pressure before and after each pump. Analyze whether cavitation may occur anywhere in the pipeline. Assume that the gasoline in the pipe has a temperature of 60°F, a specific weight of 42.5 lb/ft³, a kinematic viscosity of 5×10^{-6} ft²/s, and a vapor pressure of 7 psia.

Pump No.	Location from Pipe Inlet (ft)	Distance from Previous Point (ft)	Elevation above Inlet (ft)	Elevation Increase from Previous Point (ft)	Pump Characteristics (Q in cfs and H in ft)
Inlet	0	0	0	0	
1	30	30	10	10	$H = 2200 - 18Q^2$
2	160,000	159,970	150	140	$H = 2000 - 22Q^2$
3	360,000	200,000	250	100	$H = 1800 - 30Q^2$
Outlet	633,600	273,600	300	50	

[Solution] For this case, $D = 1$ ft, $A = 0.7854$ ft^2, $L = 120$ mi $= 633,600$ ft, $g = 32.2$ ft/s^2, and $h_s = 300$ ft. Thus, Equation 9.18 yields

$$C_4 = 15,950 f \tag{a}$$

From Equation 9.19,

$$H_1 + H_2 + H_3 = 300 + 15,950 fQ^2 \tag{b}$$

The pump characteristic curves for the three pumps are:

$$H_1 = 2200 - 18Q^2 \tag{c}$$

$$H_2 = 2000 - 22Q^2 \tag{d}$$

$$H_3 = 1800 - 30Q^2 \tag{e}$$

To solve Equations b through e, let's first set $H_1 + H_2 + H_3 = H$. Then, the four equations reduce to two equations as follows:

$$H = 300 + 15,950 fQ^2 \text{ (pipeline system)} \tag{f}$$

$$H = 6000 - 70Q^2 \text{ (pump)} \tag{g}$$

Combining Equations f and g yields

$$Q = \sqrt{\frac{5700}{15,950 f + 70}} \tag{h}$$

Equation h can be solved through iteration by first assuming a reasonable value of f. For this case, the gasoline at 60°F has a kinematic viscosity of $v = 5 \times 10^{-6}$ ft^2/s. The Reynolds number is $\mathfrak{R} = DQ/Av = 2.55 \times 10^5 Q$, and the relative roughness is $e/D = 0.00015$. From the Moody diagram, the value of f at $\mathfrak{R} = 10^5$ is $f = 0.019$. Using this as the f value for the first iteration yields $Q = 3.91$ cfs. Using this value of Q, $\mathfrak{R} = 9.97 \times 10^5$. From the Moody diagram, this gives a more accurate set of values: $f = 0.014$, $Q = 4.41$ cfs, and $\mathfrak{R} = 1.12 \times 10^6$. With another iteration, f is still 0.014. Therefore, the correct value of discharge obtained through iterations is $Q = 4.41$ cfs, and from Equation g the correct head is $H = 4640$. Now that value of Q is determined, Equations c, d, and e yield, respectively, $H_1 = 1850$ ft, $H_2 = 1572$ ft, and $H_3 = 1217$ ft. Adding these three values yields $H = H_1 + H_2 + H_3 = 4639$ ft, which checks the correctness of the calculation.

To calculate the pressure head at various key locations along the pipe, we first determine the velocity of the flow which is $V = Q/A = 4.41/0.7854 = 5.61$ fps. The velocity head is $V^2/2g = 0.490$ ft. The pressure heads at key location are calculated next.

Suction side of pump 1:

$$\frac{p}{\gamma} = -10 - \left(1.5 + f\frac{30}{D}\right)\frac{V^2}{2g} = -10 - (1.5 + 0.42) \times 0.490$$

$$= -10 - 0.94 = -10.94 \text{ ft}$$

In the above calculation, an entrance headloss coefficient of 0.5 was assumed.

Discharge side of pump 1:

$$\frac{p}{\gamma} = -10.94 + H_1 = -10.94 + 1850 = 1839 \text{ ft}$$

Suction side of pump 2:

$$\frac{p}{\gamma} = 1839 - 140 - f\frac{159,970}{D}\frac{V^2}{2g} = 1839 - 140 - 1097 = 602 \text{ ft}$$

Discharge side of pump 2:

$$\frac{p}{\gamma} = 602 + H_2 = 602 + 1572 = 2174 \text{ ft}$$

Suction side of pump 3:

$$\frac{p}{\gamma} = 2174 - 100 - f\frac{200,000}{D}\frac{V^2}{2g} = 2174 - 100 - 1372 = 702 \text{ ft}$$

Discharge side of pump 3:

$$\frac{p}{\gamma} = 702 + H_3 = 702 + 1217 = 1919 \text{ ft}$$

Pipeline outlet (check calculation):

$$\frac{p}{\gamma} = 1919 - 50 - f\frac{273,600}{D}\frac{V^2}{2g} = 1919 - 50 - 1877 = -8 \text{ ft}$$

Note that the pressure head at the pipe outlet should be zero. The computed −8 ft is within computational error. The lowest pressure head in the pipeline is at the suction side of pump 1, which is approximately −11 ft (of gasoline). This is equivalent to a suction (below atmospheric) of $11 \times 42.5 = 468$ psf = 3.25 psi. Since the atmospheric pressure is 14.7 psia, the absolute pressure of the gasoline at this point is $14.7 - 3.25 = 11.5$ psia. Because the vapor pressure of the gasoline is 7 psia, which is lower than the 11.5 psia, the gasoline is not expected to turn into vapor or cavitate in the pipe. However, since the pressure in certain regions of the pump is lower than on the suction pipe, it may cause cavitation in the pump. To examine the possibility of pump cavitation, let us calculate the available net positive suction head, which from Chapter 2 is

$$\text{NPSH (available)} = \frac{p_a}{\gamma} + h_s - h_L - h_v \qquad (i)$$

The quantities in Equation i are $p_a/\gamma = 14.7 \times 144/42.5 = 49.8$ ft, $h_s = -10$ ft, $h_v = 7 \times 144/42.5 = 23.7$ ft, and $h_L = 0.451$ ft. Therefore, Equation i becomes NPSH (available) = $49.8 - 10 - 0.94 - 23.7 = 15.2$ ft. If the first pump has a required NPSH greater than 15.2 ft, then the pump will cavitate.

9.2.1.5 Pumps in Combination

Oftentimes, it is necessary to combine more than one pump, either to achieve a large discharge Q or a high head H, or both, that cannot be achieved with a single pump. In order to achieve a large discharge, pumps are combined in parallel. In contrast, to achieve high head or pressure, pumps are combined in series.

When more than one pump is combined in series, the heads H of the individual pumps are added together to form the $H \sim Q$ curve of the combined pumps. In contrast, when more than one pump is combined in parallel the discharges Q of the individual pumps are added together to form the $H \sim Q$ curve of the combined pumps. This general principle of adding individual pump curves for combined pumps applies to all pumps, not only centrifugal pumps. Based on this principle, the combinations of two identical centrifugal pumps in series and parallel are illustrated in Figure 9.5a and b, whereas the combinations of two dissimilar centrifugal pumps in series and parallel are illustrated in Figures 9.6a and b. In both figures, point A is the operational point of the system with pump 1 only, and point B is the operational point when both pumps are on.

To combine pump characteristics curves numerically, if $H = a_1 - b_1 Q^{n1}$ is the curve for pump 1, $H = a_2 - b_2 Q^{n2}$ is the curve for pump 2 and so forth, for three pumps operating in series, the combined $H \sim Q$ curve is

$$H = \left(a_1 + a_2 + a_3\right) - \left(b_1 Q^{n1} + b_2 Q^{n2} + b_3 Q^{n3}\right) \qquad (9.23)$$

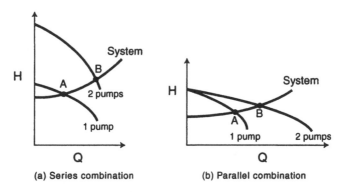

(a) Series combination

(b) Parallel combination

FIGURE 9.5 Combinations of two identical pumps.

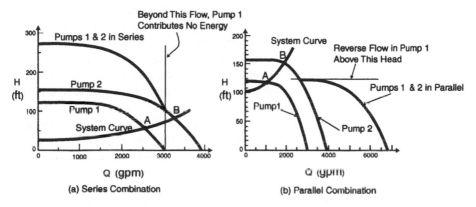

(a) Series Combination

(b) Parallel Combination

FIGURE 9.6 Combination of two dissimilar pumps.

Likewise, by rewriting the $H \sim Q$ curves of the pumps as $Q = \left(\dfrac{a_1 - H}{b_1}\right)^{1/n_1}$ and so forth, the combined $H \sim Q$ curve for three pumps in parallel is

$$Q = \left(\frac{a_1 - H}{b_1}\right)^{1/n_1} + \left(\frac{a_2 - H}{b_2}\right)^{1/n_2} + \left(\frac{a_3 - H}{b_3}\right)^{1/n_3} \tag{9.24}$$

The foregoing procedure for combining three pumps can be extended to any number of pumps.

Example 9.5 The $H \sim Q$ curve of three pumps are $H = 1 - Q^2$, $H = 2 - 1.5Q^2$ and $H = 3 - 2Q_2^2$. Determine the resultant $H \sim Q$ curves when the three pumps are in series and in parallel.

[Solution] When the three pumps are in series, the resultant $H \sim Q$ curve is, from Equation 9.23,

$$H = 6 - 4.5Q^2$$

In contrast, when in parallel, the combined $H \sim Q$ curve is, from Equation 9.24,

$$Q = \sqrt{1-H} + 0.816\sqrt{2-H} + 0.707\sqrt{3-H}$$

More about pumps used in combination can be found in Reference 2.

9.2.2 POSITIVE DISPLACEMENT PUMPS

Positive displacement (PD) pumps are those that use their prime moving part (be it a piston, a plunger or a gear) to forcibly push the fluid out the pump chamber (cylinder) and into the discharge pipe. For incompressible flow, the volume of fluid displaced (pumped) into the discharge pipe by any PD pump is the same as the volume traversed by the front face of the moving part of the pump. For this type of pump, the pressure by the pump is not generated by any centrifugal force as it is the case for centrifugal pumps. Rather, the pressure is caused by pushing the fluid with the front face of the pump's moving part. Thus, the pressure on the discharge side of the pump is practically the same as that encountered by the front face of the moving part.

There are five general types of positive displacement pumps: (a) piston pumps, (b) plunger pumps, (c) diaphragm pumps, (d) rotary PD pumps, and (e) screw pumps. The rotary PD pumps can be further subdivided into gear pumps and vane pumps. They are illustrated in Figure 9.7. All positive displacement pumps have one common characteristic: the $H \sim Q$ curve at a given pump speed is a vertical line (see Figure 9.8). This means the discharge Q is independent of head H. More details on each type of pump are provided next.

9.2.2.1 Piston Pumps

Figure 9.7a shows the main features of a piston pump:

1. **Cylinder (chamber)** — The cylinder is the casing of the pump that contains a piston and the fluid in the pump. It is open to the discharge pipe downstream, and to the suction pipe upstream.
2. **Piston** — The piston moves back and forth in the cylinder, driven by a motor or an engine either directly or indirectly. The reciprocating motion of the piston causes the fluid in the cylinder to be pushed out during the forward stroke of the piston.
3. **Rod** — The piston rod connects the piston to the reciprocating part of the machine that drives the piston.
4. **Check Valves** — Two spring-loaded check valves, one connected to the discharge pipe and the other connected to the suction pipe, control the direction of flow. When the piston is moving forward, the discharge check valve opens and the suction check valve closes. This forces the fluid in the pump to enter the discharge pipe. In contrast, when the piston is

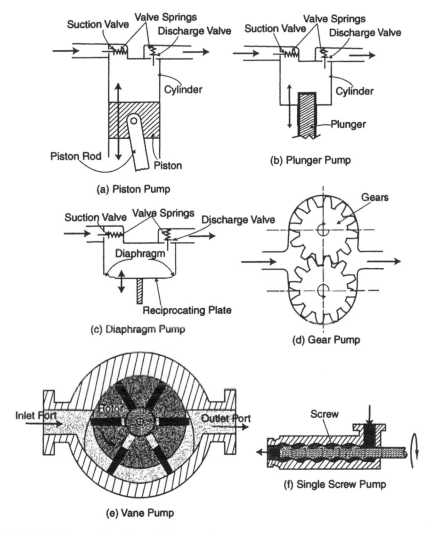

FIGURE 9.7 Types of positive displacement pump.

moving backward, the discharge check valve closes and the suction check valve opens. This causes the fluid in the suction pipe to be drawn into the pump.

5. **Packing for Piston and Rod** — The piston packing material seals the gap between the piston and its surrounding wall, preventing leakage and increasing the efficiency of the pump. Likewise, the piston–rod packing seals the gap between the piston rod and its surroundings, preventing leakage and increasing pump efficiency.

If a piston pump has a piston stroke length of L_p and a reciprocating speed of N, the average linear velocity (speed) of the piston is

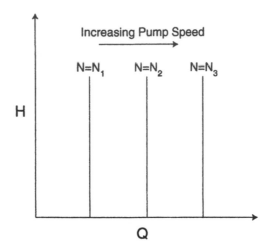

FIGURE 9.8 $H \sim Q$ curve for positive-displacement pumps (note: N_1, N_2, and N_3 are different pump speeds).

$$V_p = 2L_p N \qquad (9.25)$$

If L_p is measured in ft (m) and N is measured in revolution per second (rps), then the speed V_p will be in ft/s (m/s). If L_p is in ft and N in rpm (revolution per minute), the value of V_p will be in ft/min. If L_p is in inches, N is in rpm, and V_p is in ft/min, the above equation must be rewritten as

$$V_p = \frac{L_p N}{6} \qquad (9.26)$$

The discharge of the piston pump is

$$Q = A_p L_p N \qquad (9.27)$$

where A_p is the piston cross-sectional area in ft² (m²), L_p is in ft (m), N is in rps, and Q is in cfs (cms). Equation 9.27 shows that as long as the piston stroke L_p and the piston speed N remain constant, the discharge of piston pump is constant and independent of the pump head H or the pump pressure p.

If the net pressure on the piston by the fluid (i.e., the discharge pressure minus the suction pressure) is p_p, the average piston force on the fluid must be

$$F_p = p_p A_p \qquad (9.28)$$

Due to mechanical friction forces generated by packing, the external force to drive the piston F_e must be slightly larger than F_p.

The power delivered by the piston to the fluid is

$$P = V_p F_p = V_p p_p A_p \tag{9.29}$$

On the other hand, the power delivered by the external force driving the piston is

$$P_e = V_p F_e \tag{9.30}$$

Therefore, the efficiency of the piston pump is

$$\eta_p = \frac{P}{P_e} = \frac{F_p}{F_e} = \frac{p_p A_p}{F_e} \tag{9.31}$$

Note that a piston pump is usually much more expensive than a centrifugal pump of the same power rating. It is used only for high-head operations in which the efficiency is usually very high, of the order of 95%. In contrast, a good centrifugal pump has efficiency in the neighborhood of 80%.

Metering pumps are PD pumps, usually of the piston, plunger, or diaphragm type. A metering pump contains equipment that can measure and adjust the speed and/or stroke length of the piston (plunger), in order to produce a predetermined discharge through the pump. It combines the function of a pump and the function of a flowmeter into a single piece of equipment. Metering pumps are also called **proportioning pumps**. They are used extensive in the chemical and food industries whenever a need exists to mix different chemicals or foods in liquid or slurry forms at desired proportions. They are more expensive than ordinary PD pumps, and hence should not be used unless metering is needed — i.e., unless one needs to mix different liquids or slurries at predetermined and controllable rates.

Example 9.6 Two metering pumps are used to mix alcohol with water at a volume ratio of 1 to 2 (i.e., one part alcohol with two parts water, by volume). The pump for alcohol has a piston diameter of 2 inches and a constant stroke length of 1.5 inches, and the pump for water has a piston diameter of 3 inches and a constant stroke length of 2.5 inches. The discharge through both pumps can be adjusted through piston speed. Determine the piston speeds of the two pumps in order to produce a mixture flow rate of 60 gpm. The temperature of both liquids is the same — 20°C (68°F).

[Solution] Before one can solve this problem correctly, one needs to know an interesting phenomenon in chemistry, which shows that when a liquid is dissolved in another liquid, such as alcohol in water, the volume of the two liquids is not conserved. There is usually a slight reduction of the total volume, though the weight of the two is conserved or remains the same as before the two liquids are mixed. More specifically, if liquid 1 (say, alcohol) has a volume of V_1, and liquid 2 (say, water) has a volume of V_2, upon mixing the two the volume of the mixture (solution) will be $V_m = \varepsilon(V_1 + V_2)$, where ε, the volume reduction ratio, is a number slightly less than 1.0. The value of ε depends on not only the kinds of liquids being mixed, but also the ratio of the two liquids. It can be determined rather easily through laboratory measurements using a calibrated test tube. For a solution that is made up of 1 volume of alcohol and 2 volumes of water, the value of ε is known to be 0.97, approximately. By using this information, we are now ready to solve the problem.

From Table C3 in Appendix C, water at 68°F has a specific weight of 62.3 lb/ft². From Table C1 alcohol (methanol) at the same temperature has a specific weight of $1.53 \times 32.2 = 49.3$ lb/ft². By mixing 1.0 ft³ of alcohol with 2.0 ft³ of water, the mixture (solution) will have a volume of $\forall_m = (1 + 2) \times 0.97 = 2.91$ ft³. Thus, the specific weight of the mixture is $\gamma_m = (W_1 + W_2)/\forall_m = (49.3 + 2 \times 62.3)/2.91 = 59.8$ lb/ft³. The volumetric flow rate of the mixture is $Q_m = 60$ gpm $= 0.134$ cfs.

From conservation of mass and the continuity equation,

$$Q_1 + Q_2 = Q_m / 0.97 \qquad (a)$$

and

$$\gamma_1 Q_1 + \gamma_2 Q_2 = \gamma_m Q_m \qquad (b)$$

where the subscripts 1, 2, and m represent alcohol, water, and mixture, respectively.

Substituting the values of γ_1, γ_2, γ_3, and Q_m into the above equations yields

$$Q_1 + Q_2 = 0.138 \qquad (c)$$

$$49.3 Q_1 + 62.3 Q_2 = 59.8 \times 0.134 = 8.01 \qquad (d)$$

Solving Equations c and d yields $Q_1 = 0.0454$ cfs and $Q_2 = 0.0926$ cfs.

Using Equation 9.27 for the alcohol pump yields $Q_1 = 0.0454$ cfs, $A_p = 0.02182$ ft², $L_p = 1.5$ inches $= 0.125$ ft, and $N_1 = 16.65$ rps $= 999$ rpm. Likewise, using Equation 9.27 for the water pump yields $Q_2 = 0.0926$ cfs, $A_p = 0.0491$ ft², $L_p = 2.5$ inches $= 0.208$ ft, and $N_2 = 9.07$ rps $= 544$ rpm.

Finally, by using Equation 9.25, the average linear velocities of the two pistons are 4.16 fps for the alcohol pump and 3.77 fps for the water pump.

9.2.2.2 Plunger Pumps

A plunger pump is similar to a piston pump except for the fact that the former uses a plunger (long cylinder) instead of a piston mounted on a rod (see Figure 9.7b and

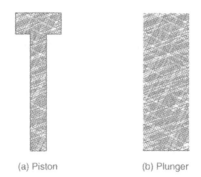

(a) Piston   (b) Plunger

FIGURE 9.9 Comparison between a piston and a plunger.

Figure 9.9). Because the two types of pumps are so similar otherwise, the entire foregoing discussion and analysis for piston pumps are also applicable to plunger pumps. Because a plunger is of the same size as the piston and is larger than the piston rod, the plunger is stronger than the piston rod and can withstand higher forces. This explains why plunger pumps instead of piston pumps are usually used in high-pressure applications.

Ordinary piston and plunger pumps are for single-action: pumping the fluid only during the first half of each reciprocating cycle when the piston/plunger moves forward. During the retreat of the piston/plunger in the second half of the cycle, the pump is not doing useful work. To remedy this shortcoming, double-acting piston/plunger pumps were developed for commercial use. By having two discharge check valves and two suction check valves as shown in Figure 9.10, fluid is pumped out continuously both during the advancement and the retreat of the piston/plunger. The double-acting piston/plunger pumps have higher discharges and better efficiencies than the single-acting ones.

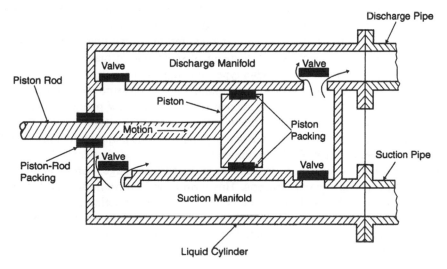

FIGURE 9.10 Double-acting piston pump.

9.2.2.3 Diaphragm Pumps

A diaphragm pump is similar to a piston or plunger pump except for the fact that the piston or plunger is separated from the fluid by a flexible membrane (diaphragm) (see Figure 9.7c). Because the piston/plunger does not make contact with the fluid, it is protected from damage by physically or chemically abrasive fluids, such as acid or abrasive slurry. The diaphragm is usually made of a chemically inert, physically strong, flexible material, and it can be replaced periodically without high cost.

9.2.2.4 Rotary PD Pumps

These are the PD pumps that have rotating parts. The most common type of rotary PD pump is the **gear pump**, which contains two gears inside the pump rotating in opposite directions (see Figure 9.7d). Due to tight packing and little clearance between the contacting opposite gear teeth and between the gears and the pump wall, fluid is displaced forcibly through the pump as in the other types of PD pumps. A second type of rotary PD pump is the **vane pump,** which, as shown in Figure 9.7e, contains a set of straight radial vanes closely packed to cause the PD action when the vanes rotate. Due to the large spacing that exists between neighboring vanes, the vane pump can pass relatively large solid particles and hence is suitable for use in hydraulic and pneumatic transports of coarse solids.

9.2.2.5 Screw Pumps

A screw pump contains a rotating screw (auger) in the center of the pump cylinder (see Figure 9.7f). Fluid or fluid–solid mixture is forced by the rotating screw to move through the pump and to enter the discharge pipe. The main motion of the fluid (or mixture) inside the screw pump is in the axial direction. Thus, both screw pumps and propeller pumps are considered to be axial-flow pumps. However, while in a propeller pump the propeller is housed in a large-diameter, disklike pump cylinder, in a screw pump the screw is contained inside a small-diameter, elongated pump cylinder. The relatively small diameter of screw pumps makes it possible to rotate at speeds higher than that of other types of pumps; some have speeds exceeding 10,000 rpm. For a given screw design, the pump head or pressure increases linearly with the screw length. Design of the screw threads of a screw pump is similar to designing an auger-type extruder, though the two machines have very different purposes or functions. Applications of screw pumps include pumping fuel oil, liquid lubricant, hydraulic fluids (for hydraulic presses), and special chemicals. Similar to other pumps, they cover a wide range of pressure and discharge. Screw pumps are also called **progressive-cavity pumps**. This term is derived from the fact that as a screw pump turns, the cavity or space between the rotor (rotating screw) and the stator (stationary wall possessing specially designed threadlike or wavy grooves) advances forward and pushes the liquid or solid–liquid mixture forward in a positive-displacement action.

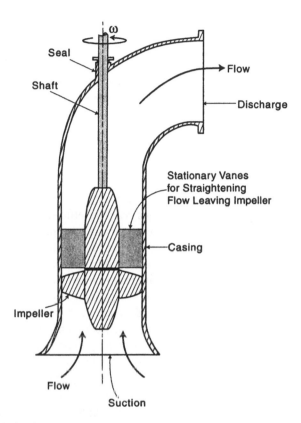

FIGURE 9.11 Axial-flow propeller pump.

9.2.3 PROPELLER PUMPS

Propeller pumps use propellers, which are designed according to the theory of airfoils described in fluid mechanics. The flow passing through the propellers of such pumps is mainly in the axial direction, making propeller pumps a special type of axial-flow pump. Due to the large diameter of the propeller and the wide space between propeller blades, large quantities (discharge Q) of fluid can be passed through the pump at low head or pressure. Thus, propeller pumps are for low head and large discharge applications, such as needed in drainage operations. Figure 9.11 shows a typical propeller pump.

9.2.4 OTHER TYPES OF PUMPS

9.2.4.1 Jet Pumps

Jet pumps are based on the conversion of fluid momentum (kinetic energy) to pressure. When a jet is issued in a pipe in the main flow direction, the momentum of the jet will be converted to fluid pressure downstream where the jet is diffused and dispersed. However, since much of the kinetic energy of the jet is dissipated into

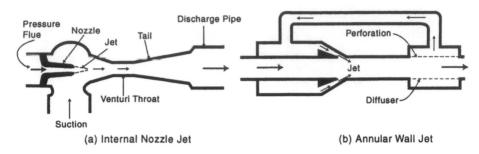

(a) Internal Nozzle Jet   (b) Annular Wall Jet

FIGURE 9.12 Two types of jet pumps.

heat during this conversion, the process of converting the kinetic energy of a jet to pressure is very inefficient. Most jet pumps have an efficiency of less than 30%, which is much lower than those of the aforementioned types of pumps. For this reason, and because the head of each stage of a jet pump is low, jet pumps are used only for special purposes such as when coarse solids must be transported hydraulically over a short distance, or when two different liquids or a liquid and a gas must be mixed — a jet pump does a good job in such mixing. Advantages of jet pumps include that they have no moving parts and hence need no lubrication, they can propel flows containing coarse solids, and the construction of a jet pump is relatively simple. Figure 9.12 shows two different types of jet pumps: (a) a jet is issued from an internal nozzle in the throat of a Venturi — the Venturi shape causes more efficient transfer of energy from momentum to pressure — and (b) an annular jet issued from the pipe wall at a small angle to the wall. Note that the annular jet pump is the least intrusive of all jet pumps. It can be used to pump fragile objects such as live fish or coal logs over short distances with little damage to the pumped objects. The efficiency of an annular jet pump is even lower than that of an internal nozzle jet pump — less than 20%. Other types of jet pumps also exist, but they all have low efficiency.

9.2.4.2 Airlift Pumps

An airlift pump consists of a long vertical pipe to convey or lift a liquid (such as water from a well) and a separate vertical pipe to carry compressed air to the bottom of the liquid pipe. By injecting air from the air pipe into the bottom of the liquid pipe, air rises in the liquid pipe and causes the liquid to rise as well. Figure 9.13 shows a typical airlift pump. The main mechanism involved in an airlift pump is the lighter density of the liquid–air mixture in the pipe as compared to the liquid outside the pipe in the reservoir or well.

As shown in Figure 9.13, the vertical distance between the pump outlet and the reservoir water level is H_1, and the vertical distance between the water level and the air pipe outlet is H_2. The ratio H_1/H_2 is a measurement of the lifting efficiency. The practical range of H_1/H_2 is between 0.4 (low head) and 3 (high head). Advantages of airlift pumps include no moving parts and hence no wear, they can be used for corrosive liquids, and they can be installed in small diameter wells that cannot accommodate an ordinary submersed pump. Disadvantages include low efficiency (less than 40%) and need for large submergence (large H_2) as compared to ordinary pumps. An

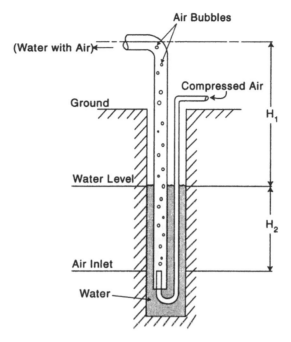

FIGURE 9.13 Air lift pump.

interesting potential application of airlift pump is in deep-sea mining, such as bringing manganese nodules from the bottom of deep seas to an anchored surface ship via a vertical pipeline. This application has been studied thoroughly in Japan.

9.2.4.3 Electromagnetic (EM) Pumps

From the Lorentz equation in electromagnetic theory, when a magnetic flux intensity B is applied across a conducting fluid carrying an electric current density J, an electromagnetic force per unit volume f_{em} is generated equal to

$$f_{em} = J \times B \tag{9.32}$$

In the above equation, the boldfaced letters J, B, and f_{em} represent vector quantities, and $\times$ represents vector cross product. When J is perpendicular to B, the force f_{em} becomes perpendicular to both J and B according to the right-hand rule of vector algebra. In such a case, Equation 9.32 can be written in scalar form as

$$f_{em} = JB = \varepsilon_m J H_m \tag{9.33}$$

where ε_m is the magnetic permeability of the fluid, and H_m is the magnetic field intensity in amp/m.

An EM pump of rectangular cross section is illustrated in Figure 9.14. The pump applies a strong magnetic field H_m (or a strong magnetic flux density B) and a strong

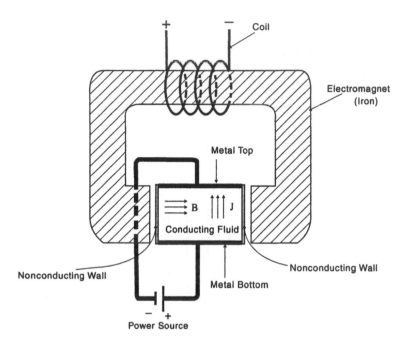

FIGURE 9.14 Electromagnetic pump — conceptual drawing.

current density J (perpendicular to B and H_m) across the liquid to be pumped. This causes a force per unit volume f_{em} to be generated on the conducting fluid in the direction of the pipe flow (perpendicular the cross section shown in Figure 9.14). When internal friction of the pump is negligible, the pump pressure generated becomes

$$\frac{\partial p}{\partial x} = JB, \quad p = JBL_{em}, \quad \text{and} \quad H = \frac{JBL_{em}}{\gamma} \tag{9.34}$$

where x is the distance in the flow direction, p is the pump pressure, H is the pump head, γ is the specific weight of the fluid, and L_{em} is the length of the EM pump (i.e., the length of the region subject to both J and B). In using Equation 9.34, SI units should be used, in which H and L_{em} are both in m, p is in N/m², which is the same as Pa (Pascal), $\partial p/\partial x$ is in Pa/m, γ is in N/m³, J is in amp/m², B is in weber/m², a unit also called **tesla**.

Equation 9.34 shows that the pump head of an EM pump is independent of the discharge Q through the pump. When H is plotted against Q, it is a horizontal straight line. This is different from positive displacement pumps in which a vertical line results when plotting H versus Q. As it is the same with other types of pumps, the $H \sim Q$ curve of the EM pump can be plotted together with the $H \sim Q$ curve of the pipeline system to determine the operation point.

From Equation 9.34, large pump pressure and high pump head can be generated electromagnetically only if both B and J are large, and L_{em} is long. To have a large

B requires a strong magnet, and to have a large current density J flowing through the liquid requires that the liquid be a good conductor. For this reason, EM pumps are practical only for special applications involving highly conductive liquids — liquid metals such as mercury or liquid sodium. Note that liquid sodium is used as the coolant in breeder reactors whose cooling systems are driven by EM pumps. EM pumps may use either permanent magnets or electromagnets, and the current can be either AC (alternating current) or DC (direct current). As discussed in Chapter 7, by using capsules with metallic walls, EM pumps can be used to pump capsule flow even though the fluid is not a conductor, e.g., air. In this case, the EM force is generated on the capsules instead of the fluid.

Example 9.7 An EM pump is used in an experiment to pump seawater having a density of 1026 kg/m² and a conductivity of 5.2 Siemens/m. The pump has the shape of a rectangular box of 2 ft (length) × 1 ft (width) × 0.5 ft (height). The top and bottom plates, each having a size of 2 ft × 1 ft, are made of aluminum and connected to 120 V of electricity. The two vertical walls of the pump are made of a strong plastic material that does not conduct electricity. The walls are perpendicular to a magnetic field generated by a magnet. The strength of the field across the walls and through the seawater in the pump is approximately uniform, having a flux density of 50,000 gauss. Find the pressure and head that can be developed by this EM pump.

[Solution] For this problem, $L_{em} = 2$ ft $= 0.610$ m, $\gamma = g\rho = 9.81 \times 1026 = 10,070$ N/m³, and $B = 50,000$ gauss $= 5$ weber/m². From Ohm's law and the definition of electric conductivity, the electrical current flowing through the pump between the top plate and the bottom plate is

$$I = \frac{V}{R} = \frac{k_e A_o V}{h_{em}} \tag{a}$$

where I is the current in amp; V is the voltage in V; R is the resistance in ohm; k_e is the electric conductivity of the liquid (seawater) in Siemens/m; A_o is the plate area, which is the same as the cross-sectional area of the electric current in m²; and h_{em} is the height of the EM pump, which is the distance between the two plates in m. Note that I is related to J, the electric current density, by

$$J = \frac{I}{A_o} = \frac{k_e V}{h_{ev}} \tag{b}$$

For this problem, $V = 120$ V, $A_o = 2 \times 1 = 2$ ft² $= 0.1858$ m², $k_e = 5.2$ Siemens/m, and $h_{em} = 0.5$ ft $= 0.1524$ m. Substituting these values into Equation a yields $I = 761$ amp. Thus, from Equation b, $J = 4,096$ amp/m². Then, from Equation 9.34, $\partial p/\partial x = 20,480$ Pa/m, $p = 12,490$ Pa, and $H = 1.24$ m.

More information on electromagnetic pumps can be found in Reference 2.

9.3 PUMP DRIVERS

There are various types of drivers (prime movers) for pumps including electrical motors, engines, gas and steam turbines, steam engines and so forth. They are described in the following subsections.

9.3.1 ELECTRIC MOTORS

Due to the widespread availability of electrical power from power distribution grids, most pumps today are driven by electric motors. Various types of electrical motors can be used to drive pumps; they include both AC (alternating current) and DC (direct current). The AC pump is used most often because the electric power from the power grid is AC. The DC pump is used when the power source is DC, such as for small pumps driven by batteries, or on board a ship or other vehicle that generates DC power. To use a DC pump when the power is AC requires rectification from DC to AC, which is costly.

Most electric motors are rotational machines that convert electrical power to mechanical power. A typical motor consists of two main components: a rotor, which is the rotating center part, and a stator, which is the stationary outer part. The stator is a set of windings made of insulated copper wire; it provides the electrical field needed to drive or rotate the rotor. Depending on their types, the rotor may be a separate set of windings or a metal structure having the shape of a squirrel cage attached to the rotating shaft. Electrical current is provided to the stator in such a manner that it creates a traveling electromagnetic field moving in the direction of the rotor rotation. This rotational field causes the rotor to rotate.

The power output of any rotating motor is

$$P_{om} = \omega_m T_m \tag{9.35}$$

where ω_m is the angular velocity of the rotating motor shaft and T_m is the motor torque.

The input power of a DC motor is simply

$$P_{im} = VI \tag{9.36}$$

where I is the supply current in amp, V is the supply voltage in V, and P_{im} is the motor input power in W.

The efficiency of any DC motor is

$$\eta_m = \frac{P_{om}}{P_{im}} = \frac{\omega_m T_m}{VI} \tag{9.37}$$

For any AC motor operating on single-phase current, the input power is

$$P_{im} = VI \cos\theta \qquad (9.38)$$

where V and I are the rms (root-mean-square) values of voltage and current, respectively, and θ is the phase angle between the voltage and the current, measured at the motor terminals. The quantity $\cos\theta$ is called the **power factor**, which is a number between 0 and 1. It depends on the motor design and operational conditions such as voltage, frequency, and the power output. Note that the quantities V and I can be separately measured with an ammeter and a voltmeter, respectively. The product of the two quantities is the apparent power rather than the real (actual) power consumed. The real power consumed is the quantity $VI\cos\theta$, which can be measured directly with a wattmeter. Therefore, the power factor $\cos\theta$ is the ratio of the actual power of an alternating current to the apparent power.

The efficiency of any single-phase AC motor is

$$\eta_m = \frac{P_{om}}{P_{im}} = \frac{\omega_m T_m}{VI \cos\theta} \qquad (9.39)$$

The overall efficiency of a pump–motor system, η, is the pump power output, P_{op}, divided by the motor power input, P_{im}. Thus, using Equations 9.7 and 9.38 yields

$$\eta = \frac{P_{op}}{P_{im}} = \frac{\rho g H Q}{IV \cos\theta} \qquad (9.40)$$

where $\cos\theta$ becomes one (unity) when the power supply is DC. From Equations 9.8, 9.39, and 9.40, it can be seen that $\eta = \eta_p \eta_m$ when $\omega_p T_p = \omega_m T_m$, or when the transmission of power from the motor shaft to the pump shaft is 100% efficient. Otherwise, the transmission efficiency, η_t, must also be included, namely $\eta = \eta_p \eta_m \eta_t$.

Equations 9.38 to 9.40 are valid only for single-phase AC. For 3-phase motors, the values of $VI\cos\theta$ must be multiplied by $\sqrt{3}$. Two wattmeters are used to measure the power input of a 3-phrase motor — one is connected between two of the lines, e.g., a and c, and the other is connected between the third line and either of the first two, e.g., b and c. The total power consumed by the motor is the sum of the two measured values. Note that large motors are mostly 3-phase; they are more efficient than single-phase motors.

Three types of motor commonly used are **induction** motors, **synchronous** motors, and **DC** motors. The first two types use AC (alternating current), which in the U.S. is supplied at 60 Hz frequency — either single phase (for small motors) or three phase (for motors of one horsepower or greater). The three types are discussed briefly after considering the following example.

Example 9.8 Water is pumped through a pipe at a discharge of 10 cfs, and the corresponding pump head is 100 ft. The pump is driven by a 3-phase AC motor. The two wattmeters connected to the motor input read a combined power of 90.3 kW. Determine the efficiency of the motor/pump system. If a voltmeter and an ammeter connected to the motor terminals read 480 V and 250 amp, respectively, find the power factor and the phase angle between the voltage and the current.

[Solution] The power output of the pump is $P_{op} = \gamma QH = 62.4 \times 10 \times 100 =$ 62,400 ft-lb/s. Since 1 ft-lb/s = 1.356 W, in SI units $P_{op} = 84,614$ W. The power input of the motor is the real power measured by the wattmeter. Therefore, $P_{im} = 90.3$ kW = 90,300 W. From Equation 9.40, $\eta = 84,614/90,300 = 0.937$ = 93.7%. Finally, from Equation 9.38, the power factor is cos $\theta = P_{im}/VI =$ 90,300/(250 × 480) = 0.7525, and the phase angle is 41.2°.

9.3.1.1 Induction Motors

The induction motor is based on the concept of electromagnetic induction from which it derived its name. The moving electromagnetic field generated by the stator causes an induced current in the rotor (squirrel cage), which interacts with the moving magnetic field to generate a torque on the rotor that causes it to rotate. Under optimal design conditions, the rotational speed of an induction motor is usually a few percent lower than the speed of the traveling magnetic field, determined by a quantity called **slip**, which is defined as follows:

$$S = \frac{N_s - N_m}{N_s} = \frac{V_s - V_m}{V_s} \qquad (9.41)$$

where N_m and V_m represent, respectively, the rotational speed and the circumferential velocity of the rotor — namely, $V_m = 2\pi r_o N_m$, where r_o is the rotor radius. The subscripts s and m in Equation 9.41 refer to synchronous and motor, respectively.

From Equation 9.41, slip is 1.0 (unity) when the rotor is stationary ($V_m = 0$), and is 0 (zero) when V_m is the same as V_s. Under steady-state operation, V_m will never reach V_s because the torque generated by an induction motor (see Figure 9.15) is at maximum when S is unity (i.e., when the rotor is stationary), and the torque reduces to zero when S becomes zero (i.e., when the rotor velocity reaches the synchronous velocity). Advantages of induction motors include (1) they cost the least to purchase; (2) they are readily available in a wide range of sizes; (3) they are self-starting; (4) they are suitable for use not only in general applications but also in specific cases where the torque needs to be controlled or when the motor speed needs to be varied.

9.3.1.2 Synchronous Motors

Synchronous motors are similar to induction motors except for the fact that the rotor rotates at a velocity equal to the synchronous velocity, so that no slip exists ($S = 0$). Unlike the induction motor, which cannot develop torque at synchronous speed, the synchronous motor can develop large torques at synchronous speed. In order to do so, the rotor of the synchronous motor contains either permanent magnets, or electromagnets connected to an external DC source of electricity. The connection

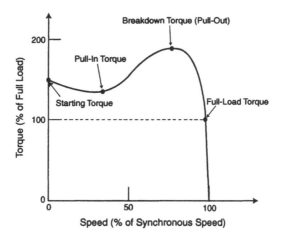

FIGURE 9.15 Typical torque curve for a squirrel-cage induction motor.

between the rotating electromagnets and the stationary outer source of DC is normally through a slip-ring and metal brushes mounted on the rotor shaft, which require frequent maintenance and cause sparks. In recent years, brushless designs have corrected this problem, but the new mechanism is not inexpensive. Generally, synchronous motors are more expensive than induction motors on the first cost (capital cost) basis. It also requires a more complicated mechanism for starting and bringing the rotor to synchronous speed. The main advantage of synchronous motors over induction motors is that the former has a higher efficiency — of the order of 95%, which is about 5% higher than that of induction motors of the same size. For this reason, synchronous motors are usually used only in places where large units (usually greater than 100 hp) are required and savings of electricity outweigh the extra cost of the synchronous motor.

For all AC motors, whether synchronous or induction types, the synchronous speed N_s is determined by the number of magnetic poles on the motor and the frequency of the power, namely

$$N_s = \frac{120 f_o}{n_p} \qquad (9.42)$$

where f_o is the frequency of the AC, n_p is the number of poles, and N_s is given in rpm. As an example, if an AC motor has two poles and runs on 60 Hz current, from the above equation the synchronous speed must be 3600 rpm. By the same token, the synchronous speed of motors with 4, 6, 8, 10, and 12 poles (note that only even number of poles are possible) is 1800, 1200, 900, 720, and 600 rpm, respectively.

9.3.1.3 DC Motors

DC motors are used normally only in places where the source of electricity is DC, small pumps operating on battery power, or when pumps are used on board vehicles such as ships, airplanes, and trains, where DC is normally the source. It is beyond the scope of this book to discuss the various designs of DC motors. Suffice it to mention that there are three different types: shunt motor, series motor, and compound motor. Their speeds can all be controlled by varying the supply voltage.

9.3.2 ENGINE AND TURBINE DRIVERS

9.3.2.1 Engines

Engines such as gasoline engines, diesel engines, and steam engines can all be used to drive pumps. Since engines are reciprocating machines, they are most suitable for driving reciprocating pumps such as the piston or plunger pumps. They are used in situations such as: (1) remote locations where electricity from the grid is not readily available, and (2) large pumps with readily available sources of fuel for which it is more economical to use the available fuel to power an engine that drives the pump than to use the electricity from the power grid as the prime source of energy for pumping. It should be realized that without electricity it is more difficult to control pumps and run the auxiliary equipment. It is always convenient to have electricity available even when the prime mover of the pump is an engine or turbine.

9.3.2.2 Turbines

Turbines are rotating machines driven by the motion of fluids (liquids or gas) going through the turbine — the opposite of pumps. Depending on the source of energy, there are three general types of turbines: **steam turbines, gas turbines,** and **hydraulic turbines**. They are normally used to drive generators, which in turn generate electricity. Sometimes they are also used to drive pumps directly without first converting to electricity. The prime advantage of using turbines to drive pumps is the relatively low cost as compared to using electric motors. This is especially true when the pumps are large (i.e., when the energy-saving potential is high), and when there is a readily available source of fuel or energy to power the turbines needed at the location of pumps, as in the case of using natural gas available from a natural gas pipeline to power a gas turbine that in turn drives the compressors used to pump the gas through the pipeline. Control of pumps driven by turbines is also relatively simple, and it has little effect on the performance of the prime driver — the turbine. This is so because turbines are protected by governors, which can adjust automatically to load changes.

The fuel to power a steam turbine is usually a fossil fuel — coal, oil, or natural gas. Nuclear power plants use nuclear fuel to power steam turbines. Renewable and cleaner-burning fuels such as biomass (combustible solid wastes such as wood-processing wastes and municipal solid wastes) can also power steam turbines.

Gas turbines differ from steam turbines in that they do not use water and do not generate steam. Instead, the combusted gas expands and is forced through the impellers or propellers of the turbine to make the latter turn. Again, they can be used to generate electricity or to drive pumps directly, as in the case of steam turbines.

Hydraulic turbines are those that derive their energy from hydropower — water head generated by dams. They will be discussed in more detail in Section 9.5. Suffice it to mention here that although hydraulic turbines are most commonly used for generating electricity, they are also used occasionally to drive pumps directly.

9.4 COUPLING PUMPS TO DRIVERS

The coupling between a pump and its driver transmits the shaft power from the driver to the pump. Depending on the need for speed control and other requirements, different types of coupling exist; they are discussed below.

9.4.1 COMMON-SHAFT COUPLING

For miniature pumps used for special applications, the pump–motor system is often designed and manufactured as a single unit utilizing a common shaft. In this case, the pump operates at the same speed, develops the same torque, and transmits the same amount of power, as that of the motor. Speed control in this case can be accomplished by controlling the speed of the motor.

9.4.2 DIRECT MECHANICAL COUPLING

In direct mechanical coupling, the motor and the pump are manufactured separately. However, their shafts are connected together in good alignment so that the two machines behave the same as a common-shaft motor pump system. Many different types of such coupling are commercially available. They may be rigid coupling or flexible coupling. The latter permits some degree of misalignment of the two shafts without causing serious problems. Generally, alignment of shafts in all types of direct mechanical coupling should be as precise as possible during pump installation. Otherwise, the pump and/or motor may soon be damaged or have serious vibration problems. As in the common-shaft case, the two shafts coupled together rotate at the same speed and transmit the same torque and power. Speed control can be done only by controlling the speed of the motor using an adjustable-speed drive.

9.4.3 GEARS

Coupling the motor shaft with the pump shaft via a gearbox not only allows transmission of power but also an opportunity to change speed and torque. For instance, the design speed of a 4-pole induction motor may be 1750 rpm. If the pump driven by this motor is to operate at 437 rpm, a gear of 4-to-1 ratio will be needed to connect the pump to the motor. With such a gear in use, the torque needed by the motor to turn the pump at any speed will be one-fourth of that of the pump. Oftentimes, the operational speed of a motor is much higher than the design speed of the pump used in conjunction with the motor. Having a gearbox connected between the two machines with an appropriate gear ratio allows controlled step-down of the shaft speed and increase of the torque. The gear functions differently from that of a variable-speed drive. While the latter allows control of the motor speed and the overall power transmitted, the gear simply provides a fixed means to match the pump with the motor. Therefore, both a gear and a variable-speed drive may be needed in a given pump application. Another function of the gear is it allows pumps to be set up in a direction such that the pump shaft is either parallel to or perpendicular to the shaft of the drive (motor). The latter may be needed, for instance, when the pump is horizontal while the motor is mounted vertically above the pump.

9.4.4 BELTS

Belts serve the same function as gears and are often used in lieu of gears for speed reduction. Usually, a belt system consists of a V-belt made of rubber connected between two sheaves (grooved wheels) mounted on the two parallel shafts of the motor–pump system. The difference in the diameter of the two sheaves at which the belt is mounted determines the speed ratio to be accomplished. The belt can be tightened easily by using spring-loaded sheaves. As compared to gears, belts are much lighter and costs less. They are also more flexible in the adjustment of the speed ratio. Some of them are capable of having the speed ratio adjusted continuously while the belt and the shafts are turning. This is done within a certain range by using adjustable-pitch sheaves. However, belts have a much shorter life span and need more frequent replacement, adjustment, and maintenance than gears do.

9.4.5 FLUID COUPLING

Fluid coupling is a rotating machine whose casing contains both a pump impeller and a turbine runner of similar size and shape mounted next to each other on separate shafts for the purpose of smooth transmission of the torque from the impeller to the runner. The fluid contained in the coupling is usually an oil, which not only transmits the torque but also lubricates the machine and conducts and dissipates heat. It is the same device used in automatic transmissions in automobiles. When installed between a rotating pump and a rotating motor, the pump-side shaft of the coupling is connected directly to the motor or drive shaft, whereas the turbine (runner) side of the coupling is connected directly to the pump shaft. Use of such a device not only prevents fluctuations of the torque and speed in the drive shaft from being transmitted to the pump shaft, it also prevents the build-up of large torques in the motor (drive) shaft during start-up of pumps. The extra torque needed to accelerate the pump during start-up is minimized as the pump slowly picks up its speed during start-up with a fluid coupling. Note that soft couplings such as fluid coupling and eddy-current coupling are not necessary for starting up a centrifugal pump if the valve that controls the flow through the pump is completely closed during start-up. This is so because without flow going through the pump, only a small quantity of fluid is moved (circulated) within the pump casing when the centrifugal pump is on. The pump itself is now acting as a fluid coupling, limiting the torque that can be generated during start-up.

9.4.6 EDDY-CURRENT COUPLING

Eddy-current coupling serves the same purpose as fluid coupling except for the fact that it is based on a totally different principle — **eddy current** — for providing a soft coupling. Note that the term *eddy current* used by electrical engineers has a very different meaning from the same term used in fluid mechanics. The electrical engineer's eddy current is the electrical current induced on a conductor by a moving magnetic field. It is the same principle that induction motors are based on. In the case of eddy-current coupling, the conductor is an outer metal cylinder mounted on a shaft that is connected (directly coupled) to the drive (motor) shaft which needs to be at constant speed. A slightly smaller concentric inner cylinder, the source of magnetic field, is mounted on a separate shaft that is connected (directly coupled) to the load (pump) shaft. The inner cylinder contains a field winding excited by direct current (DC) supplied from outside the coupling. This in turn generates a magnetic field of an intensity dependent on the field current strength. When the two concentric cylinders rotate relative to each other, a moving magnetic field is generated by the inner cylinder, and an eddy current is induced in the outer cylinder. This generates a torque on the inner cylinder, causing it to rotate in the same direction as the rotating outer cylinder. Because the torque generated is limited, no dangerous large torque is possible during pump start-up and load variation. This makes the eddy-current coupling a soft coupling. The torque on the pump shaft in this case can be controlled by the strength of the field current.

9.5 PUMP CONTROL, OPERATION, AND MAINTENANCE

9.5.1 PUMP CONTROL

9.5.1.1 Discharge Control by Valves

Oftentimes, it is necessary to alter the flow rate (discharge) going through a pipeline, Q, which under steady-state operation is the same Q going through the pump. It should be kept in mind at the outset that the discharge Q through a centrifugal or propeller pump is related to the pump head, H, through the pump characteristic curve as shown in Figure 9.2 for a typical centrifugal pump. For such a pump, altering the discharge Q through the pump also changes the pump head H and vice versa. It is not possible to adjust the discharge and the head of such pumps independently. In contrast, because the $H \sim Q$ curve for a positive displacement (PD) pump is a vertical line for any pump speed N and stroke length L_p (for piston or plunger pumps), varying H over a wide range will not affect Q. One must either change the pump speed or the stroke length, or both, in order to change the discharge. This difference between PD pumps and centrifugal/propeller pumps must be understood in controlling pumps.

A common way to control the discharge going through a centrifugal or propeller pump is to use a valve normally located on the discharge side of the pump. The pump may even be allowed to continue to run for a short time (a few minutes) after the valve is completely shut off and the discharge is zero. However, for positive displacement pumps, throttle of the flow by a valve must be accompanied by other adjustments such as providing a bypass line and directing the flow to the bypass. Even though closing or partially closing a valve is the most effective way to regulate the discharge through a centrifugal pump, it can produce many problems when it is not properly executed. For instance, if the valve is closed too rapidly, large pressure surges (water hammer) may be generated to damage the pump and other pipeline components. Generally, the water hammer effect can be minimized by closing the valve slowly, as analyzed and discussed in Chapter 2. Another problem associated with valve closure is the following. If the valve is completely or essentially closed while the pump continues to run over an extended period, not only is energy wasted, but the pump and/or the motor may also overheat. A third problem associated with the closure of a discharge valve when the pump is running is the high pressure generated in the pump casing. This not only can cause excessive leaks through the packing around the shaft but also may result in the weakening of the pump housing (casing) especially if the valve closure is frequent. Fatigue of materials is generated, which shortens the life of the pump. All these and other concerns affect the way a pump is to be controlled in a given situation.

9.5.1.2 Controlling Pump Speed by Motor Speed

A way to adjust the discharge Q without relying on valve action is to control the speed of the pump, which in turn is controlled by the speed of the motor that drives the pump. Unlike valve closing, which wastes energy, reducing the flow by reducing the motor speed does not waste energy. The power consumption is automatically less when the motor speed is reduced and so forth. Therefore, using motor speed control instead of valve control to alter discharge saves energy.

Speed control for electric motors is usually done in one of two general ways: **voltage control** or **frequency control**. Depending on the type of motor used, different speed controllers are available commercially. They include (1) **AC adjustable-voltage** type — for relatively small (less than 100 hp) induction motors that possess high slip (about 10%), (2) **wound-rotor induction motors with secondary controls** — for wound-rotor instead of squirrel-cage type of induction motors, (3) **adjustable-frequency** type — which uses a rectifier to change 60 Hz AC to DC and then uses an inverter to change the DC back to AC at a different frequency, and (4) **DC motors with silicon-controlled rectifier (SCR) power supply**. It is beyond the scope of this book to discuss each type. Suffice to mention that the motor manufacturer/distributor should be asked to select the proper speed controller for specific motors. In fact, in most cases the motor and the speed controller can be purchased from a single source with the speed controller as a part of the motor system.

9.5.1.3 Motor Starters

The main function of a motor starter is to connect the needed supply current to the motor so that the motor can start — the same as an electric switch but for large current. The starter often also includes an overload protection relay, which can be tripped whenever the circuit is overloaded, and/or a fuse or circuit breaker to protect the equipment from short circuit. Both manual and magnetic starters are available commercially. In the case of manual starters, a single handle provides the manual control needed. In automatic starting, the electric contact to the motor is made by a magnetically operated contactor. Depending on the motor type and control equipment, the voltage applied to start a motor may be either full voltage (i.e., the same voltage as needed for steady-state operation) or reduced voltage. Reduced voltage is often used to start a pump because of the large transient current going to the motor during its start. The transient current may be several times as high as the steady-state current at the same voltage, and it can overheat and damage a motor unless protection is provided by using a circuit breaker, protection relay, or reduced starting voltage. Use of reduced starting voltage prevents frequent tripping of the relay or circuit breaker. Since the torque generated by a pump is almost zero at start up when Q is zero, and since the torque increases as Q increases, the torque needed from the motor to drive the pump during start-up is less than what is needed during steady-state operation. This shows why a reduced voltage can start a pump. After the pump has accelerated to a sufficient speed, the full voltage is then applied. This can be done most conveniently with an automatic starter.

Other measures are also needed for pump control and protection. For instance, to prevent pumps from being damaged by transient overpressure (water hammer), either a pressure relief valve or a pressure damper such as a surge tank or an air chamber must be used near the tank on the discharge line. Chapter 2 contains more information about such devices. Large pumps also must be protected by temperature and vibration sensors, whose signals are used to turn the pump off automatically whenever serious overheating and/or vibration problems occur. Nowadays, large pumps are automatically controlled by PLCs (programmable logic controllers), which communicate electronically with a computer that controls the operation of the entire pipeline system.

9.5.2 PUMP OPERATION

Pump operation varies greatly with the type of pump and driver used, and the associated equipment and system design. Only certain general and important operational issues can be addressed here. They should be kept in mind by pump operators.

9.5.2.1 Priming

When a pump is started for the first time after installation, or restarted after shutdown, the pump casing may be empty of the liquid to be pumped and filled with air instead. Turning on this empty pump will not generate enough suction (vacuum) to draw the liquid into the pump — because a pump is not a compressor. The pump impeller will be spinning without liquid flowing through it, which will cause overheating of the pump and the motor. Damage will then occur if the pump and the motor are not quickly turned off. To prevent this from happening, all pumps except a few special types (the self-priming type and the piston/plunger pumps) must be filled with water first before starting, and this is called **priming**.

The best way to prime a pump is to place the pump at an elevation below the liquid level of the intake reservoir. This will bring water in from the reservoir by gravity, causing the pump to remain primed after the pump is stopped. If this is not practical and the pump must be placed at an elevation above the surface elevation

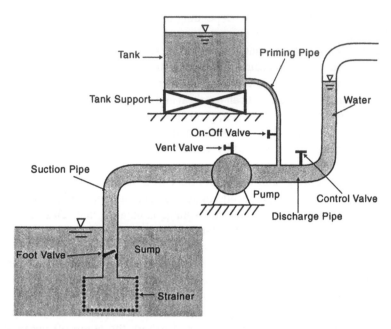

FIGURE 9.16 A typical priming system for a pump located above reservoir level.

of the liquid in the reservoir, then a priming system must be incorporated into the design of the pumping system. Many alternatives are available to prime pumps. Their selection depends on local situations and practical considerations. A common type used is illustrated in Figure 9.16. It consists of an elevated tank of liquid (the same liquid to be pumped); the liquid in the tank is used to fill the pump and the suction line completely by gravity before the pump is started. To prevent the liquid from draining back into the sump (intake reservoir), a foot valve (which is a check valve) is mounted near the bottom of the inlet pipe. The foot valve allows the liquid in the pipe to move in one direction only — the direction that the liquid is to be pumped. During the priming period, the on-off valve on the prime pipe is opened to allow the liquid to fill the pump, the intake pipe, and a small portion of the discharge pipe (up to the level indicated in the figure). Meanwhile, the vent valve on the top of the pump casing remains open to allow air to escape from the system. As the air is drained out completely and the liquid starts to emerge from the vent, the vent is closed and the system is ready to be started. For a centrifugal or axial-flow pump, the control valve on the discharge pipe should now be closed to start the pump. If priming is not done correctly and thoroughly, a large amount of air may be trapped in the pump and the inlet pipe, and the pump may fail to engage the liquid. From the noise generated and pressure gauge readings, one can usually tell whether the liquid has entered the pump during startup. As soon as this priming process is completed, the control valve should be opened gradually, and the flow be allowed to increase until it reaches the desired steady state value. If the liquid fails to flow in spite of the priming, the pump should be stopped immediately, and a more thorough priming be conducted. Even with successful priming, the pump will still have considerable amount of air in it during its initial stage of operation.

This generates intense noise, which is normal during start-up. However, as soon as the liquid starts to move through the pipe, less and less air will remain in the pump and the inlet pipe, the noise will subside, and the flow will approach steady state. For water to be pumped, in lieu of an elevated tank, it is often more convenient and effective to connect the priming pipe to the water supply line (tap water line), provided that tap water exists at the location of the pump.

9.5.2.2 Cavitation

Cavitation is a special type of damage caused by liquid flow on solids, including pumps, turbines, pipes, and valves that come in contact with the flow. It is caused by the low pressure generated by liquids, causing the liquid to vaporize (boil at room temperature). The vaporized liquid forms bubbles that are carried by the flow to downstream locations. When such bubbles enter regions of higher pressure, they collapse back into liquid. These collapsing bubbles can generate localized high-pressure fluctuations of the order of 100,000 psi. Any solid that comes into contact with such bursting bubbles can become severely damaged in a relatively short time, on the order of days or weeks. The damage is usually in the form of pitting or cavities — hence the name cavitation. When cavitation occurs in centrifugal or propeller pumps, it usually damages the running blades (vanes) and parts of the pump casing. Pumps damaged by cavitation not only lose efficiency but also may develop vibration and other problems. This is a serious matter that must be corrected. Normally, the problem can be corrected by increasing the pressure of the flow in the regions of high suction, such as on the suction side of the pump. Careful consideration of the possibility of cavitation during pump selection and system design, such as making sure that the available NPSH is at least a few feet greater than the required NPSH, will prevent pump cavitation. Further discussion on cavitation can be found in fluid mechanics texts such as Reference 3.

9.5.2.3 Vibration and Noise

All pumps vibrate and emit some noise. However, large vibration can quickly damage a pump and hence must be corrected before the pump is allowed to continue to run. There are various causes of pump vibration; the most common cause is misalignment of the shaft. Other common causes include having a loose part rotating in the pump, cavitation, and resonance. Resonance is a general cause of vibration, which happens whenever the frequency of vibration of any structure (i.e., the natural frequency) matches the frequency of the excitation force or torque. For instance, a pump may have a natural frequency N. If the pump is running at or near this frequency, resonance will occur to cause large vibration. In such a case, running the pump at a speed significantly higher or lower than N will usually solve the vibration problem. As with other problems that face us humans, the most effective way to solve a vibration problem is to identify the cause. Once the cause is identified, removing the cause will usually solve the problem. Pump noise is most often a result of vibration; thus, reducing pump vibration reduces noise. Noise is also often a result of cavitation and air entrainment in the pump. Oftentimes, from the location, intensity, and the pitch of the noise, an experienced pump operator can find the cause of the noise. Special diagnostic tools are also available commercially to study vibration and noise problems related to pumps. The foregoing discussion on the vibration and noise of pumps is also applicable to turbines, compressors, blowers, fans, motors, and other rotating machines, except for the fact that without liquid flow, dry machines such as motors, compressors, blowers, and fans can not cavitate and hence have no vibration or noise problem associated with cavitation.

9.5.2.4 Overheating

Even though it is permissible to operate a centrifugal or axial-flow pump when the control valve is closed and no flow moving through the pipe (i.e., under shut-off condition), the pump can overheat and get damaged if the pump is allowed to run under such conditions exceeding a permissible time or period. This is due to the fact that under such conditions the pump is running at high speed, which causes the same body of liquid in the pump to rotate and generate heat. Without the liquid flowing through the pump, heat cannot be dissipated fast enough and the liquid temperature rises in the pump. A temperature increase of more than 10 to 20°F may damage the pump. From fluid mechanics considerations, the heat generated is expected to be proportional to the third power of the pump speed. Therefore, doubling the pump speed will increase the heat generated by eight times! This shows that the overheating problem is much more serious for high-speed pumps than for low-speed pumps. While a low-speed pump may be allowed to run under no-load condition for a few minutes, even 30 s may too long for a high-speed pump. A simple formula to determine the rate of temperature rise caused by operating the pump under no-load condition is the following:

$$T = \frac{0.707Pt}{cW} \tag{9.43}$$

where t is the time in seconds since the pump has started to spin under no-load condition; P is the pump input power (brake horsepower) under shut-off condition; c is the specific heat capacity in Btu/lbm; W is the weight of liquid in pounds (lbm) contained inside the pump; and T is the temperature rise in °F in time t. Note that Equation 9.43 is derived by steady-state energy balance, which assumes that the entire heat dissipated by the rotating pump is used to raise the liquid temperature. The equation is only a rough estimate because the heat losses through the pump casing and the heat used for heating the pump impeller and the casing to a higher temperature are not included. Thus, the equation errs on the safe side in estimating the permissible time to operate any centrifugal and axial-flow pump under no-load condition.

Example 9.9 A centrifugal pump is used in a pipeline system to pump water. If the pump is started under shut-off (i.e., $Q = 0$) condition, and it can tolerate a maximum temperature rise of 10°F, what is the permissible shut-off time of the pump when started at full speed? It is known from the pump characteristic curves that this pump has a shut-off brake-power of 120 hp. The amount of water inside the pump casing is estimated to be 155 lb. If the pump starts at a reduced speed equal to 60% of the full speed, what is the corresponding permissible shut-off time?

[Solution] For this problem, $P = 120$ hp, $W = 155$ lb, $c = 1.0$ Btu/°F (for water), and $T = 10$°F. From Equation 9.43, this yields a value of $t = 18.3$ s. This shows that in order to limit the temperature rise to 10°F, this pump should not be allowed to operate under full-speed and no-load condition for longer than approximately 18 s.

When the same pump is started at 60% of full speed, the brake horsepower is expected to be reduced by a factor of $(0.6)^3 = 0.216$. (This assumes that the shut-off brake horsepower is proportional to the third power of pump speed.) Consequently, now the permissible time to allow no-load operation will be increased to $t = 11.6/0.216 = 84.7$ s.

9.5.3 Maintenance

Due to the high cost and important function played by pumps in a pipeline system, and due to the fact that pumps without proper maintenance can easily be damaged, it is highly important that they be well maintained. One should carefully read and strictly follow the pump manufacturer's advice as given in the operation/maintenance manual.

Generally, pumps require close monitoring and frequent maintenance. Except in remote locations where it is difficult to station operators, pumps should be under constant care and attention of a human operator. During working hours, the operator should observe and inspect the running pumps on an hourly basis. Any irregularity discovered (such as increase of noise or sudden loss of pressure or flow rate) should be reported immediately to the engineer in charge of the pump. Necessary corrective measures must be taken to remedy any irregularities observed.

In addition to such daily attention, pumps should also receive semiannual and annual inspections. The semiannual inspection includes stopping the pump and checking the alignment of the pump and driver shafts, checking the stuffing box packing (for pumps with stuffing boxes instead of mechanical seals) and replacing with new packing, checking to see if additional or replacement grease (for grease lubricated) pumps is needed, and changing the oil (for oil-lubricated pumps). Then a more thorough inspection should be conducted annually. This includes (1) removing the bearings and checking the bearings for possible damage, replacing damaged bearings, and careful cleaning of the bearing housing; (2) inspection of the shafts and shaft sleeves (if any) for possible damage, and realignment of the pump shaft with the drive shaft; (3) checking all auxiliary piping, including the drain pipe, sealing water pipe, and cooling water pipe, and flushing them; and (4) recalibration of instruments and flowmeters. The annual inspection should also include all the items of the semiannual inspection, so that there will be only one semiannual inspection each year. In addition to the daily, semiannual, and annual inspection, pumps also may need complete overhaul when cavitation or other problems have caused a significant deterioration of the pump performance, as demonstrated by a significant loss of pump efficiency, loss of pressure, damage of impeller vanes (for centrifugal pumps) or propellers (for propeller pumps), etc. The overhauls usually involve major repairs and/or replacing the damaged parts.

For pumps that are in remote locations, it may not be practical to have an operator stationed there. In such a case, it is not possible to have daily inspections as described earlier. Still, daily attention should be given to the pump through remote monitoring of the pump performance, such as the pump and motor speed, the discharge, the pressure at critical locations, the level of pump vibration and noise, etc. In addition, the same semiannual and annual inspections described earlier also apply to remotely located pumps.

9.6 PUMP SELECTION

A good knowledge of pumps such as provided in this chapter is needed for the proper selection of pumps for a given project. In addition to possessing the pertinent technical knowledge, it is important for the pump selector to have a clear understanding of the various requirements of the project, such as the fluid to be pumped, the capacity (discharge) to be pumped, the need for storage (reservoirs), the variation of flow rate with time, the pipeline system designed for the project, the temperature of the fluid, the climate, the degree of reliability and safety that is needed, and many other requirements. In what follows, a set of key factors are listed for consideration in the selection of pumps.

9.6.1 FLUID TYPE

Even though the various aspects of pumps discussed in this chapter are common to pumps for all fluids, pumps that handle different fluids are usually designed and constructed differently to best suit individual needs. For instance, pumps that handle water are designed somewhat differently from those handling sewage, pumps for handling seawater are different from those for handling fresh water, and pumps for oil are different from those for water or slurry. Therefore, fluid type is the first factor to be considered in the selection of pumps. Once the fluid type is known, one needs to consider only those pumps designed specifically for such fluid. This greatly narrows the field of selection. One should also know that a pump may be designed to handle more than one type of fluid. In such a case, the characteristic curves of pumps depend not only on the pump but also on the fluid. For the same pump, the characteristic curves differ whenever the fluid is different.

9.6.2 FLOW PARAMETERS

The two most important flow parameters affecting pump selection are discharge Q and head H. First, the discharge Q through the pump and the pipeline must be determined from project need. For instance, a water supply project may need to convey 1.2 mgd (million gallons per day) of water through a pipe. This is equivalent to a discharge of 1.86 cfs. Once Q is determined and the design velocity V is specified (say 6 fps), the pipe diameter can be calculated and the head H to pump water through this pipe can be computed from the energy equation. This in a nutshell explains how Q and H are determined.

9.6.3 NUMBER OF PUMPS

For a small pipeline in which the interruption of operation due to pump failure is of no serious consequence, a single pump may suffice. In contrast, for large pipelines and for pipelines that serve vital functions, at least two pumps will be needed with one of them being a spare. The spare pump is generally connected in parallel with the operating pump but is isolated hydraulically from the operating system by one or two valves. Whenever the operating valve is out of service, the spare will be turned on to continue the work while the other pump is being repaired or maintained. For large systems where the pumps are costly, it may be less expensive to use three smaller pumps than two larger pumps for the same job. For instance, if $Q = 10$ cfs and $H = 1000$ ft, the flow power will be 1135 hp. One may use two pumps of 1200 hp each, with one being the spare. Alternatively, one may use three pumps of 600 hp each, with one being the spare. Both systems perform the same duties. In the first case one must buy two large pumps of a total of 2400 hp; in the second case one must buy three smaller pumps of a total of 1800 hp. Since the price of similar pumps is proportional to the horsepower rating of the pumps, the total purchase price of the three smaller pumps will be significantly less than the purchase price of the two larger pumps. This shows the merit of using three instead of two pumps. To further reduce cost, sometimes one may use four or five pumps, one of which is a spare. More than five pumps are not practical since they take up too much space, increase maintenance, and reduce system reliability.

9.6.4 OPERATING FREQUENCY

In certain special applications such as irrigation or drainage of a sump or wet pit, the pump used may operate on-and-off periodically. Suppose that the inflow discharge to the sump is Q_{in}, the sump volume is $\forall$, the pump discharge when operating is Q, the pump cycle time is Δt, and t_p is the duration when the pump is on. From continuity, it can be proved that

$$t_p = \frac{\forall}{Q - Q_{in}} \quad \text{and} \quad \Delta t = \frac{\forall}{Q_{in}} + \frac{\forall}{Q - Q_{in}} \tag{9.44}$$

The above equation indicates that by varying the pump discharge Q, the operating cycle properties of the sump pump system, t_p and Δt, can be varied. This provides a guide to the selection of the pump needed for such an on-and-off operating system. The pump should be sized such that both t_p and Δt will not be too short. This avoids overly frequent on-and-off and overly short on-time t_p, both of which are undesirable for the pump.

9.6.5 RELIABILITY

System reliability plays an important role in the selection of pumps. For pumps that serve vital functions and must be on all the time, one needs to include not only a spare pump powered by the same electrical system, but also a back-up pump that can be powered by an independent power source, such as a diesel generator, to cope with power failure. Top-of-the-line equipment should also be selected for pumps that must be highly reliable.

9.6.6 SAFETY

Safety should be kept in mind in the selection of pumps. For instance, for pumps that operate in places where an explosive environment may exist, as in the case of pumps used for natural gas or petroleum products, the pumps and their drives and auxiliary equipment must all be selected carefully to make sure that they do not generate sparks and are explosion resistant. For pumps submersed in water, operating on wet ground, or subject to rain, one must make sure that they are designed to operate in such wet environments. Otherwise, the operator's safety would be at stake.

9.6.7 PUMP TYPE

Whether one should select a centrifugal pump or another type depends mainly on the head (pressure drop) of the system. For a system that involves large head (i.e., long pipeline and/or large static head), one must use PD pumps. For intermediate head ranges, a centrifugal pump is the most suitable. For low-head operations, axial-flow pumps such as propeller pumps should be used. What is high head and what is low head cannot be defined precisely because there are head ranges for which both PD and centrifugal or both centrifugal and axial-flow pumps are suitable. Roughly, above 1000 ft of water is considered high head, between 10 and 1000 ft is considered intermediate head, and below 10 ft is considered low head.

9.6.8 PLOT OF $H \sim Q$ CURVES

Based on consideration of the foregoing factors, a number of candidate pumps known to be available commercially should be chosen for more detailed comparison and further analysis. An important step at this point is to plot the $H \sim Q$ curve of each candidate pump, and the $H \sim Q$ curve of the system together as in Figure 9.4, to see if the operating point yields the approximate discharge needed for the system. If it has been decided that only one pump is to be used, then the $H \sim Q$ curve of the single pump must be used in this analysis. If it has been decided that two or more pumps will be used either in parallel or series, then the $H \sim Q$ curve of the combined pumps must be used in the analysis. Through this step, several different pumps may be selected as the finalists for further comparison.

9.6.9 PUMP EFFICIENCY

The next step is to determine the efficiency η of the pump at the operational point. This can be determined from the efficiency curve as shown in Figure 9.2. If more than one pump is used, the efficiency of the combined system must be determined from those of single pumps. Note that efficiency is only one of many important factors to be considered in pump selection. The final pump selected may or may not be the most efficient.

9.6.10 CAVITATION

All pumps selected must not cavitate. This is determined by comparing the available NPSH with the required NPSH, with the latter determined from the pump characteristic curve (NPSH versus Q) provided by the pump manufacturer.

9.6.11 PUMP DRIVERS

Once the pump for a given project is selected, one can then determine the type of driver needed. The driver selection depends on (1) pump type — for instance, use a reciprocating driver for a piston or plunger pump; (2) power source to be used — for instance, use electric motors when electricity from the grid is available and cost-effective, and use a diesel or gasoline in places where such fuels are available but electricity is not; (3) characteristic curves — the characteristic curves of the motor, such as the torque curve and the power curve must match those of the pump; (4) system needs — for instance, if the pipeline system requires different flow rates at different times, a variable speed driver that meets the system needs must be selected; and (5) other practical considerations.

9.6.12 CONTROL SYSTEM

Selection of the control system for the pump and the prime mover depends on many factors such as (1) the kind of pump and driver selected; (2) operational needs — for instance, need for a high degree of automation and remote control for booster pumps, and need for greatest safety protection for pumps that handle hazardous liquids or gas; (3) characteristics of each piece of equipment and instruments used for control — it is not possible to make a good selection without a good understanding of the functional properties of the equipment or instrumentation.

9.6.13 COST

Needless to say, cost is also a highly important factor in the selection of pumps, drivers, and their affiliated equipment. However, one should not select the lowest cost equipment unless the equipment can meet all the needs of the project, is of unquestionable quality, and is expected to have a service life and maintenance record essentially as good as those more costly machines that can perform the same functions. Vendor reputation is also highly important since one must rely on the vendor to service the machines and replace worn parts over many years.

More information on pump selection can be found in References 2 and 4.

9.7 COMPRESSORS, BLOWERS, AND FANS

For pumping gases instead of liquids, pumps are generally referred to as **compressors, blowers,** or **fans**. It is common practice to call a gas pump a **compressor** when the pressure rise generated by the pump exceeds approximately 50 psi (3.45 bar), and to call it a **blower** when the pressure is in the range 5 to 50 psi (0.345 to 3.45 bar). In contrast, fans and other axial-flow gas pumps can only generate very low

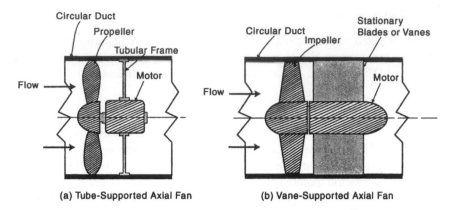

(a) Tube-Supported Axial Fan   (b) Vane-Supported Axial Fan

FIGURE 9.17 Two types of mountings of axial-flow fan.

pressure (less than 5 psi or 0.345 bar). Figure 9.17 shows two types of mounting of axial-flow fans in pipes or ducts. Because PD pumps can generate very high pressures, most compressors are of the PD type, whereas most blowers are of the centrifugal type. They all can be placed in series to increase pressure, or in parallel to increase discharge.

Most of the discussions of centrifugal, axial-flow, and PD pumps in the previous sections are applicable not only to pumps that handle liquids but also to those handling gases. The only main difference is that when treating compressors and blowers, even relatively small rise in gas pressure due to pumping action, such as a few psi, is often sufficiently high to cause significant change in the gas density. Therefore, the flow must be treated as compressible instead of incompressible. This is illustrated as follows.

Consider a centrifugal compressor or blower having a general configuration similar to that of a centrifugal pump for liquid, as shown in Figure 9.1. Under design conditions, the flow entering the impeller is mostly radial, and the power imparted to the flow by the impeller is, from Equation 9.6, equal to $\omega \rho Q r_o V_r$. This means that the work done on unit mass of the gas by the compressor is $\omega r_o V_r$. This work done on unit mass of the gas going through the compressor causes the enthalpy of the gas to rise by the amount $c_p(T_2 - T_1)$. Therefore,

$$c_p(T_2 - T_1) = \omega r_o V_r \tag{9.45}$$

Assuming that the compression process is isentropic (i.e., adiabatic and frictionless), we have from thermal dynamics

$$\frac{T_2}{T_1} = \left(\frac{p_2}{p_1}\right)^{\frac{k-1}{k}} \tag{9.46}$$

Eliminating T_2 from the two above equations yields

$$\frac{p_2}{p_1} = \left(1 + \frac{\omega r_o V_t}{c_p T_1}\right)^{\frac{k}{k-1}}$$ (9.47)

Equation 9.47 can be used to estimate the pressure rise $p = p_2 - p_1$, or the compression ratio p_2/p_1, once the quantities ω, r_o, V_t, T_1, c_p, and k are known. After p_2 is determined, Equation 9.46 can be used to calculate the temperature of the gas after compression, T_2. The result from Equation 9.47 also can be used to determine the gas density change as follows:

$$\frac{\rho_2}{\rho_1} = \left(\frac{p_2}{p_1}\right)^{\frac{1}{k}}$$ (9.48)

Example 9.10 A centrifugal compressor compresses air at atmospheric pressure of 14.7 psia and at temperature of 60°F. The compressor has an angular velocity of 23,000 rpm, an impeller radius r_o equal to 3 inches, an impeller thickness b of 2 inches, and a tangential velocity V_t equal to 500 fps. The blade angle β at the tip of the compressor vanes is 120°. Find the compression ratio p_2/p_1, the discharge pressure p_2, the discharge temperature T_2, the air density ρ_2, and the mass flow rate of air pumped by the compressor.

[Solution] For this problem, $p_1 = 14.7$ psia $= 2117$ psfa, $T_1 = 60°F = 520°R$, $\omega = 23,000 \times 2\pi/60 = 2409$ rad/s, $b = 2/12 = 0.1667$ ft, $r_o = 3/12 = 0.25$ ft, $U = r_o \omega = 602$ fps, and $V_t = 500$ fps. Furthermore, from Chapter 3, for air $k = 1.4$ and $c_p = 0.240$ Btu/lbm/°R $= 6012$ ft-lb/slug/°R. Substituting these values into Equation 9.47 yields $p_2/p_1 = 1.38$, $p_2 = 1.38 \times 2117 = 2921$ psfa $= 20.3$ psia. Then, from Equation 9.46, $T_2/T_1 = 1.096$, and $T_2 = 570°R = 110°F$. This means a rise of temperature of 50°F. From Equation 9.48, the density ratio $\rho_2/\rho_1 = 1.259$. From the equation of state of perfect gas, $\rho_1 = p_1/(RT_1) = 2117/(1716 \times 520) = 0.00237$ slug/ft³. Therefore, $\rho_2 = 0.00298$ slug/ft³. From Equation 9.10, $V_m = (V_t - U)/\cot \beta = (500 - 602)/\cot 120° = 177$ fps. Finally, the mass flow rate is calculated from $2\pi r_o b V_m \rho_2 = 0.138$ slug/s $= 4.45$ lbm/s $= 267$ lbm/min.

From the above example, it can be seen that even at very high speed, the pressure rise developed by a single-stage centrifugal compressor is rather small as compared to that developed by centrifugal pumps for handling liquid. This is due to the fact that the density of gas at or near atmospheric pressure is usually much lower than that of the liquid. Since pressure rise due to fluid motion is generally proportional to fluid density, much smaller pressure rise results from pumping gas than liquid. However, the low density of gas enables the compressor to be rotated at a much higher speed than for liquid pumps. From fluid mechanics, the pressure rise of turbulent flow is proportional to the second power of speed. This means doubling speed can increase the pressure by four times. For this reason, centrifugal compressors and blowers operate at much higher speed than centrifugal pumps do. Still, there are limitations as to how high the speed should go. One such limitation is that the speed should not produce any transonic or supersonic velocity because such flow generates shock waves and large drag and headloss. Multiple impellers are usually used (placed in series on the same shaft) in order to produce large pressure changes. Positive displacement type compressors are more effective in generating high pressure.

9.8 TURBINES

9.8.1 INTRODUCTION

Turbines have the opposite function of liquid and gas pumps: they extract energy from rather than impart energy to the flow in a pipeline. The energy extracted from a turbine may be used either to do mechanical work or to generate electricity. Turbines are used in both liquid and gas flow through pipe. Theoretical analyses of pumps using fluid mechanics, with minor modifications, are generally applicable to turbines as well. The main difference is that the direction of flow inside a turbine is in the opposite direction of the flow in a pump of the same type. As such, torque is produced by the flow rather than imparted to the flow, and energy is generated rather than consumed by a turbine. By definition, turbines are rotating machines. The reciprocating types are called **engines**, such as gasoline engines and diesel engines.

9.8.2 TYPES OF TURBINES

9.8.2.1 Hydraulic Turbines

Hydraulic turbines are those used in hydropower projects to convert the potential energy of water to electrical power. The main components of a typical hydroelectric system are shown in Figure 9.18. They consist of (1) a dam to back up the water and create a reservoir to store energy and create head, (2) a penstock, which is the large pipe that conveys the water from the reservoir to the turbine, (3) a gate to control the flow through the penstock, (4) a turbine that drives a generator, (5) a draft tube, which efficiently drains the water from the turbine to the river downstream, and (6) the tailing pond, which is a deep pool of water for dissipating the kinetic energy of the discharge from the draft tube. Hydraulic turbines are often the largest turbines made, some over 20 ft in diameter (see Figures 9.19 and 9.20).

9.8.2.1.1 Centrifugal Type

The centrifugal type turbine is called the **Francis turbine**. It is based on the same principle of centrifugal pumps, and the turbine has similar construction as that of a centrifugal type. The flow direction of a Francis turbine is opposite to that of a centrifugal pump. Water enters the turbine through the volute in order to create the tangential component of velocity, which is needed to turn the turbine runners. For turbines, the rotating part is called a **runner** rather than an **impeller**. A special feature of the turbine different from most centrifugal pumps is that the turbine has

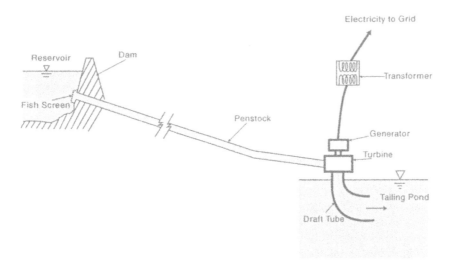

FIGURE 9.18 Typical hydroelectric system.

FIGURE 9.19 The propeller of a large Kaplan turbine being hoisted during installation at the Truman Dam power plant in Missouri. (Courtesy of the U.S. Army Corps of Engineers.)

a set of gates mounted on the casing around the runner; they are called **wicket gates**. The angle or incline of the wicket gates is adjusted to produce the most desirable inflow angle to the runner, thereby producing maximum power and best efficiency. Special governors are used to control the wicket gates automatically. The wicket gates system of Francis turbines can be seen in Figure 9.20. Most Francis turbines are medium-head turbines for heads ranging from 80 to 600 ft; they have efficiencies in the 90 to 95% range. Due to similarity between centrifugal pumps and Francis

FIGURE 9.20 Historic picture of the large Francis turbines being installed at the Grand Coulee Dam in Washington State, U.S.A., showing the wicket gates of the turbine runner. (Courtesy of U.S. Bureau of Reclamation.)

turbines, Equations 9.10 to 9.14, which are derived for centrifugal pumps, also hold for Francis turbines.

9.8.2.1.2 Propeller Type

The propeller turbine, also called the **Kaplan turbine**, is the counterpart of the propeller pump. It is an axial-flow machine in that the flow going through the turbine runner is mainly in the axial direction. A set of wicket gates are used to produce a desirable flow pattern entering the runner. The runner consists of a shaft, a hub, and several wide blades (propellers) attached to the hub. Figure 9.19 is an example. The blade angle is adjustable to attain the best efficiency for specific wicket-gate openings and heads. Kaplan turbines are low-head turbines, for heads less than 100 ft (30 m). They have efficiencies in the range similar to that of Francis turbines. Note that propeller-type windmills — those used most often nowadays for generating electricity — are similar in principle to Kaplan turbines except that their blades are longer (more slender), and not confined inside a casing. Both Kaplan turbines and Francis turbines are called **reaction** turbines because conversion of energy in both cases involves pressure changes across the turbine.

9.8.2.1.3 Impulse Type

In contrast to reaction turbines, impulse turbines (also called **Pelton wheels**) do not derive their energy from the pressure changes across the turbine. Instead, the energy is derived from the kinetic energy of a high-speed jet issued from the end of the penstock. The impulse turbine consists of a large rotating wheel with many buckets attached to the rim of the wheel. The water jet impinges on one bucket at a time as the wheel turns, to generate a more-or-less continuous torque on the wheel as it rotates. Each bucket consists of two symmetric cups to slit the jet evenly on the two cups, with the split jets leaving the bucket in opposite directions to avoid unbalanced thrust on the shaft. The buckets are designed with good shape and curvatures in order to minimize energy loss and maximize energy transfer.

From energy considerations, the power of the jet is

$$P_j = \rho Q \frac{V_j^2}{2} \tag{9.49}$$

where V_j is the jet velocity.

From the momentum equation, the mean thrust of the jet on the moving bucket is

$$F_b = \rho Q (V_j - U)(1 - \cos\theta) \tag{9.50}$$

where U is peripheral velocity of the wheel which is the same as the bucket velocity, and θ is the angle of turning of the jet.

From Equation 9.50, the power of the jet transferred to the wheel is

$$P_{th} = UF_b = \rho Q U (V_j - U)(1 - \cos\theta) \tag{9.51}$$

where P_{th} is the theoretical power transferred from the jet to the shaft.

From Equation 9.51, it can be seen that the maximum theoretical power is developed when $\theta = 180°$. In practice, when θ approaches $180°$, turbulence is generated, which decreases the efficiency of energy transfer. The optimum angle θ found in practice is between $173°$ to $176°$. Taking the derivative of Equation 9.51 with respect to U and setting the result equal to zero yields $V_j = 2U$, which yields the maximum theoretical power. In reality, due to certain energy losses unaccounted for in Equation 9.51, maximum power is produced when U is slightly less than $0.5V_j$. To achieve maximum power and best efficiency, each Pelton wheel is controlled by a governor to maintain a peripheral speed, U, slightly below $0.5 V_j$.

Pelton wheels are used only for high-head applications, when the head is at least about 600 ft. With such high heads, it is not suitable to use reaction turbines due to the high pressure that would otherwise exist inside the turbine casing. In contrast, the pressure inside a Pelton wheel is always atmospheric. The efficiency of a Pelton wheel is usually less than that for Francis turbines and Kaplan turbines. It is usually in the range of 80 to 90%, with those having the best efficiency in the highest head range. They are not as efficient as the reaction turbines because (1) there is some headloss in the needle valve to convert the total head of the flow at the end of the penstock to a jet; (2) there is considerable loss in transferring the energy of the jet to the wheel; and (3) the static head between the turbine and the water surface in the tailing pond is not utilized.

Example 9.11 A Pelton wheel of 3-m diameter is used to generate electricity in a hydroelectric project. The water surface in the reservoir is 670 m above the elevation of the jet that drives the turbine. The jet has a diameter of 18 cm. The penstock is a 1-m diameter steel pipe 6 km long. The buckets of the wheel have a turning angle of 175°. Determine: (a) the water velocities both in the penstock and the jet, (b) the optimum angular velocity of the Pelton wheel, (c) the torque and power developed on the wheel, (d) the theoretical efficiency of the wheel, and (e) the electrical power generated if the actual efficiency of the turbine is 88% and the efficiency of the generator is 95%.

[Solution] (a) Using the one-dimensional energy equation between point 1 (at the surface of the reservoir) and point 2 (at the jet) yields

$$670 = \frac{V_j^2}{2g} + h_L = \frac{V^2}{2g}\left(\frac{A}{A_j}\right)^2 + h_L = 952.6\frac{V^2}{2g} + h_L \tag{a}$$

Assume that the headloss due to local effects are negligible along the penstock. From the Darcy-Weisbach formula,

$$h_L = f\frac{L}{D}\frac{V^2}{2g} = 6000 f\frac{V^2}{2g} \tag{b}$$

Combining the two above equations and using $g = 9.81$ m/s² yields

$$V = \sqrt{\frac{13,145}{952.6 + 6000 f}} \tag{c}$$

Using the Moody diagram, Equation c yields, through iteration, $f = 0.0116$ and $V = 3.59$ m/s. Thus, the jet velocity becomes $V_j = V \times (A/A_j) = 3.59 \times 30.9 = 111$ m/s. (b) As discussed before, the theoretical optimum peripheral velocity of the Pelton wheel is $U = 0.5V_j = 0.5 \times 111 = 55.5$ m/s. The angular velocity of the wheel is $\omega = U/r_o = 55.5/1.5 = 37$ rad/s, and $N = \omega/(2\pi) = 5.89$ rps = 353 rpm. (c) $A = \pi D^2/4 = 0.7854$ m², $Q = VA = 3.59 \times 0.7854 = 2.82$ cms, $\rho = 1000$ kg/m², and $\theta = 175°$. Substituting these values into Equation 9.51 yields $P_{th} = 17,340,000$ W = 17.34 MW, which is the theoretical power imparted by the jet to the turbine. The torque is $T = P/\omega = 17,340,000/37 = 469,000$ N-m. (d) From Equation 9.49, the power of the jet is $P_j = 17,370,000$ W = 17.37 MW. Therefore, the theoretical efficiency of this Pelton wheel is $\eta_{th} = P_{th}/P_j = 17.34/17.37 = 0.998 = 99.8\%$. (e) Because the actual efficiency of the Pelton wheel is only 88%, and the actual efficiency of the generator is 95%, the combined efficiency of the turbine-generator system is $\eta = 0.88 \times 0.95 = 0.836 = 83.6\%$. Therefore, the actual electric power generated by this system is $P = \eta P_j = 0.836 \times 17.37 = 14.5$ MW.

9.8.2.2 Steam and Gas Turbines

Steam and gas turbines are similar in design and principle except for the fact that steam turbines are power by steam generated from heating water by either combustion or nuclear reaction, whereas gas turbines are powered by the exhaust gas generated from burning a fuel such as petroleum or natural gas. They are either centrifugal type or axial-flow type turbines. The centrifugal type generates much higher pressure than the axial-flow type. Whether to use the centrifugal type or the axial-flow type depends mainly on the pressure requirements of the pipeline system.

9.9 DIMENSIONLESS PARAMETERS

By using dimensional analysis in fluid mechanics, it can be proved that the following relationships between certain dimensionless pump parameters hold:

$$C_H = \Psi(C_Q), \quad C_P = \Psi(C_Q), \quad \eta = \Psi(C_Q) \tag{9.52}$$

where

$$C_Q = \frac{Q}{ND^3}, \quad C_H = \frac{gH}{N^2D^2}, \quad C_P = \frac{P}{\rho N^3 D^5} \tag{9.53}$$

In the above equations, C_H, C_Q, C_P, and η are the dimensionless head, dimensional discharge, dimensionless power, and efficiency, respectively, of the pump. As before, H, Q, P, N, ρ, and g are the pump head, discharge, power, angular speed, fluid density, and gravitational acceleration, respectively. D here denotes pump diameter instead of pipe diameter. Equation 9.52 shows that the three dimensionless parameters C_H, C_P, and η are all functions of the dimensionless discharge C_Q. By plotting the values of these three dimensionless parameters as a function of C_Q, the dimensionless pump characteristic curves are obtained. For a given pump, the dimensionless pump characteristic curves are similar in shape to the dimensional pump characteristic curves, except that the units are different — the dimensionless curves have no units.

The dimensionless pump parameters and the dimensionless pump curves are quantities and relationships that hold general truth. They are valid for different values of H, Q, N, D, and so forth, as long as the same pump is referred to. Furthermore, they are valid for a family of pumps designed to have the same shape and geometry but different sizes. This is illustrated in the following example.

Example 9.12 A centrifugal pump of 2-ft diameter running at 880 rpm develops a head of 100 ft when the discharge is 5 cfs. (a) If the pump speed is reduced to half (i.e., 440 rpm), what is the expected discharge and head? (b) If the speed remains the same but the diameter of the pump is doubled, what are the discharge and head?

[Solution] (a) Using subscripts 1 and 2 to denote the pump operation conditions before and after, respectively, the speed reduction, we have

$$\left(C_Q\right)_1 = \left(C_Q\right)_2 \quad \text{or} \quad \frac{Q_1}{N_1 D_1^3} = \frac{Q_2}{N_2 D_2^3} \tag{a}$$

For this problem, $D_1 = D_2$, and $N_1 = 2N_2$. Therefore, from Equation a, $Q_2/Q_1 = N_2/N_1 = 0.5$, and $Q_2 = 0.5Q_1 = 0.5 \times 5 = 2.5$ cfs.

From the dimensionless pump head, we have

$$\left(C_H\right)_1 = \left(C_H\right)_2 \quad \text{or} \quad \frac{gH_1}{N_1^2 D_1^2} = \frac{gH_2}{N_2^2 D_2^2} \tag{b}$$

Because $D_1 = D_2$ and $N_1 = N_2$, Equation b yields $H_2 = H_1/4 = 100/4 = 25$ ft. (b) If the pump speed remains the same but the pump diameter is doubled, Equation a yields $Q_2 = 8Q_1 = 8 \times 5 = 40$ cfs. Likewise, Equation b yields $H_2 = 4H_1 = 4 \times 100 = 400$ ft.

Note that centrifugal pump manufacturers often provide a family of pump characteristic curves for the same type of pump but of different impeller diameters. The smaller impellers are fabricated by shaving down the tip of the largest impeller, using the same casing for different impeller diameters. Such pump families do not consist of geometrically similar pumps because the spacing between the pumps and the casing is not proportionally reduced with a decrease in impeller diameter. In fact, the spacing is increased when the impeller diameter is reduced. Therefore, the dimensionless parameters discussed above should not be used to predict the characteristics of one size from another size in such a family. To be correct, the characteristic curves of each size must be individually obtained through tests.

Another useful pump parameter is the specific speed, defined as follows:

$$N_S = \frac{NQ^{1/2}}{H^{3/4}} \tag{9.54}$$

This is a quantity closely related to the design of pumps. Large specific speeds represent relatively large discharge, high speed, and low head. Axial-flow pumps fit this category. In contrast, low specific speeds represent relatively small discharge, low speed, and high head. Radial-flow pumps (centrifugal pumps) fit this category. The values of H and Q used in Equation 9.54 are those corresponding to the best efficiency of the pump.

Note that N_S as given by Equation 9.54 is not dimensionless. To be truly dimensionless, a quantity $g^{3/4}$ must be included in the denominator along with the existing quantity $H^{3/4}$. However, it is customary in North American pump practice to define specific speed without $g^{3/4}$, and to express N in rpm, Q in gpm, and H in ft. Using this convention, the ranges of specific speed for radial-flow pumps (centrifugal pumps), mixed-flow pumps, and axial-flow types (propeller pumps) are 500 to 1500, 1500 to 8000, and 8000 to 20,000, respectively.

For turbines, the North American practice defines specific speed as

$$N_S = \frac{NP^{1/2}}{H^{5/4}} \qquad (9.55)$$

Again, N_S as given by Equation 9.55 is not dimensionless. From this equation, the values of N_S for impulse turbines (Pelton wheels) are about 5; for radial-flow turbines (Francis turbines) they are in the range of 20 to 100; and for axial-flow turbines (Kaplan turbines) they are in the range of 100 to 200.

PROBLEMS

9.1 A pump causes water at 65°F to move at 6 fps through a pipe of 30-inch diameter. The pump head is 200 ft, and the pump efficiency is 85%. Determine the needed electrical power to the pump in watts, kW, MW, and hp. Solve this problem in two ways: (a) by using the ft-lb units to calculate until the power is calculated in hp, and (b) by converting all quantities in ft-lb units to SI at the beginning of the calculation. The two ways should produce the same results.

9.2 Solve the same problem given in Example 9.1 if the pipe diameter is reduced to 2 inches. Compare the results with those given in the example.

9.3 A centrifugal pump has a radius r_o of 3 ft, a thickness b of 1 ft, a blade angle β of 130°, and a rotational speed of 1770 rpm. The fluid is water at 70°F. Find the theoretical pump characteristic curves for torque, head, and input power. Plot the results.

9.4 Solve the problem in Example 9.3 if a different pump is used with $H = 300 - 200Q^2$. Examine the adequacy of this new pump for this system.

9.5 Solve Example 9.4 with the pipe diameter reduced to 10 inches.

9.6 When two pumps are combined in series or parallel, what is the efficiency of the combined pump system if the individual pumps have efficiencies η_1 and η_2.

9.7 Plot the three $H \sim Q$ curves given in Example 9.5 separately and in combination (both in series and parallel combinations). Use Equations a and b in the example to plot the combined pump curves.

9.8 Three identical pumps are to be used both in series and parallel, in two separate applications. The pump characteristic of each individual pump is $H = a - bQ^n$, where a, b, and n are positive constants. Find the equations for the combined $H \sim Q$ curves for the three pumps staged both in series and parallel. Express the equations in their simplest forms.

9.9 Water at 70°F is pumped through a steel pipe of 12-inch diameter and 100-mi length, against a static head rise of 500 ft. The required discharge through the pipe is 4.5 cfs. A piston pump is used with stroke length of 4 inches and reciprocating speed of 3000 rpm. Determine: (a) the average linear speed of the piston, (b) the cross-sectional area of the piston, (c) the pump head needed, (d) the average force on the piston during operation, and (e) the power input to the motor that drives the piston, assuming that the pump–motor combined efficiency is 90%.

9.10 From information available on the Internet, select a commercially available PD pump that can be used to supply the flow of 4.5 cfs through the system described in the previous problem. If one pump is not enough, use multiple pumps in series or parallel combination. Do the same by using commercially available centrifugal pumps.

9.11 For the same system described in Example 9.6, if the mixture flow rate is to be increased to 80 gpm without changing the piston speeds, how much should the individual stroke lengths of the two pumps be adjusted?

9.12 If the same EM pump in Example 9.7 uses fresh water instead of seawater in the experiment, what is the expected current consumed and the pressure and head generated? Assume that the conductivity of the fresh water is one-hundredth that of the seawater.

9.13 A centrifugal pump is running under shut-off condition with no flow going through the pump. The shut-off power is 20 hp, and there is 240 lb of water in the pump casing. What is the pump temperature rise in 60 s? If the pump speed is increased by 50%, what is the expected temperature rise during the same period?

9.14 A sump pump used to drain a building basement is observed to operate on 15-min cycles, being on for 5 min each time. The discharge through the sump pump is measured to be 0.1 cfs. Find the rate of leakage inflow to the sump and the volume (capacity) of the sump.

9.15 A Pelton wheel drives a generator at 600 rpm. The water jet diameter is 8 cm and the jet velocity is 120 m/s. For best efficiency, peripheral velocity of the wheel is 47% of the jet velocity. The jet deflection angle is 170°. Determine: (a) the diameter of the Pelton wheel, (b) the theoretical power, head, and torque developed by this Pelton wheel, (c) the real torque, head, and power, assuming the machine to be 88% efficient, and (d) the specific speed.

REFERENCES

1. Olson, R.M. and Wright, S.J., *Essentials of Engineering Fluid Mechanics*, 5th ed., Harper & Row, New York, 1990.
2. Karassik, I.J. et al., Eds., *Pump Handbook*, McGraw-Hill, New York, 1976.
3. Streeter, V.L., *Fluid Mechanics*, 5th ed., McGraw-Hill, New York, 1971.
4. Walker, R., *Pump Selection: A Consulting Engineer's Manual*, Ann Arbor Science Publishers, Ann Arbor, MI, 1972.

10 Instrumentation and Pigging

10.1 FLOWMETERS

10.1.1 TYPES OF FLOWMETERS

Flowmeter usually refers to the instrument for measuring the flow rate through a pipe, either the volumetric flow rate (discharge Q, in units such as cfs and m^3/s), or the mass or weight flow rate as in the units of lb/s, or kg/s. Flowmeters play a central role in pipeline instrumentation and monitoring. There are many types of flowmeters. Some fluid mechanics books, such as Reference 1, have whole chapters covering the principles and practice of flowmeters. Special books on flow measurements, such as Reference 2, are available. This chapter briefly describes the important types of flowmeters commonly used in measuring pipe flow.

10.1.1.1 Venturi

Discussion of the Venturi meter (simply called the *venturi*) can be found in most texts in fluid mechanics. The meter, as shown in Figure 10.1, is a short segment of pipe consisting of a convergent cone followed by a divergent cone. To minimize energy loss, the divergent cone is much longer (having a much smaller cone angle) than the convergent part. The throat of the venturi is the narrow part connecting the convergent section to the divergent section. The meter is tapped both at the throat and at the upstream straight section for measuring the pressure difference between these two taps. This pressure difference can be used to determine the discharge Q as will be discussed next.

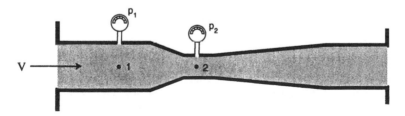

FIGURE 10.1 Venturi meter.

289

The Bernoulli equation in fluid mechanics states that when a flow is steady and incompressible, and when frictional loss is negligible, the total energy of a fluid particle of unit mass is constant along a streamline, namely,

$$\frac{V^2}{2} + \frac{p}{\rho} + gz = \text{constant (along a streamline)} \qquad (10.1)$$

where V, p, ρ, and z are the velocity, pressure, density, and elevation, respectively, of the fluid particle at any given location along a streamline, and g is the gravitational acceleration. From this equation, it can be proved that the discharge of incompressible flow through a venturi is

$$Q = C_v A_2 \sqrt{\frac{2[(p_1 - p_2)/\rho + g(z_1 - z_2)]}{1 - (A_2/A_1)^2}} \qquad (10.2)$$

where subscripts 1 and 2 represent the upstream tap and the downstream tap (the throat), respectively. Note that a factor C_v is included in Equation 10.2 to take into account the energy loss not included in the Bernoulli equation. For turbulent flow through a venturi with smooth interior, C_v is approximately 0.98. When the venturi is horizontal, the quantity $(z_1 - z_2)$ is zero, and Equation 10.2 reduces to

$$Q = C_v A_2 \sqrt{\frac{2[(p_1 - p_2)/\rho}{1 - (A_2/A_1)^2}} \qquad (10.3)$$

Equation 10.3 is the counterpart of Equation 3.53 for compressible flow treated in Chapter 3. For gas flow through a venturi at relatively high speed, the density of the gas changes significantly from point 1 to point 2 of the venturi, and the compressible flow solution given by Equations 3.53 and 3.54 should be used to determine the mass flow rate, which is $\rho_2 A_2 V_2$.

For a given venturi in a given pipe, the values of C_v, A_1, A_2, z_1, z_2, g, and ρ are all known and constant, and so Equation 10.2 (or Equation 10.3) is reduced to

$$Q = C\sqrt{\Delta h} \quad \text{where} \quad \Delta h = \frac{(p_1 - p_2)}{\rho g} + (z_1 - z_2) \qquad (10.4)$$

where C is a constant.

The above equation shows that the discharge Q through a venturi is proportional to the square root of the piezometric head difference, Δh. This relationship is true for most flowmeters operating in the turbulent-flow range.

The venturi is widely used for both liquid and gas. Its advantages include accuracy in measurements and low headloss. Its disadvantages include high cost, and the possibility of causing cavitation at the throat if insufficient pressure exists in the pipeline at the location of the meter. Furthermore, as is the case with most flowmeters, a venturi should not be used for flows that contain solid particles — the types of flows discussed in Chapters 5, 6, and 7. Otherwise, it could result in flow blockage or serious wear (erosion) to the meter. Also, Venturi meters cannot be used in pipelines that must pass pigs (scrapers) (see Section 10.3).

Example 10.1 Water is flowing upward through a vertical venturi of 4×2-inch size (i.e., the throat diameter is 2 inches and the end diameter is 4 inches). The elevation change between the two pressure taps is $z_2 - z_1 = 1.5$ ft, the pressure difference read by the meter is $p_1 - p_2 = 4.2$ psi, and the velocity coefficient is 0.98. Find the discharge through the venturi.

[Solution] For this problem, $D_1 = 4$ inches $= 0.3333$ ft, $D_2 = 2$ inches $= 0.1667$ ft, $A_2 = 0.02182$ ft², $A_2/A_1 = (D_2/D_1)^2 = 0.25$, $\rho = 1.94$ slug/ft³, $z_1 - z_2 = -1.5$ ft, $p_1 - p_2 = 4.2$ psi $= 605$ psf, and $C_v = 0.98$. From Equation 10.2, $Q = 0.530$ cfs.

10.1.1.2 Orifice

An **orifice meter** is a disk with a circular central opening inserted across a pipe to measure the flow rate. As shown in Figure 10.2, the opening of the orifice may have a sharp crest, a round-entrance crest, or a nozzle-shaped crest. These different crest shapes affect the degree of contraction of the jet through the orifice, characterized by the contraction coefficient $C_c = A_j/A_o$, where A_j is the cross-sectional area of the jet at the **vena contracta**, and A_o is the area of the orifice opening. Using the same approach used for the derivation of Equation 10.2, the discharge equation of an orifice can be obtained as follows:

$$Q = C_c C_v A_o \sqrt{\frac{2[(p_1 - p_2)/\rho + g(z_1 - z_2)]}{1 - C_c^2 (A_o / A)^2}} \tag{10.5}$$

The value of C_c for both the round-entrance and nozzle types is 1.0, which means no contraction of the jet. For the sharp-crested type at high Reynolds number (i.e.,

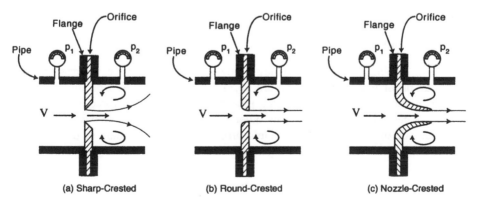

(a) Sharp-Crested   (b) Round-Crested   (c) Nozzle-Crested

FIGURE 10.2 Orifice meter of different types.

for turbulent flow), C_c can be determined approximately from the following empirical formula:

$$C_c = 0.595 + 0.29 \left(\frac{A_o}{A} \right)^{5/2}$$ (10.6)

From the above equation, the value of C_c varies from about 0.595 when A_o/A is zero, to 0.714 when A_o/A is 0.7. The equation should not be used for A_o/A larger than 0.7. When used within the allowable range, and when the Reynolds number of the flow in the pipe is greater than 10^5, the equation is expected to yield results within 5% of experimental values. As in the case of the venturi, the quantity C_v in Equation 10.5 is the velocity coefficient, which, for practical purposes, can be treated as a constant equal to 0.98. The product $C_c C_v$ is the discharge coefficient C_d. For a given orifice in a given pipe, Equation 10.5 reduces to the same form as Equation 10.4.

Orifice meters are widely used for both liquid and gas. They have the advantage of being less costly than the venturi for the same pipe, and they can easily be fabricated in a machine shop. However, they generate greater disturbance to the flow than does the venturi, and hence have higher headloss. They are not to be used for flows that contain solids, or in pipelines that must pass pigs or capsules.

Example 10.2 Assume the same conditions as in the previous example, except that a sharp-crested orifice of 2-inch diameter is used. Calculate the discharge and compare the result with that for the venturi in the previous example.

[Solution] The only significant difference between the equation for the venturi and the equation for the orifice is the existence of the contraction coefficient C_c for the orifice. For this orifice meter, $A_o = A_2 = 0.02182$ ft^2, $A_o/A = 0.25$, $p_1 - p_2 = 605$ psf, $z_1 - z_2 = -1.5$ ft, $\rho = 1.94$ slug/ft^3, $C_v = 0.98$. From Equation 10.6, $C_c = 0.604$. Then, from Equation 10.5, $Q = 0.310$ cfs. This shows that for the same size opening and under the same pressure, the discharge through the orifice is considerably less than that through the venturi. This is due to a much larger headloss generated by the orifice than the venturi.

10.1.1.3 Elbow Flowmeter

An elbow flowmeter uses an existing elbow or bend of a pipe to measure the discharge through the pipe (see Figure 10.3). The meter is based on the principle that whenever a flow passes through a pipe bend, a centrifugal force is generated. This force causes the pressure on the outer side of the bend to rise beyond that on the inner side of the bend, by an amount proportional to the square of the velocity or discharge. More specifically, the pressure rise is

$$\Delta p = \rho V^2 \frac{D}{R_b} = \rho \frac{Q^2}{A^2} \frac{D}{R_b}$$ (10.7)

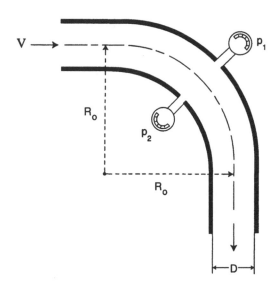

FIGURE 10.3 Elbow flowmeter.

where Δp is the pressure rise; ρ is the density of the fluid; V is the mean velocity across the bend; A is the cross-sectional area of the bending pipe (same as in the straight part of the pipe); Q is the discharge through the bend ($Q = VA$); D is the pipe diameter for the bend part, which is the same as for the straight part; and R_b is the mean radius of the bend. Equation 10.6 holds for the case that R_b is much larger than D. Note that Equation 10.7 is of the form $\Delta p = CQ^2$, where C is a constant for a given pipe bend or elbow flowmeter. This shows that, as is the case with the venturi and orifice meters, the pressure rise of an elbow flowmeter is proportional to the square of the discharge Q.

Elbow flowmeters have the advantage of low cost when an elbow already exists in a convenient location for flow measurement. In such a case, all one needs to do is to provide two taps, one on the outer side of the bend and one on the inner side, and to connect the taps to a device that measures the differential pressure, such as a manometer or a differential pressure transducer. A major shortcoming of elbow flowmeters is that the pressure difference generated, Δp, is rather small at low velocity. Therefore, it is not accurate for flow measurement when the discharge is small.

Example 10.3 An elbow flowmeter is used to measure the discharge of water through a 12-inch-diameter pipe. The radius of the elbow is 2 ft. If the differential pressure transducer connected to the elbow measures a pressure difference of 0.1 psi, what is the discharge and velocity through the pipe? Can the flow be measured accurately by this elbow meter if the discharge in the pipe is as small as 0.1 cfs?

[Solution] For this elbow, $D = 12$ inches $= 1$ ft, $A = \pi D^2/4 = 0.7854$ ft, $R_b = 2$ ft, $\rho = 1.94$ slug/ft^3, and $\Delta p = 0.1$psi $= 14.4$ psf. Substituting these values into Equation 10.7 and solving for Q yields $Q = 3.03$ cfs. Therefore, $V = Q/A$ = 3.85 fps.

When the discharge decreases to 0.1 cfs, Equation 10.6 yields $\Delta p = 0.0157$ psf = 0.000109 psi. It is highly doubtful that such a low differential pressure can be accurately measured.

10.1.1.4 Rotating Flowmeters

Many types of rotating flowmeters are available commercially. They include propeller type, turbine type, vane type, gear type, and cup type. They are unsuitable for flows that contain solids. They are used widely in both liquid and gas pipelines in branches (distribution lines) where pigs do not go through. Water meters used for rate charges to the customers most often use rotating flowmeters. Most of them are the positive-displacement type, with the volume of flow passing through the meter being directly proportional to the number of turns of the meter.

10.1.1.5 Vibratory Flowmeters

Vibratory flowmeters have a loose part exposed to the flow, vibrating at a frequency proportional to the fluid velocity and the discharge. A common type is a wobbling disk. The disk wobbles at a frequency proportional to the volumetric discharge of the liquid through the pipe. Another type is based on the vibration of a circular cylinder held perpendicular to the flow. The vibration is caused by vortex shedding, which is a phenomenon discussed in elementary fluid mechanics texts. As in the case of the wobbling type, the frequency of vibration of the vortex shedding type is proportional to the fluid velocity and hence the discharge. Vibratory flowmeters can be used for both liquid and gas that do not contain solids and where pigs or capsules do not pass through.

10.1.1.6 Rotameters

Rotameters use a float inside a transparent vertical tube to determine the flow rate (see Figure 10.4). The meter must be mounted in a vertical part of the pipe with the flow going upwards. The transparent tube is tapered, having an expanded diameter with height. As the flow rate increases, the float moves to a higher location in the meter tube. The meter is calibrated and graduated to determine the flow rate (discharge Q) from the location of the float. Floats of different densities are needed for use with the same rotameter in different ranges of discharge. In such a case, calibration must be done for different floats. Rotameters have many limitations; for example, they can only be used in vertical position with upward flow and cannot be used in fluids that contain solids, and so on. However, the rotameter has a special advantage that most other flowmeters do not have: it can accurately determine flows of low velocity. It is only used in small pipes with relatively low flow, and in places where a section of pipe is vertical with upward flow.

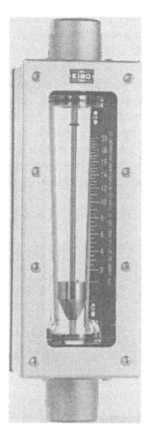

FIGURE 10.4 Rotameter. (Courtesy of King Instrument Company.)

10.1.1.7 Magnetic Flowmeter

The magnetic flowmeter has a bore normally of the same diameter as the inner diameter of the pipe. The meter is connected to a straight section of the pipeline by flanges. A conducting fluid such as water is forced to flow through the flowmeter either by gravity or by a pump. The electromagnet of the flowmeter generates a strong magnetic field having a flux density B across the flow. From Faraday's law of electromagnetic induction, the conducting fluid cutting across the magnetic field generates a voltage E_o across the bore in a direction perpendicular to the magnetic field. By placing two electrodes on opposite sides of the circumference of the meter bore, this voltage E_o can be measured with a high-impedance voltmeter. Application of Faraday's law yields

$$E_o = C_1 BVD_b = C_2 BQ \quad \text{or} \quad Q = C_3 E_o / B \qquad (10.8)$$

where C_1, C_2, and C_3 are constants that can be found from calibration; V is the mean velocity across the pipe; Q is the discharge; and D_b is the diameter of the magnetic flowmeter (namely, the bore size). From Equation 10.8, the discharge measured by a magnetic flowmeter is directly proportional to the voltage measured, E_o, and inversely proportional to the magnetic flux density B, which in turn is proportional to the intensity of the magnetic field.

Equation 10.8 is independent of the pressure, temperature, density, viscosity, and conductivity of the fluid flowing through the meter. As long as the fluid has a minimum conductivity close to that of tap water, the meter reads the same voltage for a given discharge. However, because the meter requires a minimum fluid conductivity, many low-conductivity liquids and gases, including oil and natural gas, cannot use this type of meter. Unless such liquids and gases are seeded with ions (i.e., ionized) immediately upstream of the flowmeter, their flow rates cannot be determined by conventional magnetic flowmeters. Figure 10.5 shows a commercially available magnetic flowmeter.

FIGURE 10.5 Magnetic flowmeter. (Courtesy of Krohne, Inc.)

Even though magnetic flowmeters are more expensive than most other types of flowmeters, they have unique properties unmatched by other types, including (1) high accuracy (within 1% error), (2) insensitivity of the meter reading to fluid property changes (e.g., the readings are unaffected by the change of fluid pressure, temperature, density, viscosity, and conductivity of the fluid), (3) instantaneous reading that enables the meter to measure both steady and unsteady flows, (4) can be easily connected to modern computer-based data acquisition systems because the output is a voltage signal that is linearly proportional to discharge, (5) can have the same diameter (bore) as the pipe inner diameter, thereby eliminating any disturbance to the flow, creating no extra headloss (local headloss coefficient equals zero), and allowing free passage of solids, pigs, and capsules, and (6) minimum wear by solids contained in the flow. Due to these advantages, magnetic flowmeters are currently widely used in various industries for measuring the discharge of flows in pipes of various sizes.

10.1.1.8 Acoustic Flowmeter

Most acoustic flowmeters use high-frequency sound (of frequencies higher than 20 kHz) to measure discharge, and hence they are also called ultrasonic flowmeters. There are two general types of such flowmeters: the transit-time type and the Doppler type.

The transit-time flowmeter uses two probes, one upstream and the other downstream, mounted diagonally across the pipe as shown in Figure 10.6. Each of the two probes contains both a transmitter that transmits (emits) the high-frequency sound, and a receiver that receives (senses) the sound. Both sounds (i.e., ultrasonic pulses) are emitted at the same time but received by the two probes at different times. The downstream probe receives the sound sooner because sound waves travel faster downstream than upstream. If t_{12} is the travel time of the wave from the upstream probe to the downstream probe, t_{21} is the travel time of the wave from the downstream probe to the upstream probe, V is the mean flow velocity in the pipe, C is the celerity of sound (i.e., the speed of sound in stationary fluid), L is the

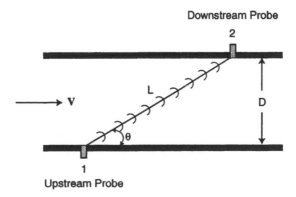

FIGURE 10.6 Transit-time acoustic flowmeter principle.

straight-line distance between the two probes, and θ is the angle between the pipe wall and the line connecting the two probes, it can be proved that

$$t_{12} = \frac{L}{C + V\cos\theta} \quad \text{and} \quad t_{21} = \frac{L}{C - V\cos\theta} \tag{10.9}$$

By assuming that C is much larger than V, the above equation yields

$$V = \frac{C^2 \Delta t \tan\theta}{2D} \quad \text{where} \quad \Delta t = t_{21} - t_{12} \tag{10.10}$$

This shows that for a given pair of probes mounted on fixed locations of a given pipe, the velocity V and the discharge Q are both linearly proportional to the measured time difference Δt. The celerity C of the fluid can be determined from Equation 2.78.

The Doppler type acoustic flowmeter also uses ultrasonic pulses. However, it relies on solid particles and/or air bubbles entrained in the liquid flow to reflect the waves back to the receiver that is located either separately across the pipe or in the same probe. What the meter measures is the average velocity of the entrained particles, which is assumed to be the same as the mean velocity of the fluid across the pipe. The flowmeter is based on the Doppler effect, which says that when a high-frequency wave emitted from a source hits a solid particle moving away from the source, the reflected wave will have a frequency different from the frequency of the emitted wave. The difference, i.e., the frequency shift, is proportional to velocity of the particle away from the source. From the Doppler effect, the mean velocity across a pipe and the discharge can be determined by measuring the Doppler shift electronically, and by relating frequency shift Δf to the discharge Q through theory or calibration.

Note that the transmitters and the receivers of ultrasonic flowmeters of both types may either be flush-mounted through holes in the pipe wall, or clamped on the outside of the pipe wall. The clamp-on type is especially convenient and does not weaken the pipe wall. It is the most commonly used today. The transit-time type is used in cases where the liquid is clean (free from solid particles and bubbles). In contrast, the Doppler type is used in pipes entrained with many small particles of solids and/or air bubbles. For accurate measurement of discharge without calibration, the transit-time ultrasonic flowmeter may need several pairs of probes, so that the multiple paths through the pipe can be covered to yield a good cross-sectional average velocity for discharge determination. Modern ultrasonic flowmeters use sophisticated electronics and computer technology to process data, in such a manner that noise from various sources, such as echoes from the pipe wall, is filtered out.

Acoustic flowmeters have been used in a variety of applications for pipe sizes ranging from as small as 0.5 inch to as large as 20 ft. While most other flowmeters including electromagnetic flowmeters increase in cost when the size of pipe increases, the cost of ultrasonic flowmeters remains independent of pipe size. This makes the ultrasonic flowmeter the logical choice for large pipes, such as for penstocks and aqueducts. It is also the logical choice for certain liquids and gases that cannot be measured with a magnetic flowmeter due to insufficient electric conductivity. Generally, for cases in which a magnetic flowmeter is applicable and costs about the same as an acoustic flowmeter, the former is likely to be a better choice because it is more reliable and accurate.

10.1.2 CALIBRATION OF FLOWMETERS

For accurate flow measurements, all flowmeters must be calibrated prior to use. Calibration is usually done in the manufacturer's laboratory before the instrument is shipped to the customer. When the calibration of a new or used flowmeter is in doubt, recalibration may be needed. This is usually done either in the user's own laboratory if the user has such a facility, in a commercial laboratory, or in the hydraulic laboratory of a university. The U.S. National Institute of Standards and Technology (NIST) also offers this service. For a fee, the agency calibrates both gas and liquid flowmeters. Depending on their degree of sophistication and required accuracy, three alternate systems of flowmeter calibration can be used. They are discussed in the following sections.

10.1.2.1 Constant-Head System

The most basic and proper setup for flowmeter calibration is the constant-head system shown in Figure 10.7. It is usually performed in a well-equipped hydraulic laboratory. The system consists of (a) a water sump, which is usually under the ground level of the hydraulic laboratory in order to save space, (b) a pump that sends the water from the sump into a constant-head tank, (c) the constant-head tank that supplies, by gravity, a steady flow of water for testing or calibrating the flowmeter, (d) a piping system that takes the water flowing out of the constant-head tank by gravity and brings it to the flowmeter for the calibration test, and (e) a weighing tank or a calibrated volumetric tank to receive the water through the flowmeter, and to determine the discharge either by weight or volume.

The constant-head tank is needed in order to maintain a steady flow of water through the flowmeter for calibration. Without it, the discharge coming from a pump would be oscillating to a certain extent, making accurate calibration impossible. The flowmeter must be mounted in a horizontal section of the pipe with a straight section of at least ten diameters in length connected to both upstream and downstream of the flowmeter. The straight sections are needed in order to maintain a fully developed velocity profile immediately upstream and downstream of the flowmeter. The diameter of the test section must be the same as that of the pipe in which the flowmeter is to be used. The weighing tank is needed for accurate determination of the discharge. By weighing the amount of water accumulated in the tank over a measured period of time, the weight flow rate in lbs/s or kg/s can be determined accurately. Knowing the density or the specific weight of the water at the calibration temperature, the discharge Q in cfs or other units can then be found from the weight flow rate measured. This explains how flowmeters are usually calibrated in hydraulic laboratories.

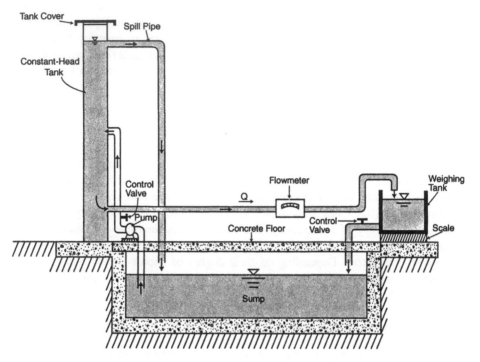

FIGURE 10.7 Constant-head system for flowmeter calibration.

In lieu of a weighing tank, a volumetric tank may be used to determine the discharge from the volume of water accumulated during the test over a certain period, in a way similar to the determination of flow rate with the weighing tank. The choice between a weighing tank and a volumetric tank is a practical matter. When the discharge is small, a weighing tank often provides the most practical means to determine the discharge accurately. In contrast, when the discharge is high, a huge weighing tank may not be practical or available, and a volumetric tank is used instead. The tank is usually an underground reservoir with its volume calibrated against its height, so that the volume of the water in the tank can be determined easily from the water height with a calibration curve or a simple formula. When properly used, both methods can determine discharge accurately, within 0.2% error.

When a flow is first started for a calibration test, the discharge may not be steady and the system may suffer from air entrainment. The experimenter must wait at least 5 min to allow steady flow to establish and air bubbles in the system to escape. Then the system will be ready for testing. Needless to say, great care must be exercised during every step of the test in order to insure accurate calibration.

10.1.2.2 Meter Prover

A meter prover is a specially designed short loop of pipe used for calibrating or checking the accuracy of flowmeters used in pipe. The flowmeter to be tested or calibrated is mounted in a test section of the prover. The system is designed in such a way that elastomer spheres (balls) can be dropped into the pipe downstream of the test section, called the *prover barrel*. For an incompressible flow, the discharge through the flowmeter is the same as that through the prover barrel. Therefore, by measuring the discharge through the prover barrel, the discharge through the flowmeter can be determined. The discharge through the prover barrel is determined from the average travel speed of the ball through the barrel between two points spaced at a sufficiently large distance. The ball diameter is slightly larger than the inner diameter of the barrel, so that there is little leakage of flow around the ball. Consequently, the ball will travel through the barrel at the same speed of the liquid in the barrel. The discharge is determined from the average speed of the ball multiplied by the average cross-sectional area of the barrel interior. This discharge is then used to calibrate the flowmeter. To minimize error, the barrel uses a pipe of very round and smooth interior, and care is exercised to prevent any leakage around the ball. Also, more than one ball run is made to yield an average speed. In spite of all such care, the accuracy of calibration with a prover is never as good as the constant-head method described in Section 10.1.2.1. This is due to the fact that the barrel often contains bends, which provide larger resistance to the ball than in the straight section. Consequently, some fluctuations of discharges exist in the prover during the test, affecting the calibration results. The relatively short test section of the prover also affects calibration accuracy. For provers, a 1% error should be considered good.

10.1.2.3 Meter-to-Meter Calibration

This involves using a flowmeter known to be accurate to calibrate another flowmeter of questionable accuracy. It is often done *in situ*, for calibrating flowmeters that are already used in a pipeline, but are suspected of being damaged or yielding false or inaccurate readings. In such a case, another flowmeter of known accuracy, such as a calibrated venturi, orifice meter, or magnetic flowmeter, can be temporarily inserted into the pipeline to calibrate the questionable flowmeter with the latter remaining in its usual position.

Most flowmeters purchased from a commercial source come with their sensors and readout equipment, such as the pressure transducers and the electronic circuits, to provide the output signal. In such a case, the flowmeter and the readout equipment should be calibrated as a single package in order to minimize calibration error, and to simplify the calibration task. In cases where the flowmeter is supplied without the readout equipment, as in the case of a plain venturi or orifice meter, then a separate set of sensors or equipment, such as a manometer or a pressure transducer, can be used to measure the pressure differential. Such sensors and equipment will be discussed in the next section.

10.2 SENSORS AND EQUIPMENT

Various sensors and equipment other than flowmeters are used in monitoring and controlling pipelines. The most important ones are discussed here briefly.

10.2.1 MANOMETERS

Manometers provide the most economical, reliable, and accurate way to measure pressure — both gage pressure and differential pressure. Unfortunately, because the practical range of pressure that can be determined with a manometer is limited, and because manometers can only measure steady-state and slowly varying pressures, in many cases a pressure transducer instead of a manometer is used. Still, manometers are used widely in hydraulic laboratories for research, teaching, model-testing, and flowmeter calibration. They are also used widely in water plants, chemical plants, power plants, food processing plants, and many other facilities.

The principle of the manometer, based on hydrostatics, is treated thoroughly in introductory fluid mechanics, and hence is not treated here. Only a few practical aspects of using the manometer are mentioned briefly here. One aspect is the choice of fluid to be used in a manometer. Any liquid that does not evaporate under the temperature and pressure that the manometer is exposed to can be a candidate. This includes liquids such as water, oil, alcohol, and mercury, and excludes volatiles such as gasoline. Due to its availability and nonpolluting nature, water is the first liquid that should be considered for use as manometer fluid. When water is used, a small amount of a surfactant should be added to reduce the surface tension of the water. The amount added is usually so small that it will cause negligible change of the water density. Because the sensitivity of a manometer is inversely proportional to the density of the liquid used in the manometer, for low-pressure cases (water head less than 4 inches) it may be advisable to use a liquid having a density lower than that of water, such as an oil or ethanol. In contrast, in cases where the pressure or pressure differential to be measured with water is too high (say, when the water head is higher than 4 or 5 ft), use of mercury may be more practical. Since mercury has a density that is 13.6 times the density of water, use of mercury instead of water can reduce the size (height) of a manometer by more than ten times.

In recent years, there has been increasing public concern about mercury as a health hazard and pollutant. However, with proper design and due care, the use of mercury with a manometer is very safe. It seldom, if ever, causes a serious health hazard or pollution to the environment. The general threat of mercury that genuinely exists in some industries that use mercury for processing does not exist in using mercury in manometers because of the small amount of mercury used in a manometer, the small contact area between water (or air) and the mercury in the manometer tube, and the infrequent contact between people and the mercury in the manometer. Still, when using mercury as the manometer fluid, care must be taken to ensure that the mercury does not spill on the floor or into the flow (and then be carried into the pipe, or drained back into the sump of a hydraulic laboratory). Such minor accidents can be prevented by proper design of the manometer system including a mercury trap, and by exercising care in operating manometers that use mercury.

Another aspect worth mentioning is the size of the tube used to make a manometer. Because capillary action due to surface tension is inversely proportional to the tube diameter, a small tube can cause significant errors in manometer measurements. To prevent capillary effect from causing errors in pressure readings, manometer tubes should have a minimum diameter of $1/2$ inch or 6 mm.

10.2.2 PRESSURE TRANSDUCERS

Pressure transducers are classified in many different ways. From a functional standpoint, pressure transducers are often classified as (1) absolute-pressure transducers that measure the absolute pressure instead of the relative or gage pressure, (2) relative-pressure transducers that measure the gage pressure, (3) differential-pressure transducers that measure the differential pressure between two taps, (4) static-pressure transducers that measure both the steady-state pressure and a slowly varying pressure that varies at a frequency within the frequency range of the transducer capability, (5) dynamic-pressure transducers that are used only for high-frequency variation of pressure and do not measure or under-represent the low-frequency part of the signal, (6) high-temperature pressure transducers that are capable of operating in a high-temperature environment, and (7) miniature pressure transducers for measuring pressures connected to small taps and in situations where the measuring equipment must be compact, etc.

From operational principles, pressure transducers are classified as (1) the capacity type in which the pressure changes the spacing between the two plates of a capacitor, which in turn changes the signal, (2) the strain-gage type, which uses a strain gage to measure the force that the pressure exerts on a membrane, and (3) the piezoelectric type, which uses a piezoelectric material that generates a voltage in proportion to the pressure exerted on the material, etc.

In the selection of the pressure transducer for a given task of a pipeline system, one must carefully consider many factors including the pressure range to be measured, the type of pressure to be measured (whether absolute, relative, or differential pressure), the accuracy of the transducer for the range of pressure to be measured, the temperature range that the transducer must function in, the frequency response required, the size of the transducer, and the compatibility of the transducer output with the data acquisition system (computer), etc.

10.2.3 TEMPERATURE SENSORS

There are two general categories of temperature sensors — the contact type and the noncontact type. The noncontact type is also called a *pyrometer*. The contact-type temperature sensors can be further divided into two general groups — thermometers and temperature transducers.

By definition, thermometers are basic instruments that register the temperature of any material or object in direct contact with the thermometer without having to use an electronic circuit to obtain the temperature reading. Most thermometers are based on the principle of thermal expansion of liquid, which forces the liquid in the thermometer to rise in a narrow glass tube — the liquid-in-glass thermometer. Another type uses a bimetallic coil that moves a needle as the temperature changes. Whatever the type, thermometers can be used for measuring fluid temperature in a pipeline system at convenient locations such as the water in a reservoir, and for sensing the temperature of pipeline components and machines, such as the temperature of an operating pump or motor. Selection of thermometers must be based on the range of temperatures to be measured, convenience in mounting and reading, and other practical considerations.

Temperature transducers come in three general types: (1) resistance temperature detectors, (2) thermally sensitive resistors, and (3) thermocouples. The first type, resistance temperature detector, is based on the principle that the electrical resistance of metals increases linearly with the temperature of the metal. Therefore, by running a current through a small metal wire piece and measuring the voltage across the wire, the resistance of the wire and the corresponding temperature are determined. The second type, thermally sensitive resistor (called *thermistor* for short), is similar to the first type except for the fact that the material used is such that a small change in temperature can cause a large change in resistance. Finally, the thermocouple is based on the principle that when two dissimilar metals are joined together to form a thermocouple, a natural voltage of the order of a millivolt is generated across the junction. This junction voltage increases as the temperature of the thermocouple increases. Therefore, by measuring the voltage across the thermocouple, the temperature of the thermocouple can be determined. The thermocouple appears to be the type of temperature transducer most widely used. It has the advantages of being rugged and compact, and the signal generated can be recorded easily and continuously by data acquisition systems such as a strip chart or a computer. The compactness makes it possible to record the temperature at a given point, be it in a liquid, a gas, or a solid in contact with the thermocouple.

Finally, in recent years, remote sensing of temperature by using infrared sensors (i.e., pyrometer) has been developed for medical and various industrial uses over a wide range of temperature. They can be used in areas of a pipeline system where measurement by direct contact is impractical, as in regions of high temperature, when dealing with a corrosive liquid or gas, or when measuring the temperature of a rotating part such as the shaft of a motor or pump.

10.2.4 Velocity Sensors

Sometimes, it is necessary or desirable to measure the fluid velocity at specific locations in a pipe, such as the centerline of the pipe. When that happens, one should carefully choose the most appropriate velocity sensor for the measurement. A good understanding of the various types of velocity sensors is important in making the proper selection. In what follows, a brief discussion of the important types is presented. For more detailed information, one should read a fluid mechanics book such as Reference 1, or a book on flow measurement such as Reference 2.

The most fundamental and accurate method to measure the velocity of fluid, be it a liquid or a gas, is by using a Pitot tube. Two common types of Pitot tubes are the simple Pitot tube (also called *stagnation tube*) and the Prandtl-type Pitot tube. The simple Pitot tube is simply an L-shaped small tube, usually made of stainless steel, which can be inserted into the pipe for velocity measurements. As shown in Figure 10.8a, the tube is inserted into the pipe at the location (desired radial distance from the centerline) where the velocity is to be determined. The tube must be aligned with the flow, with the tube opening facing the flow and becoming the stagnation point. The end of the stagnation tube must be connected to either a manometer or a pressure transducer to determine the pressure at the stagnation point p_s. The wall pressure (static pressure p_o) must be measured separately with a pressure gage,

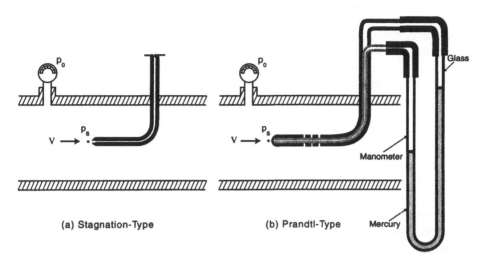

(a) Stagnation-Type   (b) Prandtl-Type

FIGURE 10.8 Two types of Pitot tubes: (a) stagnation type, (b) Prandtl type.

transducer, or manometer. Knowing the difference between these two pressures, $\Delta p = p_s - p_o$, the velocity of the flow at the location of the tube, undisturbed by the presence of the tube, is

$$u = \sqrt{2(\Delta p)/\rho} \qquad (10.11)$$

The Prandtl tube is a modification of the stagnation tube in that the L-shaped tube is actually a double tube, one placed inside the other (see Figure 10.8b). While the front opening (the stagnation point) is connected to the inner tube, a set of side openings is placed around the outer tube at appropriate distances from the front. This arrangement enables the stagnation pressure p_s to be measured by the inner tube, and the static pressure p_o to be measured by the outer tube. By connecting the two ends to a differential manometer or a differential pressure transducer, the pressure difference $\Delta p = p_s - p_o$ is measured. This eliminates the need for the static-pressure tap shown in Figure 10.8a. Note that Pitot tubes are simple, inexpensive, and accurate. They can be made of small tubes of 1-mm diameter, and hence cause the least disturbance to the flow in the pipe. One shortcoming of the Pitot tubes is that their accuracy decreases as the fluid velocity decreases. This is due to the fact that the pressure difference Δp in Equation 10.11 is proportional to the square of the velocity u. It becomes very small when the velocity is small. The accuracy often becomes unacceptable when velocity is below 1 fps for liquid, and below 30 fps for gas.

At low velocities, other velocity sensors may be more accurate, such as current meters, propeller or turbine meters, hot-wire or hot-film anemometer, and the laser anemometer. The current meter consists of a set of cups mounted on the circumference of a wheel, similar to the anemometer used outdoors at weather stations to measure wind speed. The speed of the flow is proportional to the rotational speed of the current meter. Current meters are calibrated to read the velocity of the flow for a fluid of a given density, usually water. Miniature current meters are available from commercial sources to measure the velocity in pipe for both liquids and gases. Propeller or turbine anemometers are miniature propellers and turbines, similar to toy windmills. They operate in a manner similar to the current meters, except that their axis of rotation is horizontal — the current meter rotates about a vertical axis. The hot-wire anemometer uses a tiny tungsten or platinum wire as the sensing element. A small electric current is run through the wire. As the velocity of the flow increases, heat transfer between the hot wire and the flow increases, causing the temperature of the hot wire to fall and the resistance to decrease. For a constant current the decreased resistance causes the voltage across the wire to drop. This principle is used to operate hot wires and hot films; the latter is a variant of the former, but more rugged and more suitable for use in liquid flow.

Finally, the laser Doppler velocimeter (LDV) is now available commercially to measure the velocity of liquid inside a tube or pipe. This method differs from all the other types of velocity sensors in that it is nonintrusive and hence does not disturb the flow. The principle of LDV is identical to that of the Doppler-type ultrasonic flowmeter; it is based on the Doppler effect and frequency shift of a wave reflected from particles contained in the flow, except for the fact that an optical wave (laser) instead of an acoustic (ultrasonic) wave is used. Because optical waves are used, the pipe containing the liquid flow must be transparent or have transparent windows to allow the laser beam to enter and leave the pipe. The instrument has been developed to the point that with more than one laser, it can be used to measure the three-dimensional turbulence in pipe. However, due to its high cost — on the order of $100,000 per unit — this instrument has been used to date only in research projects where the small velocity in a pipe and/or the turbulence must be measured without disturbing the flow, or in special applications that can justify such cost.

10.2.5 VIBRATION SENSORS

In critical parts of a pipeline system where vibration may be a problem, vibration sensors should be used to monitor the vibration of such parts. The two most common types of vibration sensors are accelerometers and displacement sensors. The accelerometer, as its name suggests, measures the acceleration of any structural part. An accelerometer consists of a short tube that contains a small object of a given mass attached to a coil spring, and a displacement sensor. The mass is free to move back-and-forth along the tube, responding to acceleration and deceleration along the tube. The sensor (i.e., the tube) is attached to a vibrating part of a machine or structure. As the sensor vibrates, the mass in the tube presses against the spring and causes a vibratory displacement, which is measured by a displacement sensor. A computer then converts the displacement signal to yield the acceleration as a function of time.

Another way to measure vibration is to use a displacement sensor without the mass and spring. This requires a two-part sensor with one part attached to the vibrating element of a machine, whereas the other part is attached to a neighboring stationary element. Vibration causes a change in the spacing between these two parts, which in turn generates a signal that reflects the displacement. By taking the first and second derivatives of the displacement, the velocity and the acceleration are found.

10.2.6 STRAIN GAGES

For safe operation of a pipeline system, one must make sure that all parts of a pipeline including valves and pumps are operating within the elastic limit of the material. When there is doubt as to whether a pipe or any other part is stressed to beyond the elastic limit of the material, strain gages can be used to find the answer. Strain gages are transducers that measure directly the strain (i.e., unit elongation) of any part of a structure or machine that is under stress. Once the strain is measured, the stress can also be calculated from the strain by using Hooke's law. This requires knowledge of the Young's modulus of the material, and the assumption that the deformation is within the elastic limit.

Strain gages are based on the fact that the electric resistance of a metal wire is altered by stretching the wire. Modern strain gages are made of a tiny wire looped into a grid of the shape of the radiator tube in a car or air-conditioning unit. The wire loop is then sandwiched between two pieces of thin plastic sheet to form a chiplike transducer. The transducer is a rectangle of a few millimeters in length and width, with a thickness less than 1 mm. To use it, the strain gage is normally glued to the surface of the structure for which the strain is to be measured. A small electric current is run through the wire and the voltage is measured to determine the resistance of the wire. The strain gage is calibrated to read zero when the structure is not under stress. Then, the signal measured when the structure is under stress is proportional to the strain. By attaching two strain gages near each other in mutually perpendicular directions, two components of strain and stress can be measured. Strain gages can be used not only for stationary structures but also moving structures. For instance, it can be glued to the shaft of a pump to measure the strain and stress of the shaft when it is rotating. Strain gages can also be attached to the inside of a structure to measure the strain and stress inside the structure. However, because this requires cutting a slot in the structure, it should not be done unless it is necessary.

10.2.7 DENSITY MEASUREMENTS

The density of fluid in a pipe can be measured by different means. Whenever possible, the density of liquid in a pipe can be determined by taking a sample of the liquid, and then using a volumetric container or hydrometer to determine the density of the sample. A volumetric container is a glass or metal container that has a precisely determined volume for measuring the density of liquid. By knowing the precise volume and the weight of the sample, the density, specific weight, and specific gravity can all be determined easily. A hydrometer is a calibrated glass float used to determine the specific gravity of liquid. It is simple, inexpensive, and accurate. It is the same familiar device used for checking the antifreeze concentration of a car. For gas, direct measurement of density by weight and volume is impractical. The common method to determine the density of gas in a pipe is to measure the pressure and temperature of the gas, and then use the equation of state to calculate the gas density in a manner discussed in Chapter 3.

Related to the density measurement is a technique to measure directly the mass flow rate through pipe. The method is based on the Coriolis principle, which states that if a particle of mass m moves at a velocity $\vec{V}$ relative to a system that rotates at $\vec{\omega}$, a Coriolis force equal to $2m\vec{V}\times\vec{\omega}$ is generated, where the arrows above V and ω designate them as vector quantities, and the multiplication sign within the term represents vector cross-product. This principle can be used to determine the mass flow rate through a pipe if the pipe is vibrated in one plane. The vibrating flow causes a Coriolis force in the direction perpendicular to the vibrating plane. By measuring the displacement of the pipe caused by the Coriolis force, the mass flow rate through the pipe can be determined. Once the mass flow rate is determined by this method, dividing the mass flow rate by the volumetric flow rate, Q, yields the average density of the fluid, or of the solid–fluid mixture, going through the pipe. Q can be measured by any of the flowmeters discussed before, such as the magnetic flowmeter. Though simple in principle, flowmeters constructed based on the Coriolis concept are rather complicated and expensive, especially for large pipes.

10.2.8 PIG AND CAPSULE SENSORS

Modern long-distance pipelines such as natural gas pipelines and oil pipelines use pigs (scrapers) for cleaning the pipe interior and many other purposes (see next section). It is important for the operator to know the location of the pig in a given pipeline at any given time, at least approximately. This requires special sensors either carried by the pig or located outside the pipeline, so that the operator can use these detectors to locate the pig. The type carried by the pig is a radio transmitter that emits low-frequency electromagnetic waves, which can penetrate the steel pipe and be picked up by a ground receiver. From the change of intensity of the waves received, the operator can sense the approach of the pig. The most commonly used pig detector, on the other hand, is a simple mechanical device (a small lever) intruding into pipe. As the pig hits the lever, the latter is deflected and a switch is turned on. This in turn triggers a warning flag or sends a signal to the operator at a distant place. Instead of using a mechanical detector, nonintrusive devices such as proximity sensors are also often used. Finally, a dielectric sensor has been developed by the Capsule Pipeline Research Center, University of Missouri-Columbia, to measure the passage of capsules through pipe [3]. As shown in Figure 10.9, the

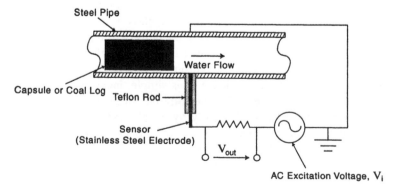

FIGURE 10.9 Dielectric sensor for detecting pigs and capsules.

sensor consists of a stainless steel electrode imbedded in a Teflon rod, with the rod flush-mounted on the pipe. An AC current of a few kHz is applied between the electrode and the ground, which is the metal pipe. This sensor has been tested successfully for detecting not only capsules of various materials, but also commercial pigs. It has the advantage of being able to recognize the type of capsules and pigs, and can record unlimited number of passages by capsules and pigs. The disadvantage is that it requires wall mounting (tapping the pipe), which must be leak-proof. Also, the signal is sensitive to the conductivity of the liquid in the pipe, which may change due to temperature change. For this reason, it is important to use a temperature compensating circuit when this sensor is used. In spite of these problems, at present, this is the most suitable sensor for monitoring and control of hydraulic capsule pipeline (see Chapter 7).

10.3 PIGS (SCRAPERS)

10.3.1 PURPOSES AND TYPES

The term **pig** is commonly used in the pipeline industry to refer to special devices sent into a pipeline for cleaning the pipe interior or inspection, etc. They are also often referred to as scrapers. The use of pigs for pipeline operation is commonly referred to as **pigging**.

There are many different types of pigs designed for different purposes. Perhaps the most commonly used pigs are **cleaning pigs**, which are used for cleaning the pipe interior — removing debris left in the pipe right after construction, removing scale that builds up on the pipe interior surface from years of operations, removing paraffin and sand build-up in a crude oil pipeline, removing water and other fluids which settle out of gas, crude, and product pipelines due to insufficient flow velocity for entrainment, etc. Removing these materials from the pipe by pigging not only results in a cleaner fluid or product going through the pipeline, it also increases the pipe diameter and reduces the pipe roughness. Consequently, the pipeline becomes more efficient — being able to transport a larger quantity of product with the same or less energy. Furthermore, with the removal of these undesirable materials from the pipe, pipeline corrosion can also be reduced and the pipeline will have a longer life. This shows that cleaning pigs serve many purposes including increasing pipeline efficiency and reducing corrosion. They should be used periodically, at least annually, as a part of regular pipeline maintenance. Cleaning pigs come in various designs. For instance, a large manufacturer in the U.S. produces two types of cleaning pigs — one that uses a set of steel brushes for cleaning (Figure 10.10), and the other using a set of urethane blades. The former is for removing hard objects adhering to the pipe wall, such as mill scale; the latter is for removing gummy or loose deposits. A unique feature of such cleaning pigs is that they contain, or have options to include, bypass ports (holes) through the front disk. Because the pig in use experiences a large pressure drop across its front disk, jets are issued through the ports and between the pigs and the pipe wall. These jets create intense turbulence in front of the moving pig, causing debris to be suspended for easy removal. Another type of cleaning pig commonly used in the water supply industry is a bullet-shaped pig made of polyurethane.

FIGURE 10.10 A cleaning pig containing steel brushes. (Courtesy of T.D. Williamson Co.)

FIGURE 10.11 A gaging pig. (Courtesy of T.D. Williamson Co.)

Another type is **gaging pigs**, also called **caliper pigs**, which are used for gaging the pipe interior diameter to determine whether there is any collapsed or kinked pipe or deep protrusions of welds into the pipe. An ordinary gaging pig contains a metal disk (a so-called *gaging flange*) having a diameter slightly (about 5%) smaller than the supposed inner diameter of the pipe. Figure 10.11 is an example. A pipe passes the gaging or caliper test if such a pig passes through the pipeline unhindered. A more sophisticated type of gauging pig contains instruments that record the size and location of pipe diameter reduction. It can detect dents, buckles, flat spots, and construction debris that change the pipe diameter significantly. Such a caliper survey can detect a change in pipe wall thickness of at least $1/8$ inch.

A third type is the **batching pig**, which serves the purpose of separating different batches of liquid products, such as gasoline and jet fuel, being transported simultaneously through the same pipeline. One type of batching pig is a seamless, liquid-cast polyurethane ball, the same as those used in meter provers discussed previously. The ball, when inflated with water and/or glycol to reach an appropriate pressure,

FIGURE 10.12 A spherical pig for batching/displacement. (Courtesy of T.D. Williamson Co.)

is slightly larger than the inner diameter of the pipe (see Figure 10.12). The spherical shape allows the pig to go around sharp bends. However, any spherical pig may get stuck in a check valve if the length of the bowl in the check valve is longer than the diameter of the sphere. In such a case, the flow can go around the sphere through the clearance between the sphere and the check valve, unable to push the sphere pass the check valve. Therefore, care must be exercised in the selection of check valves when spherical pigs are used. In fact, in the selection of valves one must always consider the valve capability to pass pigs unhindered. A minor problem in using liquid-inflated pigs for batching is that whenever the pig breaks in a pipe, the liquid released may contaminate the product in the pipe. To solve this problem, multi-cup urethane pigs are often used. Such pigs have the general shape of the cleaning pigs except that they do not have brushes or blades. Note that batching pigs are often used for displacing the water and/or air trapped in a pipeline, such as during a hydrostatic testing of a pipeline, or a part of a pipeline. Thus, they are also often referred to as *displacement pigs.*

A fourth type of pig is the **inspection pig** — the most sophisticated of all pigs. An inspection pig carries highly developed electronic circuits, sensors, and/or scanners that allow the pipe interior to be observed and pipe diameter and wall thickness to be measured continuously and 360° around the pipe as the pig travels down the pipeline. Pipe thickness and damage to the pipe wall are measured either by electromagnetic induction or ultrasonic sensors. The data recorded can be stored either in memory chips or computer disks carried on board, or transmitted outside the pipe through a transmitter. Such pigs are particularly useful for inspecting old pipes, and sometimes not-so-old pipes, to determine how much the pipe has been affected by corrosion, erosion, and other possible damage. They play an important role in modern pipeline integrity monitoring systems.

10.3.2 PIGGING SYSTEM

The pigging system includes the pigs, the pig launcher, and the pig receiver (trap). A typical pig launcher for a gas pipeline is shown in Figure 10.13. It includes (1) an end closure that can be opened to insert a pig into a launch barrel, (2) the launch

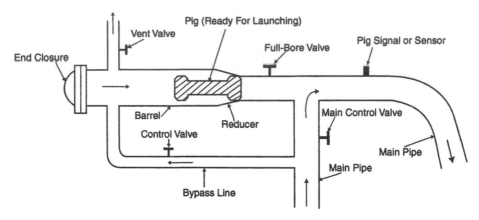

FIGURE 10.13 A typical pig launcher for gas pipeline.

barrel, which is an oversized pipe to facilitate insertion of the pig, being longer than the pig by 1.5 to 2 times, (3) a reducer to connect the barrel to a pipe of the same size as that of the main pipe, (4) a bypass line of a size 1/3 to 1/4 that of the main pipe, (5) four valves to control the gas flow through various parts of the launcher system (one of the four valves being a full-bore on-off valve, such as a ball valve, to allow passage of the pig through the launcher), (6) a blow-down line (vent line) to allow the system to reduce to atmospheric pressure after the pig is launched, and (7) a pig signal or sensor at the end of the launcher to detect the passage of the pig. A similar system can be used for launching pigs into liquid pipelines, except that the blow-down line may not be needed.

The procedure for launching pigs can be summarized as follows. (1) Close the control valve on the bypass line and close the full-bore valve of the launcher. Then, open the vent valve to vent the barrel to atmospheric pressure. (2) Insert a pig into the barrel until the front of the pig has entered the reducer and the pig can no longer be pushed forward easily due to increased friction. (3) Close the end closure, and open the control valve slowly to purge the air through the vent valve. (4) After the air is purged, close the vent valve to allow the system pressure to equalize. (5) Open the full-bore valve, and open the bypass control valve fully; the pig is now ready for launching. (6) Partially close the main control valve, which will cause more gas to flow through the bypass and behind the pig. Continue to open the main control valve until the pig moves out of the launcher. (7) As soon as the pig has passed the pig sensor, the main control valve is opened fully, and the two valves connecting the main pipeline to the pig launcher can be closed to isolate the launcher from the main pipeline. The gas flow is now in its normal state without going through the launcher. The system will operate like this until another pig needs to be launched. The pig receiving system, called the *pig receiver* or *pig trap*, is similar to the pig launcher structurally except that the bypass connections and pig sensor are placed at somewhat different locations (see Figure 10.14). It operates in reverse order of that of a pig launcher. More detailed discussion of pigs, launchers and receivers can be found in References 4 and 5.

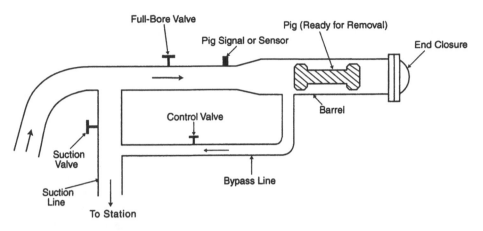

FIGURE 10.14 A typical pig receiver (trap) for gas pipeline.

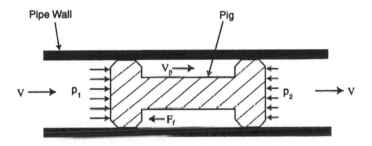

FIGURE 10.15 Pig motion in pipe.

10.3.3 PIG MOTION ANALYSIS

A simple analysis can be made to understand the motion of pigs in pipe. Referring to Figure 10.15, a pig moves through the pipe at constant velocity V_p. Due to the existence of large contact friction between the pig and the pipe, the pig moves through the pipe at a velocity V_p smaller than the mean flow velocity V, in a manner similar to the motion of capsules in pipeline during regime 2, discussed in Section 7.3.2.2. The flow creates a drag force F_D on the pig equal to

$$F_D = C_D A\rho\left(V - V_p\right)^2 / 2 \tag{10.12}$$

During steady-state motion, the drag force F_D is equal in magnitude but opposite in direction to the contact friction force F_f, namely,

$$F_D = F_f = \eta N \tag{10.13}$$

where N is the total (scalar sum of) normal forces that the pig exerts on the pipe in the radial direction, and η is the contact friction coefficient. Eliminating F_D from the two above equations yields

$$V_p = V - V_d \quad \text{where} \quad V_d = \sqrt{\frac{2\eta N}{C_D \rho A}} \tag{10.14}$$

Equation 10.14 shows that the pig velocity is smaller than the fluid velocity by an amount equal to V_d as given by the above equation. The magnitude of V_d is directly proportional to the square root of the contact friction coefficient η and the normal force between the pig and the pipe, and is inversely proportional to the drag coefficient C_D, the fluid density ρ and the cross-sectional area of the pipe A.

Furthermore, if we model the pig as a capsule having two end disks of equal diameter, Kosugi's equation in Chapter 7 reduces to

$$C_D = \frac{4k_d^4}{\left(1 - k_d^2\right)^2} \tag{10.15}$$

where k_d is the diameter ratio of the disk, namely, $k_d = D_d/D$.

Equation 10.15 holds only for large values of k_d, such as $k_d > 0.95$, which is the case for pigs. Note that the disk diameter used here for calculation is not that of the pig when it is outside the pipe, which is greater than the pipe diameter. Instead, D_d is the diameter of the disk after the pig has been squeezed into the pipe, and when the disk is moving in the pipe. Due to the existence of a thin film of fluid flowing between the disks and the pipe (i.e., the leakage flow), the diameter D_d is always slightly smaller than D. In cases where there are holes through the disk, such as provided in the front disk of a cleaning pig, an equivalent disk diameter should be used, which can be determined from the leakage flow ratio to be discussed later. In general, the diameter D_d used in Equation 10.15 is the effective diameter determined from leakage flow.

Example 10.4 A pig with two end disks having an effective disk diameter of $D_d = 0.98D$ is used in a 10-inch steel pipe for cleaning the pipe interior. The fluid is water flowing at 6 fps. The pig being squeezed into the pipe exerts a total normal force of 400 lb on the pipe wall, and the contact friction coefficient between the pig and the pipe is 0.6. Find the velocity of the pig moving through the pipe.

[Solution] For this case, $\eta = 0.6$, $N = 400$ lb, $\rho = 1.94$ slug/ft^3, $V = 6$ fps, $A = 0.545$ ft^2, and $k_d = 0.98$. From Equation 10.15, $C_D = 2353$. Then, from Equation 10.14, $V_d = 0.439$ fps, and $V_p = 5.56$ fps.

Another useful relation can be derived from capsule flow analysis. From continuity equation of incompressible flow,

$$Q = Q_p + Q_L \tag{10.16}$$

where $Q = VA$ is the fluid discharge, $Q_p = V_p A_p$ is the pig discharge, and Q_L is the leakage flow discharge.

Defining $N_L = Q_L/Q$ as the leakage flow ratio, Equation 10.16 yields

$$V_p = VA(1 - N_L)/A_p \cong V(1 - N_L) \quad \text{or} \quad N_L = 1 - (V_p / V) \tag{10.17}$$

Example 10.5 What is the leakage flow ratio for the pig in Example 10.4?

[Solution] Because $V_p = 5.56$ fps and $V = 6.0$ fps, from Equation 10.17, $N_L = 1 - (5.56/6.0) = 0.073$ which is a 7.3% flow leakage through and around the pig.

PROBLEMS

10.1 A venturi of 8 × 4-inch size is used to measure the water flowing through an 8-inch-diameter horizontal pipe. A U-tube manometer is connected to this venturi and the elevation difference for the mercury on the bottom of the U-tube is 18 inches. (1) Find the pressure difference $(p_1 - p_2)$. (2) Find the discharge Q through the pipe.

10.2 A venturi of 8 × 4-inch size is used to measure the air flow through an 8-inch steel pipe. The pressures measured by Bourdon gages connected to the two taps of the venturi, one at the inlet and the other at the throat, are 40 psig and 10 psig, respectively. The temperature of the air in the pipe at the upstream tap is 70°F. (a) Find the discharge of the airflow by assuming incompressible flow. (b) From the discharge found, determine the weight flow rate in lb/s and mass flow rate in slug/s. (c) Find the mass flow rate by assuming isentropic compressible flow. (d) From the result of part (c), find the density and temperature of air at the throat. (e) Find the Mach number of the flow at both the venturi inlet and the throat. (f) Compare the results of part (c) with part (b) and discuss the difference. Which of the two solutions yields more realistic results at high speed?

10.3 A sharp-crested orifice of 6-inch opening is used to measure the air flow through a 12-inch diameter horizontal pipe. The air density is 0.0022 slug/ft³. The two pressure taps of the orifice meter are connected to the two ends of a U-tube manometer and water is used as the manometer fluid. If the manometer reads a water level difference of 4.2 ft, what is the discharge through the pipe? Does it make any difference if the pipe is not horizontal? Why?

10.4 For many types of flowmeters such as venturi and orifice, the discharge Q is determined by measuring the piezometric head difference Δh. In calibrating such a flowmeter, when the experimental data of Q are plotted against the data of Δh on a log-log paper, the result yields a straight line with a slope equal to 2.0. Explain why. (Hint: use Equation 10.4.)

10.5 A magnetic flowmeter reads 8.2 V when the discharge through it is 0.5 cfs. When the meter reads 2.2 V, what is the anticipated discharge?

10.6 An orifice meter reads a pressure differential of 2.5 psi when the discharge is 1.5 cfs. When the discharge increases to 3.0 cfs, what is the anticipated meter reading?

10.7 A Prandtl-type Pitot tube is inserted into the centerline of a pipe to measure the air speed at the centerline. The density of the air in the pipe is 0.0021 slug/ft^3. (a) If the Pitot tube reads a pressure differential of 0.1 psi, what is the air speed measured? (b) If water is used in a U-tube manometer to measure the pressure differential of the Pitot tube, what is the manometer reading in terms of the water head?

10.8 If the fluid in the previous problem is water instead of air, what is the water speed with the differential pressure measured as 0.1 psi? Explain why the result is so different when the fluid is water instead of air.

10.9 A transit-time acoustic flowmeter is used to measure the discharge of water in a pipe of 1 ft diameter. The discharge is 5 cfs, and the two transducer probes are located diagonally from each other having an angle of 45° with the pipe centerline. Find the time shift determined by the electronic circuit of the sensor.

10.10 A weighing tank is used to calibrate a flowmeter in a water pipe. If 2500 lb of water accumulate in the weighing tank in 80.4 s, what is the discharge Q sensed by the flowmeter in cfs. Assume the water temperature to be 70°F.

10.11 A sphere (ball) used in a meter prover for calibrating a flowmeter travels a distance of 120 ft in 86 s. The inner diameter of the prover barrel is 1 ft, and the inner diameter of the pipe in which the flowmeter is calibrated is 6 inches. What are the velocity and discharge through the flowmeter?

10.12 Inclined-tube manometers are sometimes used to amplify the readings when the pressure encountered is low. Suppose one uses a manometer with 30° incline angle instead of using a vertical-tube manometer to measure the small pressure differential encountered by the Pitot tube in Problem 10.7. What will be the manometer reading in inches of water along the tube?

10.13 A pig with two end disks is used to gage a pipe of 8-inch inner diameter. The effective diameter of the pig in the pipe is 7.95 inches. The fluid is water, flowing at 5 fps. The pig in the pipe encounters a contact friction of 300 lb. Find the time that it takes the pig to travel 100 mi through the pipe.

REFERENCES

1. Streeter, V.L., *Fluid Mechanics*, McGraw-Hill, New York, 1971.
2. Baker, R.C., *Flow Measurement Handbook*, Cambridge University Press, Cambridge, U.K., 2000.
3. Du, H.L. and Liu, H., Dielectric sensor for detecting capsules moving through pipelines, *Transactions on Mechatronics*, IEEE/ASME, 5(4), 2000, pp. 429–436.
4. Pigging Products and Service Association, *An Introduction to Pipeline Pigging*, Pipes & Pipelines International, Beaconsfield, U.K., 1995.
5. *Guide to Pigging*, T.D.W. Pigging Products, Tulsa, OK.

11 Protection of Pipelines against Abrasion, Freezing, and Corrosion

11.1 LINING, COATING, AND WRAPPING

Lining is the application of a protective coating on the inside surface of pipes, whereas **coating** refers to the same except for its application to the pipe exterior. Both lining and coating are intended to reduce corrosion and abrasion of pipes. Lining also serves the purpose of forming a smooth pipeline interior, which reduces frictional loss, and it helps reduce damage to pipes by cavitation in some situations.

Various materials are used for pipe lining and coating. They include the following:

Bitumastic materials — Bitumastic materials such as coal tar, asphalt, or bitumen are used — as both lining and coating — of steel, cast-iron, concrete, and wood pipes.

Cement — Cement is often used as a lining for steel and cast-iron pipes. It is a preferred type of lining for transporting saline or brackish water.

Lead lining — An old practice no longer in use at present for fear of lead poisoning.

Glass — Glass is used as the lining for steel pipe in special applications. Glass lined pipes have the strength of steel pipe and the corrosion resistance of glass pipe. They are especially suited for conveying acids.

Rubber — Soft rubber lined steel pipes are sometimes used in slurry transport because they are abrasion resistant. Hard rubber-lined pipes are sometimes used for transporting strong acid and alkali solutions.

Brick lining — Bricks were once used to line large diameter conduits but are rarely used today.

Fluorocarbon lining — Fluorocarbons, such as Teflon, Kynar, etc., are used when transporting high-temperature and/or corrosive fluids.

Thermoplastic lining — PVC, polyethylene, etc. are used when transporting loads at normal temperature. High density polyethylene (HDPE) is used for slurry pipelines to reduce pipe wear.

Thermosetting lining — Epoxies, polyesters, etc. can be sprayed on and baked to produce a hard glasslike pipe interior.

Galvanized lining and coating — Used on steel pipes. Note that galvanized steel is not suitable for welding.

Cladded piping — First, stainless steel or nickel is cladded onto steel plates. Then, the plates are rolled into pipes. Cladding is formed by homogeneous bonding or spot-welding.

Wrapping — Tape or encasing is applied around a pipe to increase its resistance to corrosion and abrasion. It can be done on pipes with or without coating. Steel pipes are often coated with tar or bitumen and then wrapped with one or more layers of plastic or kraft paper.

11.2 INSULATION, TRACING, JACKETING, AND ELECTRIC HEATING

The best way to prevent the liquid in a pipeline from freezing in cold weather is to bury the pipe below the frostline, which depends on geographic location. In places where a pipeline or a portion of the pipeline must be above ground, consideration must be given to preventing the liquid in the pipe from freezing. Insulating the pipe with an insulating material is often sufficient to solve this problem. When insulation alone is insufficient, the pipe must be heated in one way or another.

The purpose of tracing and jacketing is to heat pipelines to prevent freezing in cold weather. While **tracing** involves attaching one or more than one small steam pipe to a large liquid pipe for heating the latter (Figure 11.1), **jacketing** involves placing a pipe larger than the liquid pipe around the latter. The steam used in heating the liquid by jacketing moves through the annular spacing between the two pipes (Figure 11.2). Hot water may also be used in tracing and jacketing.

A special type of tracing, called **internal tracing,** uses steam tubing in the center of the liquid pipe (Figure 11.3). Although this is the cheapest type of tracing, it is difficult to repair and becomes troublesome when rupture occurs inside the liquid

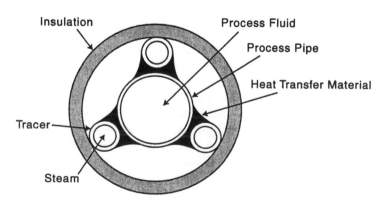

FIGURE 11.1 Steam-traced pipe (cross-sectional view).

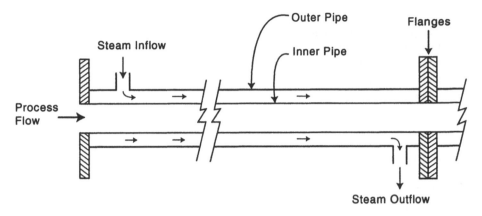

FIGURE 11.2 Jacketed pipe (profile).

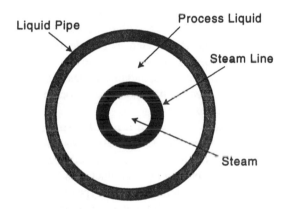

FIGURE 11.3 Internal tracing (cross-sectional view).

pipe. Furthermore, the tubing inside the pipe reduces the effective cross-sectional area of the pipe for the liquid flow, and increases headloss.

Finally, electric heating by placing or wrapping resistance heating elements, such as a heating ribbon, around a pipe is another way to heat pipelines. Although the method is the simplest, it may not be the most economical because of the high cost of electrical energy as compared to steam. This is especially true in places where the steam used may be a waste product or by-product, such as at a power plant.

11.3　PROTECTION AGAINST CORROSION

11.3.1　TYPES OF CORROSION

Corrosion is the second largest cause of pipeline damage. Corrosion is defined as being the gradual damage of pipe due to chemical or electrochemical reactions of pipes with their environment. The environment includes the fluid in the pipe, the soil, water, and atmosphere around the pipe, and other metals attached to or in contact with the pipe. Unlike the erosion damage to pipe, which is caused by the physical process of abrasion or wear such as encountered by slurry pipelines or pneumatic conveying of solids in pipes, and unlike cavitation, which is caused by vapor pockets in liquids generated by low pressure, corrosion is caused by chemical or electrochemical reaction. The following is a brief discussion of the various types of corrosion.

11.3.1.1 Chemical Corrosion

This is due to the contact of a pipe with a corrosive substance such as an acid, which attacks or reacts with the surface of the pipe to cause damage. It can happen to both metallic and nonmetallic pipes, and it can damage either the pipe interior or exterior surfaces depending on whether the corrosive chemical is inside or outside the pipe. A number of chemicals exist that can damage (corrode) pipes (see, for instance, Reference 1). Metal pipes can be damaged by various acids, halogens, and salts; concrete pipes can be corroded by various acids and salts; and plastic pipes can be damaged by various acids, hydrocarbons, and chlorine. Generally, steel pipes require no lining for transporting hydrocarbons such as petroleum, but they do require a lining for transporting water unless oxygen is removed from the water. Corrosion-resistant metals include copper, brass, nickel, stainless steel, titanium, and many alloys. Plastic pipes cannot be used for transporting hydrocarbons, but they are inert to water even when containing oxygen.

Chemical corrosion happens only when a pipe transports or otherwise comes into contact with a corrosive chemical. Chemical corrosion can be controlled by either avoiding contact with corrosive chemicals, selecting a pipe material inert to the corrosive chemical that the pipe will be in contact with, or by using a pipe lining or coating inert to the corrosive chemical. A special type of chemical corrosion takes place when a metal pipe is in contact with water containing dissolved oxygen. For steel or cast iron pipe, the metal (iron) reacts chemically with the water and oxygen to form various hydrated ferric oxides, which form rust. This type of corrosion can be prevented by using deoxygenated water, which is a common practice for protecting boilers of power plants.

11.3.1.2 Electrochemical Corrosion

This is the most common type of corrosion that occurs in metal pipes, and it is of electrochemical origin. Such corrosion can be subdivided into two types: galvanic corrosion and electrolytic corrosion. Each type is discussed below.

11.3.1.3 Galvanic Corrosion

Galvanic corrosion is based on the same principle that galvanic cells (batteries) are based on. The following is a brief description of how a galvanic cell works.

Referring to Figure 11.4, a piece of zinc (Zn) and a piece of copper (Cu) are immersed in an electrolyte, and the two metals (electrodes) are connected by a wire that is a good conductor, such as copper. From Table 11.1, zinc has a higher electrical potential than copper. Consequently, a current will flow from the copper to the zinc

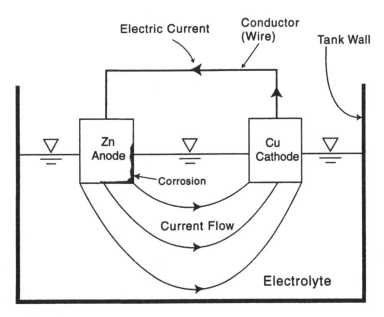

FIGURE 11.4 Galvanic cell.

TABLE 11.1
Galvanic Series of Various Metals and Alloys in Seawater

[ANODIC] — Magnesium, zinc, aluminum, cadmium, mild steel, cast iron, ductile iron, stainless steel (active), lead, tin, nickel (active), brass, copper, bronze, nickel (passive), stainless steel (passive), silver, titanium, graphite, gold, platinum — [CATHODIC].

Note: 1. The materials listed above are in descending order of electrode potential. Those listed first corrode the most, and those listed near the end corrode the least in seawater. In other electrolytes, the order may alter somewhat. 2. When any two materials in the list are connected by a wire and immersed in an electrolyte, as shown in Figure 11.4, the one with higher potential (i.e., listed earlier) will become the anode and corrode, whereas the one with the lower potential (i.e., listed later) will become the cathode and be protected. The farther the two materials are separated from one another in the electrochemical series, the greater the potential difference between the two electrodes, and the stronger the corrosion current will be.

through the connecting wire. To complete the electric circuit, a current of the same magnitude will flow through the electrolyte, from the zinc to the copper. This electrical current through the electrolyte removes molecules from the surface of the zinc electrode, causing galvanic corrosion to the zinc.

The foregoing galvanic cell is one used commonly in car batteries. The same type of reaction can happen when any two different metals, such as iron (Fe) and Copper (Cu), come into contact with each other, and when they are immersed in an electrolyte. Furthermore, the connecting wire between the two metals serves only one purpose, which is to bring the two metals into electrical contact. Thus, the same phenomenon will happen without the connecting wire, as long as the two metals are in direct contact with each other. From the foregoing, it can be seen that for a galvanic cell to form, two conditions must be satisfied: (1) there must be two dissimilar metals in contact with each other or connected together by a conductor, and (2) the two metals must be immersed in an electrolyte.

Galvanic corrosion is common in metal pipes due to the existence of many dissimilar metals in a pipeline, such as a valve made of a different type of steel than that used for the pipe, or a pump having an impeller made of or coated with bronze. The severity of galvanic corrosion depends not only on the potential difference between the two connecting metals but also on the contacting electrolyte. The higher the conductivity of the electrolyte, the more current flows through the galvanic cell, and the more severe corrosion becomes. Therefore, wet soil and soil that contains salts are highly corrosive to pipelines and other metal structures.

A modified type of galvanic corrosion also exists. In this modified type, the entire pipe can be a single metal of uniform potential. However, the electrolyte (such as soil) in contact with one part of the pipe is different from that contacting another part of the pipe. Due to the difference in the contacting electrolytes (soil), the pipe-to-soil potential for one part of the pipe is different from the pipe-to-soil potential for another part of the pipe. This also creates a galvanic cell that corrodes the pipeline.

From the foregoing discussions, it can be seen that whenever a metal pipeline contains different metals, or whenever the soil conductivity differs along a pipe, galvanic cells take place and they corrode the pipe.

11.3.1.4 Electrolytic Corrosion

Electrolytic corrosion is based on the same principle of electrolysis (i.e., electric plating). As shown in Figure 11.5, if two metals, A and B, are immersed in an electrolyte, and if an outside direct current (DC) source, such as a battery, is connected between the two metals, a current will flow through the electrolyte, causing metal to be lost from metal A (which is the anode) and be transported to and plated on metal B (the cathode). This causes metal A to corrode. This type of corrosion can occur in metal pipes for one of two reasons: (1) using a pipeline to ground any DC source of electricity such as a battery, a rectifier, a DC generator or an electric welding unit, or (2) stray currents are generated through the soil in the neighborhood of an underground pipeline. For example, experience shows that the stray currents generated from cathodic protection systems serving neighboring pipelines or neighboring utilities and the stray currents generated from cable cars (trolleys) powered by DC can all cause serious corrosion and rapid deterioration of underground metal structures including pipelines.

For both types of electrochemical corrosions (i.e., for both galvanic and electrolytic corrosions), it is always the anode that is corroded, and the cathode that is protected. This represents the underlying principle of cathodic protection of metal pipes and other metal structures to prevent or minimize corrosion.

A phenomenon closely related to electrochemical corrosion is polarization, which is the attraction of hydrogen ions, H^+, from the electrolyte to the close proximity of

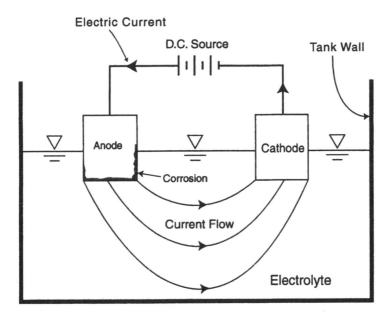

FIGURE 11.5 Electrolytic cell.

the cathode. As the hydrogen ions lose their charge at the cathode, hydrogen gas is generated near the surface of the cathode, which acts as an isolating layer to oppose or retard the current through the circuit. This retardation action by the hydrogen gas is called **polarization**, which impedes corrosion. Flowing water in the pipe disturbs or destroys polarizations, which in turn increases the corrosion rate.

11.3.1.5 Other Types of Corrosion

Other types of corrosion include bacterial corrosion, pitting corrosion, crevice corrosion, thermal galvanic corrosion, stress corrosion, erosion corrosion, cavitation corrosion, etc. They are briefly discussed below.

Bacterial corrosion is caused by the presence of certain bacteria and/or algae, which can produce substances that corrode pipe. For instance, it is known that anaerobic sulfate-reducing bacteria thrive in poorly aerated soils such as clay. During their life cycles, they produce acid that corrodes pipe and other underground structures.

Pitting corrosion is a special type of electrochemical corrosion that occurs at certain points along a pipe where the paint or coating is damaged, and where metal is exposed to the environment (electrolyte). Because the exposed areas are small, current densities caused by electrochemical corrosion cells going through these areas are high, causing high corrosion rates and rapid damage of pipe.

Crevice corrosion happens in small gaps that exist between metals, such as in crevices under bolt heads and rivet heads, inside bolt holes, and at any overlapping metals. It is caused by the existence or entrapment of materials that behave as dissimilar electrolytes with the surrounding environment. Prevention of crevice corrosion involves elimination of crevices where possible, such as by using welding rather than riveting or bolting for connecting steel pipes, closing crevices by welding, or caulking, etc.

Thermal galvanic corrosion is caused by the contact of two pipes of different temperatures, such as a hot-water pipe and a cold-water pipe. Such contacts cause galvanic corrosion with the hot pipe being the anode and the cold pipe being the cathode.

Stress corrosion happens in places experiencing high stress and strain, and in metals subjected to fatigue (i.e., repeated stresses or repeated stress reversals). Such high stresses or repeated stress reversals promote cracks of various sizes, which expose the metal to the corrosive environment.

Erosion corrosion is the combined effects of erosion and corrosion. Without erosion, corrosion will produce a crust to prevent or retard further corrosion. However, with erosion the crust is removed by physical forces of erosion, and new metal is exposed to continue the corrosion process. Thus, erosion aggravates corrosion. An example of erosion corrosion is the corrosion that takes place in certain slurry pipelines, other than in coal slurry pipelines. For coal slurry pipelines, the oxygen in water is adsorbed by the coal particles in the pipe, resulting in a low level of corrosion. Erosion is also low when the coal particles are fines. Erosion corrosion in slurry pipelines is especially prominent in areas affected by particle impacts, such as on the outer bends of elbows transporting the slurry.

Cavitation corrosion is the combined effects of cavitation and corrosion, causing rapid deterioration of the metal including steel pipe. Cavitation and corrosion reinforce each other. For instance, cavitation causes the removal of the protective layer (such as rust) caused by corrosion. This in turn accelerates corrosion. On the other hand, corrosion causes pitting and other surface irregularities on the pipe interior, which in turn causes localized flow separation and low-pressure spots to promote cavitation.

11.3.2 CORROSION-RELATED MEASUREMENTS

11.3.2.1 Corrosion Coupon Test

This is the test to determine the corrosion rate of any metal in a given liquid, due to chemical corrosion. The metal is cut into small pieces of a standard size — the **coupons** or **chips**. The coupons are then immersed in and rotated through a bath of the test liquid in a manner specified by a certain standard such as ASTM D4627-92 (1997). The corrosion rate is determined by the loss of weight of the coupon with time. Stress corrosion is tested using a different standard, such as ASTM G30-97.

11.3.2.2 Soil Resistivity Measurement

The electrical resistivity of the soil around a pipe affects the corrosion of the pipe. It is often measured as part of the work related to cathodic protection.

TABLE 11.2
Corrosion Classification of Soil

Classification	Soil Resistivity (ohm-cm)
Noncorrosive	>10,000
Mildly corrosive	2000–10,000
Moderately corrosive	1000–2000
Corrosive	500–1000
Very corrosive	<500

From Ohm's law in physics, when a voltage V is applied across the two ends of a conducting cylinder of length L and cross-sectional area A, a current I will flow through the conductor such that

$$V = \left(\frac{\rho L}{A}\right)I = RI \tag{11.1}$$

where ρ is the **resistivity** of the conductor, and the quantity $R = \rho L/A$ is the **resistance** of the conductor. When R, L, and A are given in ohm, cm, and cm², respectively, the unit of resistivity is ohm-cm.

The foregoing concept can be applied to soil. Dry soil has high resistivity and is noncorrosive, whereas wet soil (especially those containing salts) has low resistivity and is corrosive. Depending on their resistivity, soil can be classified as in Table 11.2.

The most common way to measure the resistivity of soil is the **Wenner method** (also called **four-pin method**), which involves placing four electrodes into ground, separated at a equal distance L as shown in Figure 11.6. A current source such as a battery is connected to the two outer electrodes to generate a current I through the

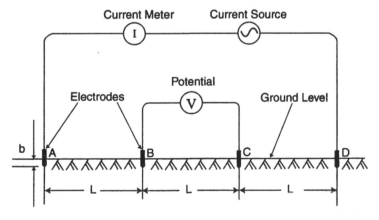

FIGURE 11.6 Four-terminal (Wenner) measurement of soil resistivity. (Note: depth of electrodes, b, must be small compared to spacing, L.)

two electrodes. If a voltage V is measured between the two inner electrodes, the resistivity of soil is

$$\rho = 2\pi L \frac{V}{I} \qquad (11.2)$$

When using the above equation to calculate soil resistivity, the distance L must be much longer than the buried length of each electrode, i.e., $L >> b$. The current source can be either a DC (direct current) or AC (alternating current). The electrodes can be made of various materials such as iron or copper sulfate. The value of ρ measured is the average resistivity of the soil layer over a depth approximately the same as the distance L between neighboring electrodes. Although different manufacturers make Wenner-type resistivity meters of different designs, the fundamental principle used is the same.

Example 11.1 An engineer used the Wenner method to determine the resistivity of the soil along a proposed pipeline right-of-way. He inserted the four pins (electrodes) of the measuring apparatus into the soil along a straight line, with equal distance of 10 m between neighboring pins. Then he connected a 12-volt power source to the two outer pins, and measured a current of 18 amp through the circuit. Determine the soil resistivity, and whether the soil is corrosive.

[Solution] For this case, $V = 12$ V, $I = 18$ amp, $L = 10$ m = 1000 cm. Substituting these values into Equation 11.2 yields $\rho = 4189$ ohm-cm, which is the soil resistivity. Then, from Table 11.2, the soil resistivity falls in the range of mildly corrosive soil.

11.3.2.3 Pipe-to-Soil Potential

The potential difference between a buried pipe and the soil around it is very important in the study of corrosion. As shown in Figure 11.7, this potential is measured by connecting a high-impedance voltmeter or potentiometer between the pipe and a special electrode (half cell) placed in contact with the soil above the pipe. The electrode used can be one of many types, such as the copper sulfate electrode, hydrogen electrode, calomel electrode, zinc electrode, etc., with the copper sulfate being the most commonly used.

The copper sulfate ($CuSO_4$) electrode, as shown in Figure 11.8, is made of a copper rod immersed in a saturated water solution of copper sulfate. The electrode has a porous plug bottom, which must make a good contact with the soil when the electrode is used. As shown in Figure 11.7, while the copper rod of the electrode must be connected to one end of a voltmeter, the other end of the voltmeter is connected to the pipe through a welded (or soldered) lead — the **cadweld**. A high impedance voltmeter is required to draw insignificant current from the pipe to the electrode. The wire used to connect the pipe to the voltmeter should be well insulated. The pipe-to-soil potential should be measured at various locations along a pipeline. The pipe is considered in good shape (noncorroding) if the pipe-to-soil potential is

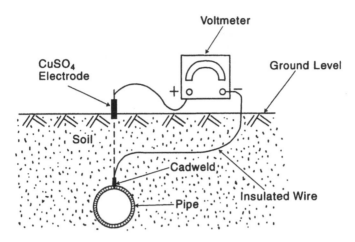

FIGURE 11.7 Voltmeter measurement of pipe-to-soil potential.

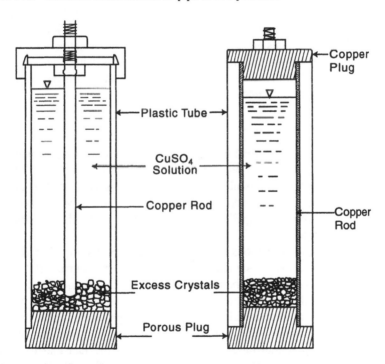

FIGURE 11.8 Copper sulfate electrodes.

higher than −0.85 V. In places where the potential is lower than −0.85 V, cathodic protection is needed.

11.3.2.4 Line Current Measurement

Another quantity that is often measured in corrosion detection is the **line current,** which is the current measured between two points along a pipeline. Figure 11.9

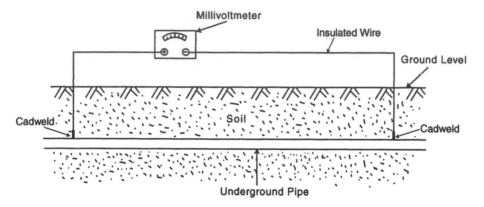

FIGURE 11.9 Line current measurement.

shows the electrical circuit for measuring the line current. It is normally measured between points separated at distances a few hundred feet apart. The existence of line current indicates that corrosion is in progress in the pipe. With line currents measured at regular intervals along a pipe, the location of corrosion cells along the pipe can be determined. Note that line current measurements cannot detect corrosion cells that are of small dimension, as in the case of an anode on the bottom of a cathode in the same cross section of a pipe.

11.3.3 CATHODIC PROTECTION

Cathodic protection is an electrical method for combating corrosion in metal structures, including steel pipes, both on ground and in water. The method requires the use of an electrical current to counter or cancel the current generated by corrosion going between a steel structure and the surrounding ground or water. By making the protected metal structure (pipeline) a cathode instead of anode, the structure is protected from corrosion.

There are two general methods to provide cathodic protection. The first is the **impressed current** method, which requires the use of a direct current (DC) source, most often a rectifier — similar to the commonly used battery charger. By connecting the negative terminal of the rectifier to the metal structure (steel pipe) to be protected, and connecting the positive terminal to the ground through an electrode, the pipe or structure becomes a cathode. The connection is identical to that shown in Figure 11.7 for pipe-to-soil potential measurements, except for the fact that the rectifier replaces the voltmeter. When the connection to the ground is through a copper sulfate electrode, to stop corrosion the voltage of the rectifier should be adjusted to maintain a pipe-to-soil potential of -0.85 volt. This must be done at various points along the pipe so that the pipe-to-soil potential at all points becomes the same, -0.85 V.

Another way to provide cathodic protection is the method of **sacrificial anode**. It involves connecting a zinc or magnesium electrode to the pipe and its environment (ground or water). This will create a galvanic cell as shown in Figure 11.4, except that now the pipe instead of the copper becomes the cathode. In protecting the pipe, the zinc or magnesium electrode becomes the anode, which gets corroded. This shows why the zinc and magnesium electrodes are called **sacrificial electrodes.**

Note that coating and/or lining metal pipes, even if the coating/lining is imperfect and leaks currents in certain places (the leakage is called **holiday**), coating or lining will greatly reduce the surface area of the electrode (pipe) in contact with the electrolyte. The reduction in this contact area causes a proportionally smaller corrosion current, thereby saving much energy used in cathodic protection.

PROBLEMS

11.1 A soil engineer uses the Wenner method to measure the soil resistivity and determine the corrosiveness of soil at a proposed pipeline site. She places the electrodes (pins) at 10 ft spacing. With a 1.5-V power source, she measures a current of 4 amp. What are the resistivity and the corrosiveness of the soil?

11.2 A person measured a pipe-to-soil potential of +1.2 V. Does the pipe need cathodic protection at this location and why? If your answer is yes, what potential should be applied to the system and how? Provide a brief discussion to show that you understand the problem and know what to do.

REFERENCES

1. Smith, W.H., *Corrosion Management in Water Supply Systems*, Van Nostrand Reinhold, New York, 1989, pp. 56–64.
2. Nayyar, M.L., Ed., *Piping Handbook*, McGraw-Hill, New York, 1992.

12 Planning and Construction of Pipelines

12.1 PROCEDURES INVOLVED IN PLANNING AND CONSTRUCTION OF NEW PIPELINES

The procedure involved in the planning and construction of any new pipeline system depends on several factors including the material (fluid or solid) to be transported by the pipeline (whether it is natural gas, oil, water, sewage, slurry, or capsules), the length of the pipeline, and the environment (whether the pipeline is in an urban or countryside setting, whether it is on land or offshore, whether the climate is warm or cold), etc. However, there are more similarities than dissimilarities in constructing different types of pipelines. Once a person understands how a given type of pipeline is built, it is not difficult to figure out how another type should be built. The following is an outline of the procedure used for long-distance steel pipelines that carry natural gas or oil:

Step 1. Preliminary planning — Determine the origin and the destination of the pipe, the product to be transported, the approximate length, diameter and type of the pipe to be used, the velocity of flow, headloss, power consumption, capital cost, operating expenses, economics, and many other practical considerations. All the design and calculations done during this stage are preliminary and approximate. Once the preliminary investigation confirms that the pipeline project is economically feasible and practical, then the next step is taken.

Step 2. Route selection — A pipeline route should be selected from, and marked on, both a highway map and a topographical map such as the Quadrangle Map of the U.S. Geological Survey. Aerial photography and surveys of the pipeline route are undertaken to obtain data needed for the design and preparation of route maps and property plats, which are normally required for right-of-way acquisition.

Step 3. Acquisition of right-of-way — The acquisition of the right-of-way for a pipeline can come either through a voluntary process — negotiation with land owners for the purchase, lease, or easement of their land needed for the passage of the pipeline, or through condemnation, which is an involuntary legal process. For public-owned pipelines and for pipelines that are privately owned that serve the public, the state and federal governments grant the right of **eminent domain**, which is a legal term for the right to condemn land. Landowners who lose their land through condemnation are normally compensated at a **fair market value**. For interstate pipelines that must cross railroads many times, it is difficult to build such pipelines without cooperation from railroads. The railroad industry in the U.S. has traditionally disallowed pipelines to cross railroads, especially pipelines that carry freight in competition with railroads. The U.S. Congress has granted eminent domain to interstate pipelines that carry natural gas and oil, but has not yet granted the same to coal pipelines, or pipelines that transport other solids. The matter remains controversial in the U.S. due to strong lobbying efforts by the rail industry against such legislation. The lack of eminent domain discourages investment in and commercial use of freight pipelines in the U.S., depriving the public of the benefits of freight pipelines.

Step 4. Soil borings, testing of soil and other data collection — Once the acquisition of the right-of-way has been completed, the pipeline developer can undertake necessary geotechnical investigations and determine whether groundwater and/or hard rock will be encountered, and collect other data along the route that are needed for the design of the pipeline.

Step 5. Pipeline design — To be discussed in Chapter 13.

Step 6. Seek legal permits — Permits from different state and federal agencies may be needed, such as the U.S. Environmental Protection Agency, U.S. Forestry Service (if the pipeline crosses federal forest), U.S. Department of Transportation (for pipelines that carry hazardous fluids such as petroleum or natural gas), etc.

Step 7. Start construction — The construction of pipeline involves the following sub-steps:

    a. **Right-of-way preparation** — For large pipelines, this may involve clearing a path of a minimum width of 50 ft (15 m), and removing trees and flattening the path somewhat so that trucks and heavy equipment can be brought in.

    b. **Stringing** — Bringing in the pipe and setting the pipe in a line along one side of the right of way — stringing.

    c. **Ditching and trenching** — Use hydraulic backhoes or some other equipment to dig ditches or trenches of rectangular or trapezoidal cross section. The depth of the ditch (trench) should be such that the pipe will be below the frostline or at least 3 ft (1 m) beneath the land surface, whichever is greater. Staying below the frostline prevents damage to the pipe by freezing and thawing of the ground; it is especially important for pipelines that convey water. Even in a nonfreezing climate, major pipelines should be at least about 3 ft (1 m) underground to reduce the chance of damage from human activities, such as plowing and land leveling. Two problems often encountered in ditching (trenching) are groundwater and hard rock. They should be avoided during the route selection step of the process whenever possible and practical.

d. **Boring** — When passing through obstacles such as a highway, railroad, or rivers, boring may be used to get the pipe across the obstacle from underneath. Modern boring machines can bore long holes to install pipes under rivers and other obstacles. Boring methods will be discussed in more detail in Section 12.5.

e. **Tunneling** — Needed for crossing mountains or hills.

f. **River crossing** — Three methods for river crossing are **ditching** (i.e., cutting a ditch in riverbed and then burying the pipeline there), **bridging** (i.e., building a new bridge or utilizing an existing bridge to carry the pipeline across a river), and **boring** (i.e., boring a hole underneath the riverbed and then pulling a pipe through). For wide rivers of shallow water, ditching often proves to be the most economical. However, recent advances in horizontal directional drilling (HDD) have greatly enhanced the technical and economic feasibility of drilling and boring across rivers to lay pipes (see Section 12.5).

g. **Welding, coating, and wrapping** — After the ditch has been prepared, steel pipes of 40-ft length (12 m) are welded together to form a long line or string. The welded joints are radiographically inspected, and the pipeline is coated and wrapped with special protective and insulating materials before being laid in the ditch. For pipelines laid underwater, the pipe must be covered with a thick layer of concrete to prevent the pipe from floating. Welding is discussed in more detail in Section 12.4.3.

h. **Pipe laying** — The welded pipeline is lifted and laid into the ditch by a line of side-booms parked along the right of way at approximately equal intervals. Steel pipes normally do not require the use of bedding materials to support the pipes in the ditch. Iron and concrete pipes require that the ditch bottom be covered by a layer of gravel or crushed rock to facilitate drainage and reduce settling. Otherwise, such pipes may be damaged and may leak.

i. **Backfill and restoration of land** — The pipe in the ditch is then backfilled by earth, the earth is then compacted, and the land surface is restored. After the pipe is backfilled, it is hydrostatically tested with water to meet applicable code and government requirements. Restoration involves cleaning out construction waste materials and planting of grass.

For a long pipeline, the foregoing procedure is applied to a portion of the pipeline (say, a few miles) at a time. After the portion is completed, the same procedure is applied to the next portion. In so doing, disruption to the community at each place along the pipeline will be limited to a few weeks. Major pipelines in the U.S. have been built rather rapidly — completing more than 1 mi per working day per crew. For extra-long pipelines, more than one crew can be used simultaneously to shorten the construction period.

12.2 MEASURES TO ALLOW PIPELINE EXPANSION

In places where the pipeline may be affected by thermal expansion, earthquake, soil settlement, etc., measures must be taken to allow pipelines to expand (elongate) freely in order to prevent development of large stresses in the pipe. This is especially true for PVC pipes and concrete pipes, which expand several times more than steel pipes due to temperature change. Even for steel pipes that are aboveground, allowance must be made for expansion and shrinkage caused by seasonal weather change. A common way to provide allowance for thermal expansion or shrinkage is to build a zigzagged instead of straight pipeline, with the corners of the zigzags free to move either outward (during expansion) or inward (during shrinkage). This is common for both aboveground and underground pipelines. For pipelines above ground, sometimes an inverted U or loop is used instead. Special joints are also available to facilitate expansion. They include bell-and-spigot joints, slip joints, swivel joints, and certain mechanical joints. For small diameter pipe, using a joint made of flexible pipe (hose) will allow expansion.

12.3 BENDING OF PIPES

Steel pipes can be and often must be bent to follow sudden grade changes, or change in the horizontal direction of the pipeline. In most cases, bending of steel pipes can be done conveniently in the field (outdoors) by using a cold bending method. Hot bending, which usually produces better results, is more cumbersome and costly. It is done in shops rather than in the field. Figure 12.1 shows the assorted bends of steel pipes and tubes produced in a commercial shop. When a pipe is bent, not only is the cross section of the pipe deformed from a circular to an oval shape, the thickness of the pipe wall on the outer side of the bend is also reduced due to stretching. This is called **thinning**. Before one attempts to bend a pipe, one must calculate the expected ovality and the thinning of the pipe from established tables. The degrees of ovality and thinning must be kept within certain limits. Different bending methods yield different degrees of ovality and thinning. The following is an outline of various bending methods for steel pipe.

1. **Compression bending** — Bending is accomplished by pressing (compressing) a pipe with a moving roller against and around a stationary die having the same radius as that of the needed bend (Figure 12.2).
2. **Draw bending** — Draw bending is shown in Figure 12.3. It is similar to compression bending except that a mandrel is inserted through one end of the pipe. The mandrel prevents the pipe from kinking or changing shape during the course of bending. The result is a bend with less ovality and better control of thinning than the compression bending.
3. **Ram bending** — A ram (punch) with a hemispherical head is pressed against one side of a pipe supported at two neighboring points by two pivoted blocks, rollers, or clamps (see Figure 12.4). Due to its simplicity, ram bending is used most often in the field (outdoors on the construction

FIGURE 12.1 Assorted pipe and tube bending products. (Courtesy of Tulsa Tube Bending Company.)

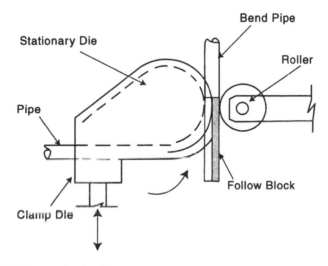

FIGURE 12.2 Compression bending.

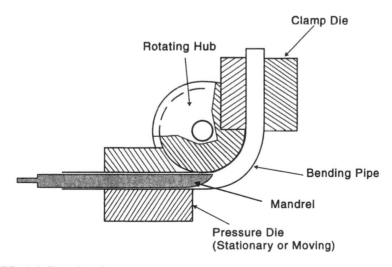

FIGURE 12.3 Draw bending.

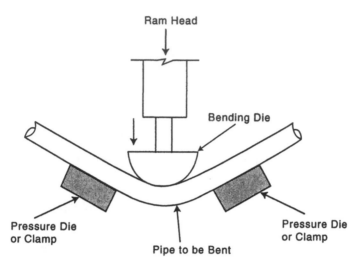

FIGURE 12.4 Ram bending.

site). However, ram bending is not as precise as the other types of bending, and produces large ovality.

4. **Internal roll bender (Rotoform)** — This is a French machine that uses a rolling head rotating inside a pipe. The rolling head presses against the pipe interior to cause the pipe to bend. Rotoform can produce compound bends or curves without heating the pipe, including bends in excess of 180°. Current technology enables bending pipes up to 12 inches in diameter.

5. **Induction bending** — In this method, the pipe passes through a special section at which a strong magnetic field is generated to heat the pipe locally. Bending is precise and control is automatic; it can form three-dimensional bends. Figure 12.5 shows an induction-type bending machine in operation. More details on pipe bending can be found in References 1 and 2.

12.4 CONNECTING PIPES

In the U.S., pipes are normally available in sections or segments* of 20-ft length. For steel pipes, the maximum section length is usually 40 ft. After pipe sections are transported to the construction site and before being laid in ditches, they must be joined (connected) together to form a long pipeline. Joining can be done in several ways including the following.

---

* Pipeliners use the term *joint* to represent a standard segment or section of pipe, which for steel pipe is either 20 or 40 ft long. This slang is confusing to the layperson and the general public, to whom *joint* means the connection between two pipe segments, or between two other objects. For this reason, this book does not refer to sections as *joints*. Rather, *joint* is used as a synonym for *connection*, which is consistent with common English usage.

FIGURE 12.5 An induction-bending machine in operation. (Courtesy of Tulsa Tube Bending Company.)

12.4.1 FLANGED JOINTS

Flanged joints or connections are bolted joints. They are more costly than welded joints, and may leak under high internal pressure. Consequently, they are used only in places where the joints must be disassembled and assembled without difficulty and without damage, such as when connecting a pipe to a pump or a fitting (such as a valve), or when connecting sections of pipes together in places where the sections must be taken apart at least occasionally. Construction of a pipeline with flanged joints involves aligning the to-be-coupled flanges together, placing a rubber gasket between the two flanges to be flanged, tightening the bolts, and welding the flanges to the pipe ends.

12.4.2 OTHER MECHANICAL JOINTS

Several other types of mechanical joints are available for connecting pipes to pipes, or pipes to fittings. They are discussed in Section 8.3.

12.4.3 WELDING

Welding is defined here as bonding sections of pipe together by fusion or melting of the pipe material, whether the material is steel, aluminum, copper, or any other material such as a plastic. For metal pipes, welding requires the application of heat at the location of the joint in order to melt the pipe material there. Welding of copper, brass, and lead pipes is often referred to as **brazing**. For plastic pipes, cold welding, often called **fusion**, is possible by using a solvent to melt the pipe at the joint, without having to apply heat. Welding is the most common way to connect sections of pipe together, and it usually results in the strongest and most lasting joint. Some basic knowledge about welding of steel pipes during pipeline construction is explained in the following discussion.

Prior to transporting the pipe sections to the construction site, both ends of each pipe section should be beveled in the shop using automatic tools. Beveling involves cutting the edge at an angle, usually 30°, so that when two sections are girth welded together, the weld material can penetrate easily into the gap between the abutting sections through a 60° groove — the V-bevel shown in Figure 8.8. Then, the pipe sections are ready for delivery to the construction site. During construction, before welding two abutting pipe sections together in the field, the beveled ends of the pipes to be welded must be thoroughly cleaned of any dirt, rust, mill scale, or solvent. This is usually done by using a power hand tool, such as a wire brush, immediately before welding. Then the two pipe sections are brought together butt-to-butt, carefully aligned, and slightly spaced — to leave a minute gap between the abutting ends of the pipes for penetration of weld into the gap. Such precision alignment is possible only through the use of heavy equipment, such as sidebooms, to suspend and position the pipe sections on the ditch side, by using highly skilled and experienced pipe welders, and by using an internal lineup clamp (for pipes with larger than 8-inch diameter). After alignment has been attained and checked, the joint is spot-welded to gain enough rigidity to maintain the alignment. This initial welding (first pass), done while the internal lineup clamp is still in place, is usually referred to as the **root bead**. If the atmospheric temperature is low (below 40°F), or if the pipe is wet (due to rain, snow, frost, or dew), the abutting parts of the pipe sections must be preheated before the initial welding. After the initial welding is completed and the weld material cooled down and hardened, the internal lineup clamp can be released and taken out of the pipe for use in the welding of the next joint. Meanwhile, the uncompleted first joint is now strong enough to allow the pipe to be supported by wooden skids, and it is ready for the second pass of welding, called **hot pass**. Before welding is complete, two more passes are done, the **filler bead**, and the **cap bead**. The lower right sketch in Figure 8.8d shows the four passes to be made.

Welding in the field can be done manually, semi-automatically, or automatically. Manual welding uses a welding rod (electrode), and a handheld welding machine. The welding machine is powered electrically by a portable welding generator, mounted on either a welding tractor or the welder's own pickup truck or rig. The welding machine produces an electric arc to melt the tip of the welding rod and the abutting pipe ends so as to fuse the pipe ends together. The welding rod has a special coating which, when melted by the electric arc, produces an inert gas that shields the welded joint, protecting it from being oxidized at high temperature by the oxygen present in the atmosphere. This results in a high-quality weld.

In semiautomatic welding, in addition to using an electric arc and a weld wire (electrode), the machine emits a continuous stream of an inert gas such as carbon dioxide, helium, and argon, which shields the joint and prevents it from reacting with atmospheric oxygen at high temperature. Automatic welding is similar to semiautomatic welding except for the fact that the process is fully automated (i.e.,

FIGURE 12.6 At the National Pipeline Welding School in Tulsa, Oklahoma, operated by American Pipeliners Union Local 798, instructor Gary Allison demonstrates how to weld a 24-in steel pipe using an electric arc and an internal lineup clamp. (Courtesy of Local 798.)

involving minimum manual intervention). It can be used in shops as well as on outdoor construction sites. It does rapid high-quality welding.

As soon as welding is done for any joint, it must be inspected visually for possible defects and examined by an x-ray machine to ensure that the weld is of high quality. Welding of the pipeline, inspection of each weld, and interpretation of x-ray pictures of each weld, all require special training, a high degree of skill, and much experience. They must be done by well-trained, experienced personnel. The Pipeliners Union in the U.S. has schools to train and certify skilled welders. Figure 12.6 shows such a training school. More details about welding can be found in References 2 and 3.

12.5 BORING AND TUNNELING TO INSTALL PIPE — TRENCHLESS TECHNOLOGIES

To minimize construction costs, the open-cut (ditching and trenching) method to lay pipe is generally used in rural and remote locations, except when pipelines must cross rivers, lakes, roads, and other obstacles. When a pipeline must cross such obstacles, it is no longer possible or practical to use the open-cut method. Instead, one must consider other alternatives such as rerouting the pipeline, using or building a bridge (for river crossing), and underground construction — boring and tunneling.

Due to recent technical advances in the boring/tunneling field, it is now increasingly feasible to install long stretches of pipelines underground, and to do it at reasonable cost without using the open-cut method. Use of such new technologies greatly reduces the havoc caused by the open-cut method in urban areas, and in environmentally sensitive locations. Such new technologies are called **trenchless technologies** by the promoters to emphasize their main advantage — avoiding cutting long trenches for installing or repairing pipes. There are two general types of trenchless technologies: those for construction of new pipelines, and those for pipeline maintenance, repair, and renovation (renewal). The sections that follow describe various underground (trenchless) technologies for installing pipelines. Trenchless technologies for maintenance, repair, and renovation are discussed elsewhere in this book. References 4 and 5 provide good coverage of trenchless technologies.

12.5.1 HORIZONTAL EARTH BORING

Horizontal earth boring (HEB) uses a machine that bores a horizontal or nearly horizontal hole (small tunnel) underground for laying pipes beneath obstacles such as a roadway. The boring and subsequent installation of pipes are done by a machine without workers being present in the borehole. Three types of HEB — horizontal auger boring, microtunneling, and horizontal directional drilling — are discussed in the following sections.

12.5.1.1 Horizontal Auger Boring

The key components of horizontal auger boring (HAB) include (1) a **cutting head**, consisting of a set of cutters mounted on the front face of the boring machine, to cut earth by the rotation of the cutters; (2) an **auger** with its front end connected to the cutting head and its tail end connected to the prime mover that drives the system, to convey the spoil (i.e., the earth or rock that has been cut loose) to outside the borehole; (3) a **nonrotating casing** around the rotating auger, which is the pipe to be installed; (4) a prime mover that provides the torque to rotate the auger and the cutters, and provides the thrust to advance the pipe (casing) along with the cutting head and the auger; (5) a system to inject bentonite slurry around the pipe to reduce friction between the pipe and the surrounding earth in order to facilitate the advancement of the pipe during the action of boring; (6) a system to receive, store, process, and recycle the bentonite slurry; and (7) control and monitoring equipment. The entire boring system is controlled from outside the borehole, and no human needs to be present inside the borehole or pipe. Therefore, the system can lay pipes of diameters as small as a few inches, and as large as several feet. Figure 12.7 is a photograph of such a machine.

The casing, the auger, and the prime mover of the HAB system are supported in one of two ways. The first way is the **track type**, which uses a stationary track fixed to the ground; the second way is the **cradle type**, which uses heavy construction equipment such as a crane or an excavator. The track type operates cyclically. Each time after a section (segment, or joint) of pipe is pushed into the earth, the system pauses to allow connection or welding of another section of pipe in the bore pit.

FIGURE 12.7 A horizontal earth-boring machine used for installing pipes. (Courtesy of Herrenknecht Company and Ruhr-University of Bochum, Bochum, Germany.)

Construction workers must be present in the pit to connect the pipe sections and to operate the machine. In contrast, the cradle type is suspended by equipment, and hence can be operated entirely from aboveground. The pipe sections also can be connected or welded aboveground, which allows continuous boring to take place. Although the control of the direction and grade by this method is not as accurate as that by using stationary track and also requires a larger right-of-way, the method is good enough for installing steel pipes, which do not need as accurate grading and directional control as installing concrete pipes. For this reason, the cradle method is commonly used for steel pipe, whereas the stationary track method is usually used for concrete pipes.

12.5.1.2 Microtunneling

Microtunneling is the high-tech version of the horizontal earth-boring system, invented and first used in Japan in the 1970s. The system uses a laser-guided and remote-controlled pipe jacking system, and it permits accurate monitoring and control of the horizontal direction and the grade of the pipe. Because no human entry into the pipe or tunnel is needed, the technology is applicable to small as well as large pipes. This technology is used most often for installing pipes of less than 1-m diameter. The system is rather versatile. It is applicable to all types of soil, and a large variety of depths (up to about 50 m below ground), either above or below groundwater table. The method is best suited for installing sewer pipes, which normally require accurate grade and alignment. One of two methods is used to remove the cut materials (spoil) from the pipe. The first method uses an auger — the same as for the HAB system discussed previously. The second method involves using slurry to mix with the spoil for hydraulic removal by pipes. The slurry also facilitates cutting, and creates enough pressure in front of the machine to prevent or reduce infiltration of groundwater into the borehole.

12.5.1.3 Horizontal Directional Drilling (HDD)

HDD is a new technology borrowed from the oil and gas industries, which have been using it for decades for deep-well drilling of oil and natural gas. In the mid-1970s, the technology was adapted to making horizontal or nearly horizontal drillings across rivers and other obstacles, for installing pipes and utility cables — including not only power lines but also lines for fiber optics. The technology is commonly referred to as **horizontal drilling**, in order to differentiate it from the vertical or nearly vertical drillings for oil and natural gas. The technology is also referred to as **directional drilling,** for the drill head is guided by an operator or a computer to maintain the predetermined drill path, and to alter the path as needed.

Prior to drilling and installing the pipe, a trailer-mounted drill rig is brought to one side of the obstacle to be crossed, such as a river (point A in Figure 12.8). Sections of the pipe to be used for crossing, and certain other equipment, are brought to the opposite side of the obstacle (point B in Figure 12.8). Drilling and pipe installation are done in two or three steps. In the first step, a pilot-hole of only a couple of inches in diameter is drilled along the desired path (i.e., a near-horizontal curved path beneath the obstacle to be crossed) of the pipeline (see Figure 12.8). Drilling is done usually with a drill bit of a few inches wide. As drilling proceeds, segments of the drill pipe are added automatically to form the **pilot string**. Through the pilot string, mud (usually bentonite slurry) is pumped through the holes of the drill bit to lubricate the drill, and to draw off the cuttings (spoil). The pilot hole continues to advance from the rig side (point A) to the pipe side (point B). Step one ends when the drill bit has emerged from the ground at point B, the pipe side.

In step 2, called **pre-reaming**, the drill bit and the hydraulic motor that drives the drill bit are both removed at point B, and a reamer assembly is attached to the

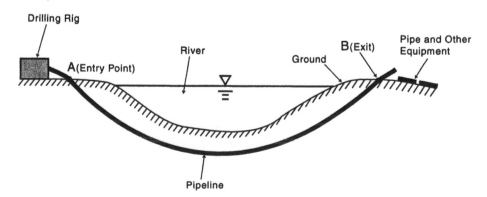

FIGURE 12.8 Horizontal directional drilling (HDD) for laying pipelines under rivers and other obstacles.

FIGURE 12.9 A horizontal directional drilling (HDD) machine in operation. (Courtesy of Ditch Witch Company.)

pilot string to enlarge the borehole. By reversing the direction of rotation of the pilot string, the rig is now used to pull the reamer into the pilot hole. Segments of the drill pipe are added at point B to the pilot string as they are being pulled back. Meanwhile, the pilot string is disassembled at point A. Step 2 ends when the reamer covers the entire path, and starts to emerge from ground at point A, the rig side. Note that step 2 is needed only for large-diameter pipes, which require large boreholes that cannot be created in a single-pass reaming.

In step 3, a larger reamer is used to enlarge the borehole and pull the carrier or product pipe across the obstacle along the borehole. Step 3 starts with the connection of the carrier pipe to the larger reamer, using a three-part connector. The connector includes a cutter, a reamer, and a swivel. The cutter, in the shape of a flywheel, cuts a hole larger than the carrier pipe diameter so that the carrier pipe can be easily pulled into ground and across the obstacle. The reamer, in the shape of a barrel, serves to open the hole and keep the pull on course. The swivel, connected between the reamer and the carrier pipe, allows the pipe to be pulled through and not to rotate with the cutter, reamer, and the drill pipe. Mud (bentonite slurry) is pumped through the hollow teeth of the reamer to lubricate the pipe being pulled across the obstacle, and to remove the spoil. Step 3 ends when the entire pipeline is pulled into the earth beneath the obstacle. Figure 12.9 shows an HDD machine (drill rig).

In large HDD systems, the control of the drill path is accomplished by steering the drill head from aboveground at the location of the drill rig, both automatically by a computer following a preset drill path, and manually by adjusting the preset path using information on the instantaneous location of the drill head. A survey system is used to pinpoint the location of the drill head from aboveground. Several such systems are available commercially, and their accuracies are improving. They can instantaneously locate the position of the drill head within a few inches. For smaller HDD systems, a **walkover** system is used where the drill head is located from signals emitted by a transmitter, housed behind the drill head. The signals are identified and interpreted on the surface by a receiving instrument, which is usually **walked over** the drill head location.

Knowing the physical properties of the soil (earth) along the proposed path of each HDD project is highly important to the success of the project. Consequently, prior to conducting any HDD, vertical boring must be conducted at many points along the proposed HDD path, and core samples (i.e., soil or rock samples) must be taken at different depths until the proposed HDD path is reached. This establishes the earth properties at selected locations along and above the HDD path. To determine the variation of the earth properties between these sampling points, other techniques, such as ground-penetrating radar, or acoustic techniques based on the reflection of high-frequency sound waves, can be used. When calibrated by the soil samples taken at discrete locations, the radar and acoustic method yield continuous information on the variation of the soil properties along and around the proposed HDD path.

Finally, the state of technology (in 2003) is that HDD can be used for installing underground pipelines of maximum diameter of about 1.5 m, maximum length of about 2 km, and maximum depth of about 100 m. Undoubtedly, as the technology further improves, such limitations will be overcome.

12.5.2 PIPE JACKING

Pipe jacking involves cutting the soil and simultaneously jacking (i.e., pushing by a machine) into the earth a pipe that is sufficiently large to allow construction workers to enter the pipe for removal of the earth inside the jacked pipe, and to operate the tunnel boring machine. It is a common technique used for installing large new sewers across existing roadways without making an open cut of the road and without interfering with the traffic on the road. To apply this method, a jacking pit or shaft is dug on one side of the road to be crossed to contain the jacking machine and sections of the pipe to be used (jacked). Directly across the road, a receiving pit or shaft is dug to receive the pipe that has crossed the road from underneath. Excavation of the earth inside the jacked pipe is done either manually, by using clay spades and/or other tools, or by a tunnel-boring machine. Excavated earth is removed using carts or other means. To reduce contact friction between the advancing pipe and the surrounding earth, and to balance the pressure around the pipe, bentonite slurry is pumped into the space immediately outside the pipe to serve as a lubricant. The pipe used for jacking must be relatively stiff and strong, such as steel pipe and reinforced concrete pipe. It must have a relatively thick wall, thicker than that needed for withstanding the internal or external pressure at the place of the crossing; otherwise, the pipe may be damaged by the large thrust developed from the jacking operation.

12.5.3 TUNNELING

Regular tunneling differs from microtunneling mainly in that the regular tunnel is larger in diameter and hence it enables construction workers and heavy equipment to enter and work in the tunnel. Another difference is that in laying a pipeline in a regular large tunnel, the pipeline is not a part of the tunnel wall that supports the earth load. A separate tunnel lining and support system is provided. The pipeline simply uses the tunnel to penetrate underground. Various methods can be used to cut the earth or rock in the tunnel, including digging (when the earth is soft), boring (with tunnel boring machines), and drill-and-blast (when hard rocks are encountered).

Removal of the spoil from a regular (large) tunnel and supplying the concrete and other materials needed for tunnel construction can be done by various means such as building a temporary railroad or conveyor belt, or by using freight pipelines such as the pneumatic capsule pipeline (PCP) discussed in Chapter 7. Using a freight pipeline for tunnel construction has a number of advantages over using rail or other vehicles, including (1) it has no moving part outside the pipe and hence is very safe; (2) it is powered pneumatically, and hence it has no exhaust gas and causes no air pollution in the tunnel; (3) the prime movers — the motor and the blower — are located outside the tunnel and hence do not take up the narrow space in the tunnel; and (4) the system transports large quantities of material at high speed. The only major limitation of the system is that the size of any article to be transported by PCP must be smaller than the capsule diameter or pipe diameter. Therefore, large construction equipment cannot be moved in and out of the tunnel by using PCP; conventional vehicles are still needed for occasional movement of large equipment in and out the tunnel. Figure 12.10 shows a large (90 m^2) and long (3 km) tunnel completed successfully in Japan using a PCP system as the main means for conveying excavated materials out of the tunnel, and for bringing construction materials into the tunnel. The PCP system used had a rectangular cross section of 1 m × 1 m, approximately, using prestressed concrete panels as walls and ceilings for ease in assembling before tunnel construction, and disassembling and removal upon completion of the tunnel. The capsules used are the rectangular type shown in Figure 7.6. Details about this PCP are provided in Reference 6. Most recently, circular cross section PCP was used successfully to excavate a vertical tunnel for waste disposal in Japan [7].

12.5.4 COMMENTS ON TRENCHLESS CONSTRUCTION

Trenchless technology for constructing underground pipelines has advanced rapidly since 1980. It is now in common use in practically all major cities around the world, and in countrysides where pipelines must cross rivers and other obstacles. The popularity of trenchless technology is growing because of the following: (1) less damage to city streets and less interruption to traffic aboveground than by open-cut construction, (2) less air pollution and less noise caused during construction than by the open-cut method, (3) deeper laying of pipes than practical for open cuts, which lessens the chance of accidental damage to the pipe from other construction activities — third party damage, (4) more economical than the open-cut method in many circumstances when damage to other structures by open cuts is taken into account, and (5) more suitable for difficult ground conditions, such as unstable soil, high watertable, or when the ground is congested with other utilities. In spite of these advantages and the growing popularity of trenchless

FIGURE 12.10 The PCP system in Japan used for constructing the Akima tunnel for bullet trains. The box-shaped long structure on the ground, both inside and outside the tunnel, constitutes the dual-conduit PCP used for supplying the concrete and removing the excavated materials during the construction period of the tunnel. (Courtesy of Sumitomo Metal Industries, Ltd.)

technology, its applicability should be kept in proper perspective. It should be realized that trenchless technology is economical and practical only in urban and other densely populated areas, and in countryside and remote locations where pipelines must cross major obstacles, such as roads, airports, major buildings, factories, power plants, rivers, mountains, national monuments, and environmentally sensitive areas. It is neither practical nor economical to use trenchless construction for new cross-country pipelines where open cuts can be done at a fraction of the cost and at many times the speed of trenchless technology, unless and until future technological advancements can result in substantial further reduction in the cost of trenchless construction to the level comparable to that of the open-cut method.

12.6 PIPELINE CONSTRUCTION IN MARSH AND SWAMP

The U.S. has extensive experience in dealing with pipeline construction in marsh and swamp. The difference between marsh and swamp is the type of vegetations encountered. While only small vegetation, such as grass, bush or water hyacinth, are present in marsh, trees are present in swamps. Trees have deep roots and are difficult to remove by ordinary construction equipment. Therefore, construction in swamp is more difficult than in marsh, and requires a somewhat different approach. In both marsh and swamp, a shallow layer of water exists, and the soil is unstable.

Swamps in the U.S. are often populated with large cypress trees having strong and deep roots. To prepare for the pipeline right-of-way, these trees are removed with backhoes and dynamite, and piled up along the sides of the right-of-way. In both swamps and marshes, a ditch having a depth of about 8 ft and a width of about 40 ft must be dug along the pipeline right-of-way. This is usually done by using draglines supported on mats made of large timbers bolted together. The mats with the dragline on them are placed on the ground beside the ditch, to form a temporary construction platform. After the dragline has completed digging the part of the ditch near where the dragline is parked, the mats behind the dragline are moved to the front of the dragline so that the dragline and other equipment and construction crew can move forward to continue digging the ditch. After ditching is completed, sections of the pipe used for the project are bought in by barges to the construction site. Welding of the pipe is done in a manner similar to welding inland pipelines. Once welded, the pipeline is coated with a thick layer of concrete so that it will sink in water even when the inside of the pipe is filled with air. The welding and concrete coating are all done on a long narrow platform. One of three kinds of platform may be used.

The first type of platform is a land-based ramp laid on one side of the ditch, with the ramp almost parallel to the ditch. After the pipes are welded together and weighted with the concrete, the weighted pipe is pushed down the ramp to enter the ditch. Floats are attached to the weighted pipe to prevent it from sinking in the ditch as it first enters the ditch. After the entire pipe has entered the ditch, the floats are removed and the pipe sinks to the bottom of the ditch. The second type of platform is a long floating platform made of several barges connected end to end. The barges are all anchored to form a stationary platform floating on the completed ditch. Welding, weighting with concrete, and launching of the pipeline are done in the same manner as the foregoing land-based ramp method. The third type of platform is the same as the second type, except for the fact that the platform (barge) is not always stationary. During launching of the pipe, the platform (barge) moves forward. The forward motion of the barge sends the pipe down the ramp. This is similar to the way pipes are laid in offshore construction. When using this third method, the lay barge moves along the entire length of the completed ditch. This requires that the ditch be deep enough and wide enough over its entire length for the barge to move through.

After the pipe is laid into the ditch by one of the three foregoing methods, the ditch should be backfilled with the same soil dug out during ditch construction. This completes the construction of pipelines in marsh and swamp. More details about pipeline construction in marsh and swamp can be found in Reference 3.

12.7 OFFSHORE CONSTRUCTION

The increasing demand for petroleum and natural gas throughout the world, coupled with the depletion of such fuel from inland wells after many years of production, has created increased need to drill and locate large reserves of oil and natural gas offshore. This has prompted the use of many offshore pipelines whose construction is both sophisticated and expensive. Many factors that need not be considered in the construction of ordinary underground pipelines must be considered in the offshore pipelines, including the corrosiveness of seawater, the dynamic action of ocean waves and undersea currents, attacks by marine life, the high external pressure caused by large depth of ocean water, and possible natural disasters such as earthquake, undersea volcanic eruptions, and the effect of hurricanes (in the Atlantic Ocean and the Gulf of Mexico), typhoons (in the Pacific Ocean), and cyclones (in the Indian Ocean and the Southern Hemisphere).

The most common way to lay offshore pipelines is essentially the same as the lay-barge method for laying pipelines in marsh and swamp discussed in Section 12.6, except for the fact that the barges for laying an offshore pipeline are much larger and much more sophisticated than those used in marsh and swamp. The lay barge of the offshore project is essentially a floating city with all the personnel, equipment, and living quarters aboard. It is also supported by several other vessels such as three or four supply ships, a tugboat, and a survey ship. With such an operation, construction of pipe proceeds rapidly — laying more than a mile of pipe per day. Speed is essential in this case due to the strong need to avoid inclement weather. Figure 12.11 shows some of the key parts of a lay barge used for laying an offshore pipeline.

After the pipe sections are welded together, coated with a protective coating material, and weighted by a thick concrete coating, it is pushed down the ramp and then down the stinger, which is mounted (hinged) on the stern of the barge. The stinger guides and supports the pipeline being lowered into the ocean, preventing the development of a large bending force caused by gravity that can damage the pipeline. The tensioner on board of the barge creates a resistance to the descending pipeline, preventing the pipeline from rapid descent into the ocean. As soon as the pipeline approaches the ocean floor, the barge moves forward to allow the pipeline to develop an S-shape curve as shown in Figure 12.11. Once the S-curve is developed, the barge can continue to move forward in order to lay the remaining pipe onto the ocean floor until the offshore platform is reached. At this point, the end

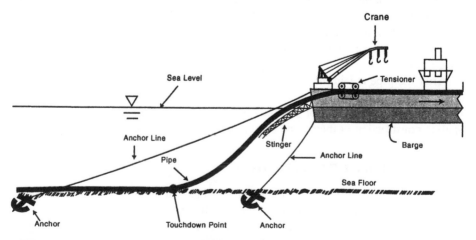

FIGURE 12.11 Lay barge used for laying pipe offshore, and its major components.

of the pipeline has been reached, and it is lowered into the ocean and then connected to the bottom of the **riser,** which is the vertical pipe that connects the off-shore pipeline laying on the ocean floor to the oil pumps aboard the offshore platform. Connection of the pipeline to the riser on the ocean floor is accomplished by using specially trained welder divers, who can weld undersea inside a dry enclosure. The dry enclosure is a pressurized chamber lowered to the ocean floor along one side of the platform. Once reaching the ocean floor, the chamber is locked onto both the pipe and the riser. Compressed air expels the water from the chamber, and the divers are able to weld the horizontal pipe to the riser. After welding is completed, the weld is inspected, both visually and by x-ray, and the connection is coated. Then, the chamber with the divers rise to the platform, and the pipeline construction is essentially complete. A submersible can then be sent to the ocean bottom to inspect and take pictures of the constructed pipeline laying on the ocean floor. Other methods of laying offshore pipelines have also been used; they are described in Reference 3.

12.8 COLD-REGION CONSTRUCTION

In regions of extremely cold weather, such as the Arctic, or near Arctic areas, special factors and issues must be considered in the construction of pipelines.

12.8.1 FREEZING

Freezing either makes pipeline construction more difficult or easier, depending on the situation. In regions where freezing occurs only during a few months in winter, such as in the central and northern parts of the U.S., pipeline construction can be much easier and much less costly if it is done in spring, summer, or fall. Winter construction should be avoided. In contrast, in Arctic regions where snow and ice start to melt in summer, there may be no roads and bridges across rivers and creeks near the construction site. This makes construction difficult in summer. In winter, temporary ice roads and ice bridges can be built to allow trucks and heavy construction equipment to move around. This makes construction easier in winter. Therefore, in cold regions, one must consider carefully which is the best season for pipeline construction, and should have the pipeline built during that season, whenever possible.

Another important consideration in cold regions is whether to bury the pipe, and if so, at what depth. All water pipelines and sewers must be buried sufficiently deeply underground in order to prevent them from freezing in winter. The rule-of-thumb for the minimum burial depth of pipelines that carry water is the frostline plus 1 ft. For other liquid pipelines that transport crude oil or petroleum products, even though they may not freeze in winter under subzero weather, the viscosity of the liquid transported may increase substantially in winter, causing excessive energy loss. The increase in energy loss is especially serious when the flow is laminar. Therefore, these liquid pipelines must also be buried to prevent a large decrease of the liquid temperature in the pipe. As with water pipes and sewers, a colder climate requires a deeper burial for crude oil and product pipelines.

12.8.2 TEMPERATURE VARIATION

The seasonal variation in the Arctic may range from $-60°F$ to $80°F$. If the pipeline is aboveground, as is the case for a large portion of the Trans-Alaska Pipeline, such a great change in temperature causes large expansion and shrinkage of the pipelines. It must be carefully considered in the design and construction of the pipeline. The solution to the problem, in many cases, is to use a zigzag path as shown in Figure 1.3 for the Trans-Alaska Pipeline. Of course, if the pipe is allowed to be buried underground, the temperature change will be much less, and the zigzag may be unnecessary. However, the builders of the Trans-Alaska Pipeline chose to have about half of its pipe aboveground in order to protect the environment, that is, to avoid melting the tundra.

12.8.3 ENVIRONMENTAL CONCERNS

Melting of the tundra in the permafrost region by a long pipeline is hazardous to the wildlife that cross the pipeline right-of-way. For this reason, pipelines in the permafrost region may not be allowed to be buried underground. Having the pipeline aboveground in such extremely cold regions causes many problems. For example, if the fluid transported through the pipeline is warm, melting of the tundra will still occur with an elevated pipeline because of heat conduction through the vertical support members of the pipe. The Trans-Alaska Pipeline solved this problem by using a refrigeration system to freeze the tundra, which is a costly method.

12.8.4 OTHER CONSIDERATIONS

Other considerations include deviations from common construction practice: no bending is necessary for pipe on flat and horizontal tundra; prewelding two 40-ft sections of pipe indoors to form 80-ft sections before bringing them outdoors to the construction site; insulating the pipe to minimize the temperature drop of the fluid in the pipe; use of glycol–water solution instead of water only for the hydrostatic pressure test of the pipe; use of construction equipment that can operate smoothly at low ambient temperature, etc.

REFERENCES

1. Fillanders, J., *Pipe and Tube Bending Manual*, Gulf Publishing, Houston, 1984.
2. Nayyar, M.L., Ed., *Piping Handbook*, 6th ed., McGraw-Hill, New York, 1967.
3. Hosmanek, M., *Pipeline Construction*, Petroleum Extension Service, Division of Continuing Education, University of Texas at Austin, 1984.
4. Najafi, M., An overview of common methods in trenchless technology, Proceedings of the International Symposium on Underground Freight Transportation by Capsule Pipelines and Other Tube/Tunnel Systems, Columbia, Missouri, Sept. 2–3, 1999.
5. Iseley, D.T., Najafi, M., and Tanwani, R., *Trenchless Construction Methods and Soil Compatibility Manual*, National Utility Contractors Association, Arlington, VA, 1999.

6. Kosugi, S., Pneumatic capsule pipelines in Japan and future developments, Proceedings of International Symposium on Underground Freight Transport by Capsule Pipelines and Other Tube/Tunnel Systems, University of Missouri-Columbia, 1999, pp. 61–73.
7. Brouwer, G., Pneumatic capsule pipeline removes soil vertically, *Civil Engineering*, 72(3), 22, 2002.

13 Structural Design of Pipelines

13.1 INTRODUCTION

To be consistent with the purpose of this book stated in the preface, this chapter presents pipeline design in the context of principles and logic, rather than codes and standards. The latter changes with time, countries, government agencies, and in some cases, with state laws and local statutes. Therefore, in addition to the information learned from this book, the designer must understand the various standards, codes, and regulations that govern the design of particular types of pipelines in particular locations in order to create an actual design for construction.

Pipeline design includes several general steps: (1) load determination, (2) critical performance evaluation such as determining the stress and/or deformation of the pipe, (3) comparison of performance with the limiting performance criteria established by codes and standards, and (4) final selection of the pipe and construction method based on the design.

Traditionally, the design of pipelines has evolved separately in different industries that use pipelines. Because different industries use pipelines for different purposes, the design requirements are different and the types of pipe materials used are different. For instance, in the petroleum industry and the natural gas industry, due to the need for long distance transportation of their products (petroleum and natural gas), these industries use primarily steel pipe with welded joints because this allows the pipeline to withstand very high pressure, often above 1000 psig and sometimes above 3000 psig. Such high pressures allow the use of long pipelines, often more than 1000 mi, with only a few booster pump or compressor stations for each pipeline. Having fewer booster stations along a pipeline reduces both construction and operating costs. This overriding interest in using high-pressure pipe has led these industries to use primarily steel pipe. Steel is selected not only for its high strength but also for its ductility, which allows the pipe to bend and even receive relatively large impacts without fracture. An inherent problem with steel is corrosion, which shortens the life of steel pipelines, especially if cathodic protection is not used or incorrectly applied (see Chapter 11). When steel pipes are designed to withstand high internal pressures generated by pumping the fluid through the pipelines and have the extra thickness for protection against the loss of thickness by corrosion, they will have more than adequate strength to withstand external loads, which are usually much smaller than the internal load for steel pipes. Consequently, the prime concern in the design of steel pipe is internal pressure and corrosion. This explains why in most cases steel pipes are laid in ditches without paying much attention to leveling and grading of the land before the pipe is laid, and without using bedding material such as gravel or sand, required for concrete or cast-iron pipes. This greatly reduces the construction cost of steel pipelines. In special cases, steel pipes are provided with paddings to protect from possible coating damage that may result from laying on hard rock.

In contrast, in the water and sewer industries, the pipes are normally under relatively low pressure, sometimes under atmospheric pressure inside the pipe, as in the case of gravity sewers. The low pressure coupled with the need for the use of noncorroding pipes has led the water and wastewater industries to use low-stress but noncorroding materials such as PVC and concrete for pipes. For such pipes, the external loads are often as important or more important than the internal load. Therefore, these industries pay great attention to external load. Often, as in the case of gravity-flow pipes, as long as the pipe is designed to be strong enough to withstand external load such as caused by the weight of earth on buried pipes, the pipes will be able to withstand the near-zero gage pressure inside the pipe.

The foregoing discussion shows that depending on their prime applications, different industries design pipes with different emphasis and using different criteria. In spite of that, there is a strong need to classify pipe design according to science and engineering principles rather than industry or types of applications. Such a need exists because even for the same industry, different design methodologies and criteria are needed at times. For instance, the water industry and the wastewater industries sometimes also must use pipes subjected to high internal pressure. In these cases, internal pressure becomes an important design consideration. The design may lead to the use of steel pipe with a lining made of an inert material such as epoxy, so that the pipe will have the high strength of steel and still be corrosion resistant. Therefore, instead of discussing the various designs used by various industries for different applications, this book chooses to classify designs based on three broad categories: (1) high-pressure pipes, (2) low-pressure pipes, and (3) intermediate-pressure pipes. High-pressure pipes are those where the internal pressure of the pipe is so high that it dominates the design. In this case, the prime attention of the designer is to ensure the integrity and safety of the pipeline from bursting or leaking caused by high internal pressure. Most long-distance petroleum and natural gas pipelines belong to this category. Low-pressure pipes are those where the internal pressure is so low, or nonexistent, that the design is governed by external loads. Most sewer pipes and culverts belong to this category. Finally, for intermediate-pressure pipes, the internal pressure load and the external loads are of similar magnitudes and hence both must be considered. This group includes pressure sewer pipes, water pipes, and certain petroleum and natural gas pipes, especially those that have large diameter and relatively thin wall, and those that are subjected to large external loads, such as underwater pipes, and pipes that are subjected to earthquakes or other natural disasters. Even though it is desirable from the learning standpoint to classify pipes according to the aforementioned three categories, it is not possible to give specific figures (numbers) for the pressures associated with each category because other factors such as pipe material and different environmental conditions also affect the design. Faced with the design of any given project, the designer must carefully weigh various factors pertaining to the project in order to determine which of the three categories the pipe falls into, and more importantly, determine what factors should be considered and evaluated in the design. The designer should not merely follow designs of similar previous projects because design differs from case to case, and, in addition, previous designs may or may not have been the best designs.

From a design standpoint, low-pressure pipes are further classified into two broad categories: (1) rigid pipes, and (2) flexible pipes. Rigid pipes are those whose deformation under load is negligible and hence the load does not depend on the pipe deformation. In contrast, flexible pipes are those whose deformations under load are significant and affect the load. This distinction is especially important for pipes buried underground. For instance, as a buried flexible pipe encounters a load such as the earth load from above it, the pipe deforms and pushes against the part of the backfill along both sides of the pipe, referred to as *sidefill*. This in turn provides a stronger external support to the pipe than before deformation. Consequently, by being flexible, the pipe can resist higher load without failure. Whether a pipe should be treated as flexible or rigid depends not only on the pipe material but also on the size and thickness of the pipe, and on the load. For instance, while steel pipes are usually treated as rigid pipes, when the pipe diameter is large and the pipe wall is thin, as in the case of large corrugated steel pipe used as culverts, the deflection under load is usually large and hence must be treated as flexible pipe. Plastic pipes are usually treated as flexible pipes. However, if the load is small, the pipe diameter is small, and the wall is thick, even a plastic pipe can be considered rigid. In the following sections, both internal and external loads will be considered, so that they can be applied to various projects and various types of pipes.

13.2 LOAD CONSIDERATIONS

Pipelines must be designed for many types of load, including but not limited to, the stress due to pressure generated by the flow (internal pressure), external pressure by fluid if the pipe is submerged underwater, external pressure generated by the weight of earth and by live loads on underground (buried) pipelines, loads due to thermal expansion, earthquakes, etc. They are considered in separate subsections.

13.2.1 STRESS DUE TO INTERNAL FLUID PRESSURE

The circumferential stress in pipe wall due to internal pressure p_i generated by the flow inside a pipe is commonly referred to as the hoop tension, which can be calculated as follows.

Figure 13.1 depicts the balance of forces on half of the cross section of a pipe. The tensile force per unit length of the pipe is $2T = 2\delta\sigma_t$, where σ_t is the hoop tension and δ is the pipe thickness. This tensile force, $2T$, is balanced by the force in the opposite direction, caused by the internal pressure p_i as follows:

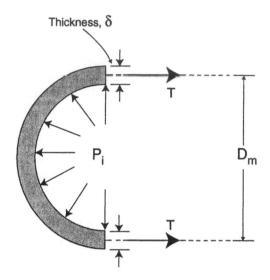

FIGURE 13.1 Analysis of hoop tension.

$$2T = 2\delta\sigma_t = p_i D_m \quad \text{or} \quad \sigma_t = \frac{p_i D_m}{2\delta} \tag{13.1}$$

where D_m is the mean diameter of the pipe, which is the average of the inner diameter D_i and the outer diameter D_o, namely, $D_m = (D_i + D_o)/2$.

From the above equation, the hoop tension stress σ_t is linearly proportional to the internal pressure and the pipe diameter, and it is inversely proportional to the thickness of the pipe. Equation 13.1 is strictly correct only for pipes with infinitely thin walls. When the pipe wall is thick, the maximum stress in the pipe due to hoop tension occurs at the inner side of the wall, having a magnitude equal to

$$\sigma_{max} = \frac{p_i(D_o^2 + D_i^2)}{D_o^2 - D_i^2} \tag{13.1a}$$

When the pipe thickness is small, it can be shown that Equation 13.1a reduces to Equation 13.1

In the above derivation, it was implicitly assumed that the internal pressure is uniform around the circumference of the pipe. In reality, due to the existence of gravity, the pressure around the circumference of a pipe is never truly uniform, unless the pipe is vertical and hence the circumference is in a horizontal plane. However, this nonuniformity of pressure caused by gravity is usually small and can be neglected, except in cases where the pipe is unusually large, the fluid is a liquid instead of a gas, and the dynamic pressure is small, as near the outlet of a large water pipe. In such a case and in certain other applications, the increase of pressure with the decrease in elevation across a pipe may need to be considered. The same can be said about external pressure generated by liquids. Because the density of gas is very small, for all practical purposes the internal and external pressures generated by gas in contact with a pipe is to be considered uniform over any cross section of the pipe.

In the foregoing treatment of hoop tension, it was implicitly assumed that there is no external pressure on the pipe. When both internal and external pressures exist, the quantity p_i in Equations 13.1 and 13.1a should be changed to the differential pressure $\Delta p = p_i - p_o$. When p_i is greater than p_o, the quantity $(p_i - p_o)$ is positive and the hoop stress is tension. In contrast, if p_i is smaller than p_o, then $(p_i - p_o)$ is negative, and the hoop stress is compression. The quantities p_i and p_o can be expressed either both in gage pressure or both in absolute pressure. The advantage of expressing both in gage pressure is that when a pipe is exposed to the atmosphere, as it is often the case, p_o becomes zero, which simplifies calculations.

When using Equation 13.1 for pipeline design, one must consider both the internal pressure p_i generated by the steady flow in the pipe, p_1, and the internal pressure generated by unsteady pressure surges, p_2, such as caused by water hammer, as long as a significantly large unsteady pressure may exist during operation, including emergency shutdown. Generally, a pipeline should be designed against both p_1 and p_2, namely,

$$p_i = p_1 + p_2 \tag{13.2}$$

13.2.1.1 Steady Pressure

The steady pressure p_1 can be calculated by using the one-dimensional energy equation given in Chapter 2. For a horizontal pipe, the maximum steady pressure p_1 occurs immediately downstream of pumps. For pipelines that dip deeply into a valley, the place of maximum p_1 may occur at the lowest point of the pipelines. In each case, the designer must use the one-dimensional energy equation to calculate the highest p_1 in the line.

Example 13.1 An 8-inch steel pipe carries water from location A to location C separated by a distance of 10 mi. The pipeline dips into a valley with the lowest elevation point B being 2 mi downstream of A. The elevations of points A, B, and C are 500 ft, 100 ft, and 520 ft, respectively. The velocity of the flow is 5 fps. Find the points of maximum pressure and design the pipe against such pressure. Assume that the maximum allowable tensile stress of the steel pipe is 20,000 psi.

[Solution] Using the one-dimensional energy equation from point A (immediately downstream of pump to point C (the pipe end) yields:

$$\Re = \frac{5 \times 8}{12 \times 10^{-5}} = 3.33 \times 10^5, \quad \frac{e}{D} = \frac{0.00015 \times 12}{8} = 0.000225$$

and

$$\frac{p_A}{\gamma} = \left(z_c - z_A\right) + \left(h_L\right)_{Ac} = (520 - 500) + f\frac{10 \times 5280}{8/12} \times \frac{5^2}{64.4} = 20 + 30745f \quad \text{(a)}$$

From the Moody diagram, $f = 0.0165$. Substituting this f value into Equation a yields $P_A/\gamma = 20 + 507 = 527$ ft, and $p_A = 32,885$ psfg $= 228$ psig.

To investigate the pressure at B, we use the energy equation between A and B:

$$\frac{p_B}{\gamma} = \frac{p_A}{\gamma} + \left(z_A - z_B\right) - \left(h_L\right)_{AB} = 527 + (500 - 100) - 101 = 826 \text{ ft}$$

Therefore, $p_B = 51,540$ psfg $= 358$ psig.

The foregoing calculation shows that the highest pressure point in this pipeline during steady-state operation is at the bottom of the valley, B. From Equation 13.1, the thickness of the pipe at or near B must be

$$\delta = \frac{p_i D_m}{2S} = \frac{358 \times 8}{2 \times 20000} = 0.0716 \text{ inches}$$

Since a standard NPS 8-inch pipe has a wall thickness of 0.332 inch (see Chapter 8), a standard 8-inch steel pipe is more than adequate for this pipeline. In actual design of pipelines, one should always use the real pipe sizes as given in Tables 8.3 and 8.4 for different pipes.

13.2.1.2 Unsteady Pressure (Water Hammer)

The unsteady pressure p_2 caused by pressure surges (water hammer) can be calculated as follows. From Equation 2.79, the celerity of pressure waves in any pipe is

$$C = \frac{C_o}{\sqrt{1 + \varepsilon \dfrac{D}{\delta}\dfrac{E}{E_p}}} \quad \text{(13.3)}$$

where E is the bulk modulus of the fluid; E_p is the Young's modulus of the pipe material; ε is a constant that depends on the type of pipe and the pipe-support system (see Chapter 2); and C_o is the celerity of pressure waves in perfectly rigid pipe, calculated as follows:

$$C_o = \sqrt{E/\rho} \quad \text{(13.4)}$$

where ρ is the fluid density. For water, $E = 300,000$ psi, and $\rho = 1.94$ slugs/ft^3. Substituting these values into the above equation yields $C_o = \sqrt{300,000 \times 144/1.94} = 4,720$ fps.

Once C is calculated from Equation 13.3, the maximum possible pressure due to water hammer is

$$\Delta p = \rho CV \qquad (13.5)$$

where V is the mean velocity in the pipe.

Note that Equation 13.5 is for rapid valve closure, i.e., valve closure time T_c shorter than $2L/C$. If the closure time T_c is longer than $2L/C$, then the water hammer pressure will be reduced according to the formula:

$$\Delta p = \frac{2L}{CT_c} \rho CV = \frac{2L}{T_c} \rho V \qquad (13.6)$$

Example 13.2 In the previous example, if a valve at the end of the pipe (location C) is closed in 10 s, what is the maximum water hammer pressure generated in this pipe?

[Solution] In this case, $C_o = 4720$ fps, $\varepsilon = 1.0$, $D = 8$ inches, $\delta = 0.322$ inch, $E = 300,000$ psi, and $E_p = 30,000,000$ psi. From Equation 13.3, $C = 4,220$ fps. Therefore, $2L/C = 25.0$ s, which is greater than T_c. This means that the valve closure must be regarded as rapid, and that Equation 13.5 is applicable, yielding $p_2 = \Delta p = 284$ psi.

The above calculation shows that the water hammer pressure is approximately 284 psi. If the pipe is to withstand both the static and dynamic pressures, then the pressure at B is $p_B = p_1 + p_2 = 358 + 284 = 642$ psig. From Equation 13.1, this internal pressure generates a hoop tension in a standard 8-inch steel pipe (thickness = 0.322 inch) of 7975 psi. Since this is lower than the allowable stress of the steel pipe, the standard pipe is adequate to handle the internal pressure. The extra thickness increases the rigidity of the pipe and extends pipeline life against corrosion.

13.2.1.3 Hydrostatic Pressure

In special cases where there is a large change of elevation of a pipeline, such as encountered in cross-mountain pipelines, a large hydrostatic pressure, p_s, may be developed in the low-elevation part of the pipe when the flow is stopped by closing a valve downstream. In such a situation, p_s may be higher than the combined steady-unsteady pressure given by Equation 13.2. Then, p_s becomes the dominating pressure to determine pipe thickness. Furthermore, if the flow is stopped by the valve while the pump is left running, then in addition to this hydrostatic pressure one must also include the pressure developed by the pump head H_0 corresponding to zero discharge. The governing pressure in this case is

$$P_o = P_s + \gamma H_0 \qquad (13.7)$$

where γ is the specific weight of the fluid, and H_0 is the pump head at zero discharge.

Depending on the situation, the designer must determine whether Equation 13.2 or Equation 13.7 governs the design.

13.2.2 STRESS DUE TO EXTERNAL FLUID PRESSURE

A pipe may be subjected to an external fluid pressure that is greater than the internal pressure, as in the case of underwater pipelines, or when a vacuum (negative pressure) is created inside a pipe. In such a situation, a net inward pressure exists on the pipe exterior, which generates a hoop compression that can be calculated in a manner analogous to the hoop tension by using Equation 13.1 or 13.1a, with p_i switched to p_o. However, before the pipe fails by compression, buckling usually occurs. The critical buckling pressure can be calculated from

$$p_b = \frac{3E_p I_t}{\left(1 - \mu_p^2\right) r_m^3} \tag{13.8}$$

where r_m is the mean radius of the pipe; μ_p is the Poisson's ratio; and I_t is the moment of inertia of the pipe thickness, which is equal to $\delta^3/12$. In terms of the mean diameter D_m and the wall thickness δ, the above equation can be rewritten as

$$p_b = \frac{2E_p}{\left(1 - \mu_p^2\right)\left(D_m / \delta\right)^3} \tag{13.9}$$

Example 13.3 Suppose an 18-inch PVC sewer pipe is laid under a lake of 80 ft of water. Before the pipe is connected to the rest of the sewer line, its interior is filled with air at atmospheric pressure. Calculate the minimum thickness of the pipe to prevent buckling.

[**Solution**] PVC has a Young's modulus of E_p = 400,000 psi, and a Poisson's ratio of μ_p = 0.38. The buckling pressure in this case, generated by the hydrostatic pressure of 80 ft of water, is p_b = 62.4 × 80 = 4992 psf.

From Equation 13.9,

$$\left(D_m / \delta\right)^3 = \frac{2 \times 400,000 \times 144}{\left(1 - 0.38^2\right) \times 4992} = 26,971$$

Therefore, D_m/δ = 30.0, and δ = 18/30.0 = 0.60 inch. This shows that the minimum thickness of this pipe must be 0.6 inch.

13.2.3 STATIC EARTH LOAD ON BURIED PIPE

13.2.3.1 Marston's Theory and Classification of Buried Conduits

According to Marston's theory [1], the earth load on an underground (buried) pipe or conduit is normally not the same as the weight of the column or prism of earth above the pipe. This load may be either greater or smaller than the weight of earth above the pipe, depending on the rigidity of the pipe, the compactness of the soil, and the construction methods used to bury the pipe and to prepare the backfill.

Marston's theory is based on the concept that when the column of backfilled soil directly above a buried pipe settles more than the surrounding soil column, some of the weight of the soil column directly above the pipe is transferred to the adjacent soil column. The result is that the load on the buried pipe in this case is less than that calculated from the weight of the soil column directly above the pipe. This condition exists when a pipe is placed in a trench and when the backfill is not compacted.

On the other hand, if the backfill directly above a pipe is compacted, the settlement of the column of soil above the pipe may be less than that of the surrounding soil column. In this situation, a reverse arch action develops, and a portion of the load from the surrounding earth column is now transferred to the buried pipe, causing a total vertical load on the pipe greater than the weight of the earth directly above the pipe.

For load computation, buried pipes are divided into two main categories: **ditch conduits (trench conduits),** and **projecting conduits (embankment conduits).** A ditch conduit is a pipe (or conduit) installed in a deep narrow ditch dug in undisturbed soil; the ditch is then backfilled. Examples of this type of conduit are sewers, drains, water mains, gas mains, and buried oil pipelines. Figure 13.2a depicts this kind of buried pipe.

Projecting conduits are further divided into two groups: positive and negative projecting conduits. A **positive projecting conduit** is a conduit or pipe installed in shallow bedding with the top of the pipe cross section projecting above the natural ground surface; the pipe is then covered with earth to form an embankment as shown in Figure 13.2b. Highway and railroad culverts are often installed this way.

A **negative projecting conduit** is a conduit installed in a relatively narrow and shallow ditch with the top of the conduit below the natural ground surface; the ditch is then backfilled with loose soil and an embankment is constructed. This method is very effective in reducing the load on the conduit, especially if the backfill above the conduit is loose soil. Figure 13.2c depicts this type of construction.

Finally, if a positive projecting conduit has a loose backfill directly above it as shown in Figure 13.2d, then the conduit behaves similarly to a negative projecting conduit, effectively transferring the load to the surrounding soil. This type of construction is called **imperfect-ditch conduit,** or **induced-trench conduit**. Although effective in reducing loads on conduits, this type of construction with loose backfill encourages channeling of seepage flow through the embankment. It should not be used in wet areas.

13.2.3.2 Rigid Conduit in Ditch

If a rigid conduit is placed in a ditch as shown in Figure 13.2a, and if the fill between the sides of the conduit and the ditch (i.e., the sidefill) is relatively loose, then most of the vertical load on the top of the conduit will be transferred to the conduit and the ditch wall rather than the side fills. In this case, the load per unit length of the conduit is

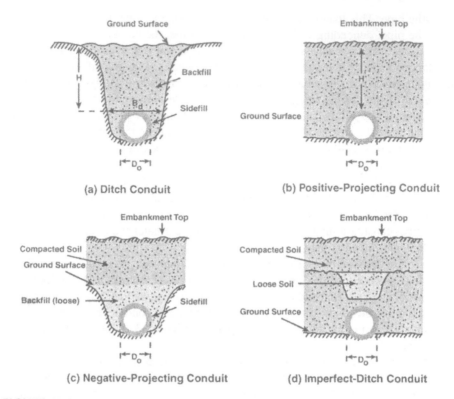

(a) Ditch Conduit

(b) Positive-Projecting Conduit

(c) Negative-Projecting Conduit

(d) Imperfect-Ditch Conduit

FIGURE 13.2 Types of buried pipes or conduits.

$$W_c = C_d \gamma_s B_d^2 \tag{13.10}$$

in which γ_s is the unit weight (specific weight) of the soil above the pipe, B_d is the width of the ditch, and

$$C_d = \frac{1 - EXP\left[-2Kn'H / B_d\right]}{2Kn'} \tag{13.11}$$

Note that H is the height of fill above the top of the conduit; n' is the coefficient of friction between fill material and the sides of the ditch, and K is the ratio of active lateral (horizontal) pressure to vertical pressure, which can be calculated from

$$K = \frac{\sqrt{n^2 + 1} - n}{\sqrt{n^2 + 1} + n} = \frac{1 - \sin A}{1 + \sin A} \tag{13.12}$$

where n is the coefficient of internal friction of the fill material, and A is the angle of repose of the material.

TABLE 13.1
Values of Kn' for Various Types of Backfill Soil

Type of Soil	Value of Kn'
Granular materials without cohesion	0.1924
Maximum for sand and gravel	0.165
Maximum for saturated topsoil	0.150
Ordinary maximum for clay	0.130
Maximum for saturated clay	0.110

Source: Data from Spangler, M.G. and Handy, R.L., *Soil Engineering,* 3rd ed., Intext Educational Publishers, New York, 1973.

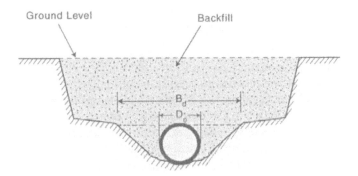

FIGURE 13.3 Determination of B_d for trench of an arbitrary cross-sectional shape.

Although the values of K and n' can be determined from laboratory tests of soil samples, in practice they are determined simply by classification of soil type as given in Table 13.1. When the characteristics of the soil is uncertain, it is usually assumed that $\gamma_s = 120$ lbs/ft³ and $Kn' = 0.150$. The value of B_d used for nonrectangular ditches should be the width of the ditch at the top of the conduit (see Figure 13.3).

From Equation 13.11, an increase in width B_d causes a decrease in the ratio H/B_d, which in turn causes a decrease in the coefficient C_d. However, since W_c is proportional to the square of B_d in Equation 13.10, a small change in B_d results in a larger change of B_d^2 than in H/B_d. Consequently, the load on the conduit in a trench, W_c, increases with increasing B_d at a power of less than 2.0. When the trench is very wide, say $B_d > 3D_o$ (where D_o is the outer diameter of the conduit), the load on the conduit W_c should approach that of an embankment conduit. Therefore, values calculated from Equation 13.10 should never exceed that of an embankment conduit of the positive projection type. If it did, the latter must be used. Calculation of the earth load on embankment conduits will be discussed later.

Example 13.4 A concrete pipe of 3-ft outer diameter is laid in a rectangular ditch of 5-ft width. The top of the ditch is 6 ft above the top of the pipe. The backfill material is saturated topsoil having a specific weight of 120 lb/ft³. What is the earth load on the pipe?

[Solution] For saturated topsoil, Table 13.1 gives the maximum value of Kn' as 0.150. But, $H/B_d - 6/5 = 1.2$, and $\gamma_s = 120$ lb/ft³. So, from Equation 13.11, $C_d = 1.008$. Substituting these values into Equation 13.10 yields $W_c = 3024$ lb/ft.

13.2.3.3   Flexible Conduit in Ditch

If the conduit is flexible and the sidefill material is tamped to the degree that the stiffness of the sidfill is essentially the same as the stiffness of the pipe, more load is now transferred to the sidefill. In this case, the vertical load on the top of the conduit is, according to Marston's theory on flexible conduits,

$$W_c = C_d \gamma_s D_o B_d \tag{13.13}$$

Comparison of Equation 13.10 with Equation 13.13 shows that the load W_c calculated from the former equation (for rigid conduits in ditches) is always greater than that from the latter equation (for flexible conduits in ditches) by a factor of B_d/D_o, which is greater than 1.0. From practice, it is known that the load on a pipe buried in a ditch lies between the values predicted from Equations 13.10 and 13.13. In his book [2], Moser reported that for flexible ditch conduits, the simple soil prism formula yields good conservative predictions of the loads:

$$W_c = \gamma_s H D_o \tag{13.14}$$

Strictly speaking, Equation 13.14 is correct only if the soil above the pipe behaves like a liquid or slurry of specific weight equal to γ_s, or if the soil above the pipe contains only pressure (normal stress) and no shear (tangential stress). However, Equation 13.14 is a good equation to use for flexible conduits in ditch not only because it yields conservative predictions but also because the predictions are rather realistic for wet soils, and soils that have settled over a long time to reach the equilibrium state of little shear. Since pipelines should be designed for a minimum of 50 years of life, being able to resist the maximum load that may occur during the life of the pipeline, assuming the condition or state of little shear in the soil is realistic from the design standpoint.

Example 13.5 Suppose that the conditions of the previous example prevail except for the fact that the pipe is flexible. Calculate the load by using both the prism formula and Equation 13.13 by Marston, and compare the results.

[Solution] For the prism formula, $\gamma_s = 120$ lb/ft³, $H = 6$ ft, and $D_o = 3$ ft. Thus, from Equation 13.14, $W_c = 2160$ lb/ft. On the other hand, when Equation 13.13 is used, $C_d = 1.008$, $\gamma_s = 120$ lb/ft³, $D_o = 3$ ft, and $B_d = 5$ ft. The equation yields

W_c = 1814 lb/ft. This shows that the load predicted from the prism formula is significantly higher than that predicted from Marston's equation for flexible pipe. The former should be used as the design load. Comparing with the result of W_c = 3024 lb/ft found in the previous example for the load on rigid pipe, the flexible pipe solutions — both from the prism formula and from Marston's equation of flexible ditch conduits — yield significantly lower values. This shows the importance of distinguishing flexible conduits from rigid conduits.

13.2.3.4 Embankment Conduit

For embankment conduits of the positive projecting type, the load transmitted to the top of the conduit or pipe will be equal to the weight of the soil directly above the pipe, plus or minus the total frictional force developed along the two vertical planes bounding the pipe. Unless the soil outside the bounding planes is compacted and the soil inside the planes is not, there is a tendency for the outside soil to settle more than that inside, resulting in a downward friction force on the center column. In this case, the total load on the conduit will be greater than the weight of the soil directly above the pipe. The opposite holds when the soil outside the plane is compacted and settles less than the soil above the pipe, or when the pipe is placed on a soft soil and settles under the load.

The load on a positive projecting conduit can be determined from

$$W_c = C_e \gamma_s D_o^2 \tag{13.15}$$

where D_o is the outer diameter of the pipe, and C_e is the load coefficient for embankment conduits, which is a function H/D_o, the coefficient of friction of the soil, n, and the product εr_s where ε is the projection ratio of the pipe (see Figure 13.4), and r_s is the settlement ratio. The settlement ratio determines the magnitude and direction of the frictional forces on the soil column above the pipe by the adjacent

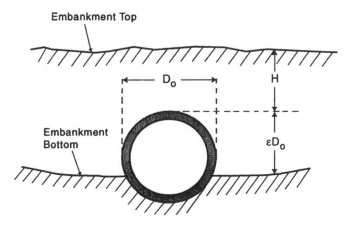

FIGURE 13.4 Determination of settlement ratio, ε, of embankment conduits.

TABLE 13.2

Variation of C_e as a Function of H/D_o and εr_s

H/D_o	-2	-1	-0.7	-0.5	-0.3	-0.1	0	0.1	0.3	0.5	0.7	1	2
0	0	0	0	0	0	0	0	0	0	0	0	0	0
1	0.80	0.80	0.82	0.83	0.85	0.87	1.00	1.21	1.34	1.43	1.5	1.57	1.76
2	1.5	1.5	1.5	1.5	1.5	1.7	2.00	2.44	2.73	2.93	3.09	3.2	3.2
3	2.07	1.81	1.9	2.03	2.18	2.51	3.00	3.67	4.12	4.43	4.68	4.95	5.62
4	2.45	2.45	2.51	2.63	2.87	3.33	4.00	4.90	5.51	5.93	6.27	6.64	7.55
5	2.75	2.75	3.00	3.25	3.56	4.15	5.00	6.13	6.90	7.43	7.86	8.33	9.63
6	3.14	3.25	3.55	3.86	4.25	4.97	6.00	7.36	8.29	8.93	9.45	10.0	11.4
7	3.23	3.69	4.10	4.47	4.94	5.79	7.00	8.59	9.68	10.4	11.0	11.7	13.3
8	3.49	4.16	4.65	5.08	5.63	6.61	8.00	9.82	11.1	11.9	12.6	13.4	15.3
9	3.72	4.63	5.20	5.69	6.32	7.43	9.00	11.1	12.5	13.4	14.2	15.1	17.2
10	4.05	5.10	5.75	6.30	7.01	8.25	10.0	12.3	13.9	14.9	15.8	16.8	19.1
11	4.39	5.57	6.30	6.91	7.70	9.07	11.0	13.5	15.2	16.4	17.4	18.5	21.1
12	4.68	6.04	6.85	7.52	8.39	9.89	12.0	14.7	16.6	17.9	19.0	20.2	23.0
13	4.97	6.51	7.40	8.13	9.08	10.7	13.0	16.0	18.0	19.4	20.6	21.9	24.9
14	5.26	6.98	7.95	8.74	9.77	11.5	14.0	17.2	19.4	20.9	22.2	23.5	26.9
15	5.55	7.45	8.50	9.35	10.5	12.4	15.0	18.4	20.8	22.4	23.8	25.2	28.8
16	5.84	7.92	9.05	9.96	11.2	13.2	16.0	19.7	22.2	23.9	25.4	26.9	30.7
17	6.13	8.39	9.60	10.6	11.8	14.0	17.0	20.9	23.6	25.4	26.9	28.6	32.6
18	6.42	8.86	10.2	11.2	12.5	14.8	18.0	22.1	25.0	26.9	28.5	30.3	34.6
19	6.71	9.33	10.7	11.8	13.2	15.6	19.0	23.4	26.4	28.4	30.1	32.0	36.5
20	7.00	9.80	11.3	12.4	13.9	16.5	20.0	24.6	27.8	29.9	31.7	33.7	38.4

soil. Spangler and Handy [1] give the following values for r_s: +1.0 for rigid pipe on the foundation of rock or unyielding soil; +0.5 to +0.8 for rigid pipe on the foundation of ordinary soil; 0 to +0.5 for rigid pipe on foundation that yields to adjacent ground; −0.4 to 0 for flexible pipe with poorly compacted sidefill; and −0.2 to −0.8 for flexible pipe with well-compacted sidefill. Once the value of r_s is found, the product εr_s can be calculated, and used to determine the value of C_e from Table 13.2. Then, Equation 13.15 can be used to determine the load on embankment conduits. Note that the prediction from Equation 13.15 also serves as an upper limit for Equation 13.10 for trenched conduits in which B_d/D_o is greater than 3. When calculating the load W_c for a trenched conduit having $B_d/D_o > 3$, both Equations 13.10 and 13.15 should be used. Whichever equation yields a smaller value of load should be used in the final design. Finally, it should be mentioned that, for values of C_e exceeding the range given in Table 13.2, predictions can be made by using a set of empirical formulas given in Reference 1.

Example 13.6 A corrugated steel pipe of 2-m diameter is used as a culvert to drain water across a highway. Suppose that the top of the highway is 6 m above the top of the pipe, and the pipe bottom is laid 1 m below the foundation (ground level) so that the projection ratio ε of the pipe is 0.5. The soil above the pipe has a density of 1925 kg/m³. The sidefills on the two sides of the pipe are well compacted to receive loads from above. Find the load on this pipe.

[Solution] The most conservative value of r_s for flexible pipe with well-compacted sidefills is –0.2. The projection ratio is $\varepsilon = 0.5$, hence, $\varepsilon r_s = -0.1$. Because $H/D_o = 6.0/2.0 = 3.0$, Table 13.2 yields $C_e = 2.51$. Then, from Equation 13.15, $W_c = 2.51 \times 1925 \times 2^2 = 19,330$ kg/m = 189,600 N/m.

13.2.3.5 Tunnel Conduit

For a pipe that forms the inside layer (lining) of a tunnel, or for jacked underground pipe (see Section 12.5.2 on pipe jacking) through otherwise undisturbed soil, application of Marston's theory yields

$$W_c = C_d B_t (\gamma_s B_t - 2C) \tag{13.16}$$

where, as before, W_c is the earth load on unit length of the conduit; C_d is the same as determined from Equation 13.11; B_t is the maximum width of the tunnel ($B_t = D_o$ for pipe jacking); and C is the cohesion coefficient, which is the cohesive force of the soil per unit area. C varies considerably with different types of soil. The U.S. Water Pollution Control Federation recommends the following approximate values for C: 0, 40, 100, 250, 300, and 1000 (in pounds per square foot) for loose dry sand, very soft clay, silty sand, medium clay, dense sand, and hard clay, respectively. Note that the values of C should never be greater than $\gamma_s B_t/2$; otherwise, Equation 13.16 is not applicable.

The static earth load considered above is often insignificant for designing high-pressure steel pipes used for long distance transportation of oil and natural gas. Design of these pipelines is often dominated by internal pressure, earthquake, earth settlement, and so forth. Such steel pipes usually have much greater thickness than required for the static earth load considered above. However, the earth load is very important for design of sewers and culverts, being the dominant load in many cases.

13.2.4 LIVE LOADS ON BURIED PIPE

When a pipeline crosses a highway, railroad, or airport from underground, not only does the earth above the buried pipe imposes a static earth load on the pipe, the vehicles (trucks, trains, and aircraft) rolling above the pipe also create a load — the live load. Unlike the static earth load, which increases with increasing buried depth, the live load generated on the pipe by vehicles decreases with the depth of earth above the pipe. This live load can be determined by using Table 13.3.

13.2.5 OTHER LOADS ON PIPELINES

Depending on individual cases, many other loads may need to be considered in the design of pipelines. For instance, when a pipeline passes through an earthquake

TABLE 13.3
Recommended Design Values for Live Loads on Buried Pipe

Live Load Transferred to Pipe (psi)				Live Load Transferred to Pipe (psi)			
Height of Cover, ft	Highway	Railway	Airport	Height of Cover, ft	Highway	Railway	Airport
1	12.50	—	—	14	NG	4.17	3.06
2	5.56	26.39	13.14	16	NG	3.47	2.29
3	4.17	23.61	12.28	18	NG	2.78	1.91
4	2.78	18.40	11.27	20	NG	2.08	1.53
5	1.74	16.67	10.09	22	NG	1.91	1.14
6	1.39	15.63	8.79	24	NG	1.74	1.05
7	1.22	12.15	7.85	26	NG	1.39	NG
8	0.69	11.11	6.93	28	NG	1.04	NG
10	NG	7.64	6.09	30	NG	0.69	NG
12	NG	5.56	4.76	35	NG	NG	NG
				40	NG	NG	NG

Note: NG = negligible.

zone, the loads induced by potential earthquakes and their effects must be considered to achieve an earthquake-resistant design. When a pipeline is elevated above ground, the effect of high winds on the pipe, both static and dynamic effects, must be considered and analyzed. When a pipeline must go undersea, loads due to ocean current and due to large forces during construction of the pipe, including those caused by the pipe's own weight while the pipe is being laid from a lay barge (see Section 12.7), must be carefully analyzed. And, when a pipeline is to be buried in an area of high groundwater elevation, the load generated by groundwater table fluctuations must be considered, etc. Due to the complexity of these analyses, it is beyond the scope and the page limit of this book to treat these loads. Readers interested in considering such loads should consult technical literature such as References 3 through 7.

13.3 PERFORMANCE ANALYSIS AND DESIGN

Structural design normally involves calculation of loads from which the maximum stresses, strains, and deformations of the structure are calculated, and then compared to the limiting states such as the allowable stresses, strains, and deformations, in order prevent structural failure from causes such as rupture, buckling, crushing, or excessive deformation. Some of these design aspects for pipes are considered herein for both high-pressure and low-pressure pipes.

13.3.1 HIGH-PRESSURE PIPES

For high-pressure pipes, analysis and design are generally focused on the stress, deformation, and failure caused by high internal pressure. Determination of hoop tension and failure caused by high pressure was presented earlier in this chapter. In cases where water hammer is a common occurrence in a pipeline, not only must the highest pressure generated by water hammer be added to the steady pressure in the calculation of the hoop tension, the dynamic effects of the water hammer, including vibration and material fatigue, must also be carefully analyzed. Many critical external loads, such as those due to earthquake (in earthquake region), high winds (for elevated pipes), ocean current (for submarine pipes), thermal stresses (for pipes welded in hot weather), etc., must be considered. In contrast, because the pressure due to earth load in this case is much lower than the internal pressure, it can be safely ignored without risk in most cases. A few special design considerations are discussed next.

13.3.1.1 Effect of Temperature Change

For pipelines with rigid supports, the pipe is restrained by the supports to expand lengthwise. If significant temperature changes occur, due to either weather change or cooling following hot welding of a restrained pipe during repair, high stresses can be generated in the pipe to cause the pipe to break, buckle, or bend excessively, or destroy the supports. This effect can be analyzed by first calculating the elongation of the pipe when it is allowed to expand freely. For a pipe section of length L, the elongation due to temperature rise of ΔT is

$$\Delta L = \alpha \Delta T L \tag{13.17}$$

where α is the thermal expansion coefficient of pipe, which is a materials property (see Table C.8 in Appendix C). While positive values of $\Delta L(\Delta T)$ represent expansion (temperature increase), negative values of $\Delta L(\Delta T)$ represent shrinkage (temperature decrease). The equation is applicable to both expansion (lengthening of pipe) due to temperature rise and shrinkage (shortening of pipe) due to temperature drop.

To determine the thermal stress generated in rigidly supported pipes, all that one needs to do next is to force the freely expanded or shrunk pipe into the original length L. This, from Hook's law, will cause a stress of

$$\sigma_T = E_p \frac{\Delta L}{L} = E_p \alpha \Delta T \tag{13.18}$$

Note that no thermal stress exists when the pipe is unrestrained. Unrestrained pipe will expand freely without stress.

13.3.1.2 Effects of Pipe Bending

Bending happens whenever a pipe is not supported uniformly, and when a lateral load exists. For instance, for pipes elevated above ground, they are seldom supported uniformly along their entire length. Any lateral load applied to the pipe, including its own weight, causes the pipe to sag or bend between adjacent supports. From textbooks on strength of materials it is known that such bending causes a moment at any cross section of the pipe, which in turn generates a flexural stress (bending stress) that is a tensile stress. The maximum flexural stress happens at the location of maximum bending moment, which most often happens at the mid-span, and at the outer face or edge of the bend. The stress is a tension on the outer part of the bend, and compression on the inner side of the bend. Such bending stresses can also happen to buried pipelines when the bedding or ground support for a portion of the buried pipe is lost due to scouring, earthquake, or ground settlement. From strength of materials, see for instance Reference 7, the maximum flexural stress $(\sigma_f)_{max}$ developed from bending is

$$(\sigma_f)_{max} = \frac{M_{max}D_o}{2I_p} = \frac{4M_{max}}{\pi\delta D_o^2} \tag{13.19}$$

where M_{max} is the maximum moment; δ is the wall thickness; and I_p is the moment of inertia of the pipe cross section. Methods to find M_{max} for different types of beam support systems under different loads are given in strength of materials texts and hence are not discussed here. Equation 13.19 is derived for thin-wall pipes. Note that $(\sigma_f)_{max}$ can be used to determine both the maximum tensile stress that exists on the outermost surface of the pipe bend, and the maximum compressive stress that exists on the innermost of the bend surface. Whichever is more critical of the two will cause failure. For instance, from Equation 13.9, it can be proved that the inner wall may buckle when the maximum compression there exceeds the following:

$$(\sigma_f)_{max} = \frac{E_p}{(1-\mu_p^2)}\left(\frac{\delta}{D_m}\right)^2 \tag{13.20}$$

where μ_p is the Poisson's ratio. In terms of bend radius R_b, Reissner [8] found that buckling occurs when the bend radius is less than the following:

$$R_b = \frac{D_n^2}{1.12\delta} \tag{13.21}$$

where D_n is the nominal diameter of the pipe.

While buckling may govern the design of steel pipes and PVC pipes in bending, for concrete or clay pipes the flexural tensile stress generated on the outer side of the bend normally governs the design. This is so because such pipe materials have much smaller tensile strength than compressive strength.

Note that maximum bending of pipe is limited not only by the stress generated but also by other factors. For instance, ductile iron pipes often contain a cement lining. To prevent damage to the lining, such pipes are not allowed to have a mid-span deflection greater than $L/120$, where L is the span.

Whenever a pipe bends, the pipe cross section will deform, changing from a circular to an elliptic shape. Since the major axis of the ellipse is perpendicular to the plane of bending, the pipe diameter increases by an amount equal to ΔX along the major axis, and it decreases by an amount equal to ΔY along the minor axis. For small deformation, $\Delta X = \Delta Y$. The following formula derived by Reissner [8] can be used to calculate such deformation, namely,

$$\frac{\Delta X}{D} = \frac{1}{16}\left(\frac{D}{\delta}\right)^2\left(\frac{D}{R_b}\right)^2 \tag{13.22}$$

where R_b is the radius of curvature of the bend, and $\Delta X/D$ is termed the **ring deflection**. One should not confuse the ring deflection with the deflection (lateral movement) of the pipe centerline due to bending of the pipe. The former can occur in a perfectly straight pipe due to earth load, as will be seen in the discussion of the modified Iowa formula — Equation 13.26.

Pipe bending affects pipeline design in another important way. As was shown in Chapter 2 (Example 2.5 and Problems 2.7 and 2.8), the flow in a pipe can produce very large forces on pipe bends, especially when the fluid pressure is high. This requires careful design of thrust blocks, which are usually heavy reinforced concrete structures, to resist such thrusts.

13.3.1.3 Seismic Design of Pipelines

Designing a pipeline to withstand earthquakes is complicated because a strong earthquake can damage the pipe in many ways. Large vibrations and differential settlement of the ground can cause large bending and shear stresses in parts of the pipe. In the case of submarine pipelines, the earthquake can cause rapid movement of the pipe relative to the surrounding water, which in turn can cause large drag and fluid-induced vibration of the pipe. It is not possible for this elementary text to treat such a complex subject. Suffice it to mention that in order to minimize vibration damage to elevated pipes, the supports of pipes must be spaced at a sufficiently small distance so that the structure's natural frequency, f_n, will be much higher than the dominant frequency of the ground movement induced by earthquakes. This prevents resonant vibration, which can destroy pipe and its supports. To be safe, the spacing between supports, L, must be within the following limit:

$$L = \left(\frac{\pi}{2f}\right)^{1/2}\left(\frac{gE_pI_p}{w}\right)^{1/4} \quad \text{where } f \gg f_n \tag{13.23}$$

where f is the design frequency of the pipe in hertz (cycles per second); E_p is the Young's modulus of the pipe material; I is the moment of inertia of the pipe wall cross section; and w is the weight of the pipe (including the fluid in it) per unit length. Note that Equation 13.23 is dimensionally homogeneous. Thus, all the units used in the equation must be consistent. For instance, if L is given in feet, g must be given in ft/s^2, w in lb/ft, and EI in lb-ft^2 (E in psf and I in ft^4).

With a span determined from Equation 13.23, the corresponding seismic stress generated due to vibration is [4]

$$\sigma_e = 0.000488 iwGD_o L^2 / I_p \qquad (13.24)$$

where i is a stress intensification factor, and G is the seismic acceleration in gs, which is dimensionless. Again, the units used in Equation 13.24 must be consistent. More details on seismic design of pipelines can be found in References 4 and 5.

13.3.2 LOW-PRESSURE PIPES

For low-pressure pipes, analysis and design are focused on soil properties, soil–pipe interaction, installation (bedding) method, and the rigidity of the pipes. They are discussed in the following sections.

13.3.2.1 Soil Classification

As mentioned before, the design of low-pressure or nonpressure pipes is focused on external instead of internal loads. Especially important to the design of these pipes is the earth load, which depends on the properties and conditions of the soil. According to the commonly use Unified Soil Classification System (USCS), soil is classified into five broad categories [2]:

Class I: This includes manufactured angular granular materials, 0.5 to 1.5 inches (6 to 40 mm) size, including materials having regional significance such as crushed stone or rock, broken coral, crushed slag, cinders, or crushed shells.

Class II: Coarse sands and gravels with maximum particle size of 1.5 inches (40 mm), little or no fines, with more than 95% materials retained on a clean no. 200 sieve. This class is further broken down into four subgroups, GW, GP, SW, and SP, in the order of decreasing amount of coarse materials retained on a no. 4 sieve.

Class III: Fine sands and clay (clay-filled) gravels, with more than 50% materials retained on a no. 200 sieve. This class is further broken down into four subgroups, GM, GC, SM, and SC, in the order of decreasing amount of coarse materials retained on a no. 4 sieve.

Class IV: Silt, silty clays, and clays, including inorganic clays and silts of low to high plasticity and liquid limits, with less than 50% materials retained on a no. 200 sieve. This class is further divided into four subgroups, ML, CL, MH, and CH, in the order of increasing liquid limit.

Class V: Organic silts, organic silty clays, organic clays, peat, muck, and other highly organic materials. This class is further divided into three subgroups, OL, OH, and PT, in the order of increasing liquid limit.

More discussion about soil classification can be found in References 2 and 9.

13.3.2.2 Soil–Pipe Interaction

The soil–pipe interaction is highly complicated by the fact that the system is structurally indeterminate. This means that the forces and stresses between the soil and the pipe cannot be determined from using only statics and dynamics (Newton's laws). The stiffness properties of the pipe and of the soil must also be included in the analysis. Further complicating the matter is the fact that the soil properties vary both with space and time; they are three-dimensional and unsteady. Due to such complexity, most analyses of soil–pipe interaction is semi-empirical relying on many simplifying assumptions and experimental data. Recent advancement in high-speed digital computers has made it possible to solve some complex soil–pipe interaction problems numerically by using the finite element method (FEM). Still, such analysis has severe limitations and is a major undertaking in each case, often requiring enormous programming time and experimental data to calibrate the numerical model. A good discussion of the use of FEM to solve such problems can be found in Reference 2.

13.3.2.3 Rigid-Pipe Analysis and Design

13.3.2.3.1 Rigid Pipe Types and Bearing Strength

Low-pressure and nonpressure rigid pipes consist of four general types: (1) asbestos cement pipes, (2) vitrified clay pipes, (3) nonreinforced concrete pipes, and (4) reinforced concrete nonpressure pipes. They are covered by four ASTM Standards: C428, C700, C14, and C76. Designs involving these pipes are based on their bearing strengths determined from a standard laboratory test called the **three-edge bearing strength test**, or simply the **3-edge test**.

The 3-edge test is conducted using the system shown in Figure 13.5, which involves a test specimen (i.e., a section of the pipe to be tested) placed on two horizontal wood blocks. Another parallel bock is placed on the top of the pipe to transmit the load from the piston of a hydraulic press to the test pipe. The load is gradually increased until the pipe fails or develops severe cracks. The load that causes pipe failure gives the 3-edge bearing strength of the pipe, in the units of lb/ft, kips/ft (kilo pounds per foot), kg/m, or N/m. The 3-edge bearing strength divided by the pipe diameter gives the D-load of the pipe, which has the units of psi, psf, kg/m^2, or N/m^2.

13.3.2.3.2 Standard Installations

The Water Pollution Control Federation (WPCF) of the U.S. gives four types of beddings (installation details) for low-pressure or nonpressure rigid pipes [10]. They are identified as class A, B, C, and D. For all the four classes, the pipe is placed in a rectangular ditch having a minimum width 8 inches wider than the outer diameter of the pipe — 4 inches of minimum clearance on either side of the pipe. Class A uses concrete (plain or reinforced) to support either the top or the bottom parts of the pipe. When the bottom is supported by concrete, the top part is backfilled with soil and carefully tamped. On the other hand, when the top is filled with concrete, the bottom is filled with compacted granular material such as gravel or crushed rock. In class B, the pipe is supported on its bottom by a thick layer of granular material,

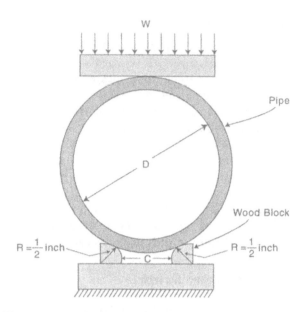

FIGURE 13.5 Three-edge bearing strength test.

and on its top by a thick layer of compacted backfill. Class C is similar to class B except for the fact that a thinner layer of compacted granular material is used on the bottom, and the backfill on the top is only lightly compacted. Finally, for class D the pipe is resting on the bottom of the ditch without using concrete or granular material. The entire ditch around the pipe is filled with loose backfill. Details of each of these four classes of construction are given in Reference 10.

13.3.2.3.3 Field Strength and Bedding Factor

The earth load that causes failure of a pipe, calculated from Marston's theory (Equation 13.10), is termed the **field strength**. Due to the support for the pipe provided by earth in the field, the field strength of a pipe is always greater than the 3-edge strength by a factor called **bedding factor** or **load factor.** Namely, the bedding factor is the field strength divided by the 3-edge strength. The bedding factors for the four classes of WPCF constructions discussed previously, A, B, C, and D, are 2.8 to 3.4, 1.9, 1.5, and 1.1, respectively.

13.3.2.3.4 Design Procedure

The design procedure for the rigid pipe systems includes the following steps: (1) determine the earth load by using Marston's theory (Equation 13.10); (2) determine the live load by using empirical data such as provided in Table 13.3; (3) combine the earth load with the live load, by adding them together; (4) select the type of construction according to the WPCF classification A, B, C, and D, and determine the corresponding bedding factor; (5) determine the safety factor from standards or codes (if no standards exist on safety factor, use a minimum of 1.5); (6) select the pipe strength by using the following formula:

$$W_3 = \frac{W_t N_s}{B} \tag{13.25}$$

where W_3 is the required 3-edge strength; W_t is the total combined load on the pipe which is the sum of the earth load W_c and the live load W_L; N_s is the safety factor; and B is the bedding factor. To illustrate this procedure, the following example is provided.

Example 13.7 A sewer made of a 24-inch (O.D.) concrete pipe is used at a large airport to drain storm water. The pipe crosses runways and taxiways at a depth of 10 ft below ground. The soil above the pipe is saturated topsoil, and the specific weight of the soil is 124 lb/ft^3. Select a pipe of appropriate 3-edge strength and define the installation method used.

[Solution] Following the six steps outlined above, we have:

1. **Determination of earth load** — For this problem, $Kn' = 0.150$, $H = 10$ ft, $B_d = 3.0$ ft (selected ditch width according to WPCF specification), and $H/B_d = 10/3 = 3.333$. From Equation 13.11, $C_d = 2.11$. Then, from Equation 13.10, $W_c = 2.11 \times 124 \times 3^2 = 2355$ lb/ft.
2. **Determination of live load** — From Table 13.3, the live load in terms of pressure at 10 ft below airport runways and taxiways is 6.09 psi. Since a linear foot (12 inches) of the 24-inch-diameter pipe covers a load area of $12 \times 24 = 288$ in^2, the resultant live load per linear foot is $W_l = 6.09 \times 288 = 1754$ lb/ft.
3. **Determination of total combined load** — The combined total load is $W_t = W_c + W_L = 2355 + 1754 = 4109$ lb/ft.
4. **Selection of installation type and determination of bedding factor** — Somewhat arbitrary, WPCF Class A installation with concrete bottom is used. For this type of installation, the bedding factor is 2.8, as given previously.
5. **The American Concrete Pipe Association (ACPA)** recommends that a safety factor N_s between 1.25 and 1.5 be used for concrete pipes. To be conservative, the value of 1.5 is used here.
6. **Finally, from Equation 13.25, $W_3 = 4109 \times 1.5/2.8 = 2201$ lb/ft.** Therefore, the pipe selected must have a 3-edge strength of 2201 lb/ft or greater.

13.3.2.4 Flexible-Pipe Analysis and Design

As a pipe of circular cross section with a diameter D is under earth load from above, the pipe deforms into an elliptical shape with a horizontal diameter increase of ΔX and a vertical diameter decrease of ΔY. Such deformation and deflections are to be referred to as **ring deformation** and **ring deflections**, respectively. In the literature, the relative ring deflection, $\Delta X/D_o$, is referred to simply as *ring deflection*. When both ΔX and ΔY are small as compared to D, the change of the circumference of the

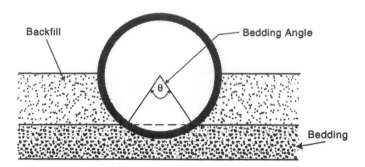

FIGURE 13.6 Bedding angle.

pipe due to this deformation is small, and ΔX is only slightly smaller than ΔY (say, $\Delta Y = 1.1\ \Delta X$). In most applications, ΔY is regarded to be equal to ΔX. The following equation holds:

$$\Delta X = \frac{KD_L W_c r_m^3}{E_p I_t + 0.061 E_s r_m^3}$$
(13.26)

This equation is usually referred to as the *modified Iowa formula* because it was first derived by Spangler in a slightly different form, and later modified by Watkins into the above form, both of whom did their work at Iowa State University. In the above equation, K is the bedding constant, which varies from 0.110 at $\phi = 0°$ to 0.083 at $\phi = 180°$, where ϕ is the bedding angle (see Figure 13.6); D_L is called the *deflection lag factor*, which is a dimensionless constant that reflects the time (normally years) that it takes before deflection will stop; W_c is load determined from Marston's formula for flexible pipe (e.g., Equation 13.13); E_p is Young's modulus of the pipe material; I_t is the moment of inertia of the pipe based on thickness, namely, $I_t = \delta^3/12$; and E_s is the Young's modulus of the soil. The deflection lag factor D_L recommended for the above equation is 1.5 if W_c is obtained from Equation 13.13. On the other hand, if W_c is obtained from the prism formula (Equation 13.14), then D_L is exactly one (1.0).

In Equation 13.26, the term $E_p I_t$ represents the stiffness of the pipe wall in resisting ring deflection, whereas the term $0.061 E_s r_m^3$ represents the stiffness of the soil around the pipe, especially the sidefill. If the sidefill is compacted as it is often the case based on good engineering practice, the soil stiffness becomes much larger than the pipe wall stiffness. Consequently, the modified Iowa formula reduces to the following simple form:

$$\Delta X = \frac{1.64 W_c}{E_s}$$
(13.27)

The above equation uses the average value of 0.10 for K, and uses the prism formula for computing W_c. The values of E_s can be found in a number of sources such as

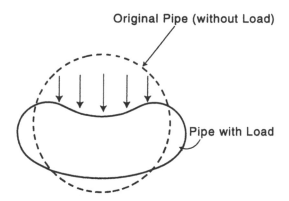

FIGURE 13.7 Reversal of pipe curvature due to over-deflection.

Reference 2. Its values for well-compacted sidefills are in the range of 1000 to 3000 psi.

Note that different pipes can tolerate different amounts of ring deflection, $\Delta X/D$. For instance, PVC pipes will not collapse into a reversal of curvature (see Figure 13.7) until the ring deflection is about 30% or 0.3. Normally, the design limit is 7.5%, which yields a safety factor of 4 from curvature reversal. For corrugated steel pipe, while the design limit is 5%, the reversal takes place at about 20%. Thus, the safety factor here is also 4. While tolerance of such large deflections is accepted in common practice for flexible pipes, such large deflections reduce the cross-sectional area of the flow, and prevents effective use of pigs for pipe cleaning and other purposes. For rigid pipes, ring deflection is normally within 2%, and is not a design concern. With such small deflection, pigs can pass through unhindered.

Ring deflection is not the only concern or limiting factor for designing flexible pipes. Many other factors that apply to rigid pipes and high-pressure pipes, such as buckling, earthquake loads, and stress caused by thermal expansion, must also be considered in the design of flexible pipes. For instance, PVC pipes expand and shrink approximately five times that of steel or cast iron. So, how to mitigate the effect of thermal expansion due to temperature change, such as by using gasketed joints, is an important design consideration for PVC pipes.

Although there are many commonalities in pipeline design, different pipe materials often require different special considerations. For example, PVC pipes become brittle at low temperature, and its strength reduces as the temperature becomes high. Such temperature effects must be carefully considered in the design of PVC and other thermal plastic pipes. Another example is fiberglass-reinforced pipe, which corrodes only when the strain in the pipe is high. To prevent such strain corrosion, designs must include calculation of the expected maximum strain to ensure that the strain is within the limit specified by the manufacturer.

Finally, it should be realized that, often, more than one factor contributes to the stress and strain in pipelines. When that happens, one must combine the stresses and/or the strain generated from different factors in order to determine the resultant stress or strain that governs the design. For example, a pipe may be subjected to not only large internal pressure but also significant external pressure due to earth load, and large thrust resulting from thermal expansion. In such a case, not only must the internal and external loads be combined properly for calculating the circumferential stress in the pipe wall (e.g., hoop tension canceling part of hoop compression), the thrust generated by thermal expansion, which is in the longitudinal direction, also contributes to the circumferential stress. From strength of materials, it is known that whenever a stress, σ_x, is generated in the x direction, it creates a strain not only in the x direction but also in the y direction, which is perpendicular to x. If the material is not free to expand or shrink in the y direction, then a stress instead of the strain will develop in the y direction, $\sigma_y = \mu_p \sigma_x$, where μ_p is Poisson's ratio. This stress is often referred to as *Poisson's stress,* which should not be forgotten in pipeline design. For any longitudinal stress σ_x generated in a pipe, Poisson's effect generates a stress $\sigma_y = \mu_p \sigma_x$ in the circumferential direction, which is an addition hoop stress. If σ_x is tension, σ_y is also tension. Likewise, if σ_x is compression, σ_y is compression. For more information on pipeline design, see References 2 through 5, 10, and 11.

PROBLEMS

13.1 A schedule 20 NPS 18-inch steel pipe is used to transport water at 6 ft/s. The pipe has a yield stress of 30,000 psi, and a Young's modulus of 3×10^7 psi. The pipeline is 12 mi long. There is a valve at the end of the pipeline and it is closed in 10 s in an emergency. (a) Determine the maximum pressure generated in the pipe due to this valve closure. (b) Determine whether the pipe thickness is adequate to withstand the calculated pressure.

13.2 A schedule 40 NPS 24-inch steel pipe is used to transport natural gas from an offshore platform to a point inland. The deepest submerged part of this pipeline is 1200 ft underwater. The density of the seawater is 2.0 slug/ft^3. (a) When pumping is stopped, the gas pressure inside the pipe is standard atmospheric pressure. Determine whether the external pressure generated by the seawater will cause the pipe to buckle. (b) During ordinary operation, the internal pressure of the pipe near the platform is 250 psig. Determine the magnitude of the net hoop stress in the pipe there, and whether it is compression or tension.

13.3 A PVC sewer pipe of 14-inch outer diameter and 0.5-inch thickness is placed in a trench. The sidefill is thoroughly tamped whereas the rest of the backfill is not. The backfill is saturated topsoil at a specific weight of 125 lb/ft^3. There is 3 ft of backfill above the pipe. (a) Find the earth load per linear foot of the pipe. (b) Determine whether this PVC pipe will buckle under this earth load.

13.4 A rigid pipe 3-ft in diameter is buried in a ditch as shown in the sketch. The backfilled soil is clay and its specific weight is 120 lbs/ft^3. The sidefill is not compacted (tamped). (a) Estimate the earth load on unit length of the pipe in lb/ft.

(b) If the pipe is flexible and the sidefill is tamped, what will be the earth load on this pipe per unit length?

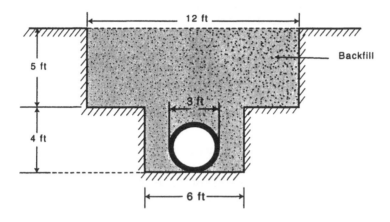

13.5 The same PVC pipe in the previous problem is used as a drainpipe to discharge water from a reservoir through an earth dam. The pipe is resting on a limestone foundation with 50% of the pipe embedded in the limestone (i.e., $\varepsilon = 0.5$). The dam height is 30 ft above the foundation (original ground surface). The earth dam is made of clay having a specific weight of 130 lb/ft^3. Find the earth load on this PVC pipe.

13.6 A concrete pipe of 50-inch outer diameter is placed under and across an existing railroad through pipe jacking. The top of the pipe is 10 ft beneath the railroad track. The soil through which the pipe is jacked is medium soft clay, and it has a density of 120 lb/ft^3. Find the earth load, the live load, and the combined load acting on the pipe.

13.7 A steel pipe of 2-ft diameter and 0.5-inch thickness is outdoors above ground, subjected to a seasonal temperature change of 110°F (from 100°F in summer to −10°F in winter). Expansion joints are used to enable a maximum of 3 inches of longitudinal movement of the pipes at each end of a joint. (a) Calculate the maximum spacing required between joints. (b) Calculate the thrust on each support if the pipe is girth welded instead of using expansion joints. (c) Repeat calculations of parts (a) and (b) if the pipe is PVC instead of steel. Compare and discuss the results.

13.8 A schedule 20 NPS 12-inch horizontal pipe carrying water at 70°F is simply supported at intervals of 20 ft. (a) Find the maximum flexural stress due to bending of the pipe under gravity. (b) Calculate the buckling stress due to such bending and determine whether the pipe will buckle. (c) Determine the maximum radius of bend of this pipe before buckling will happen. (d) Determine the ring deformation of this pipe before buckling takes place.

13.9 Suppose the pipeline in the previous example has a natural frequency of 20 Hz, and prospective earthquakes are expected to have a spectrum of acceleration with a peak at about 2 Hz. (a) Determine whether the 20-ft span is adequate for earthquake resistance. (b) If the maximum acceleration from the earthquake is expected to be 0.8 g, determine the maximum seismic stress generated in this pipeline due to vibration caused by earthquakes.

13.10 A concrete sewer pipe of 3-ft outer diameter and 1-inch thickness is installed using WPCF class B construction. The 3-edge bearing strength of the pipe, given by the manufacturer, is 3200 lb/ft. The safety factor to be used for design is 1.5. Determine the maximum combined earth load and live load that this pipe can be designed to support.

13.11 A PVC pipe of 3-ft outer diameter and 0.5-inch thickness is buried 8 ft underground. Well-compacted sidefill was used so that the value of E_s is expected to be about 2000 psi. The specific weight of the soil is 123 lb/ft³. Calculate the ring deflection of the pipe, and determine whether it is within the 7.5% required for PVC pipes.

13.12 Suppose that the steel pipe in Problem 13.2 is subjected to not only the external and internal pressures stated in part (b) of the problem, but also a thermal expansion force (thrust) resulting from an increase in the ocean temperature by 10°F. Determine how this thermal thrust affects the hoop stress and the safety of the pipe.

REFERENCES

1. Spangler, M.G. and Handy, R.L., *Soil Engineering*, 3rd ed., Intext Educational Publishers, New York, 1973.
2. Moser, A.P., *Buried Pipe Design*, McGraw-Hill, New York, 1990.
3. Uni-Bell Plastic Pipe Association, *Handbook of PVC Pipe Design and Construction*, Dallas, TX, 1979.
4. Nayyar, M.L., Ed., *Piping Handbook*, 6th ed., McGraw-Hill, New York, 1992.
5. American Lifeline Alliance, Guidelines for the Design of Buried Steel Pipe, FEMA/ASCE, Washington, D.C., 2001.
6. Liu, H., *Wind Engineering: A Handbook for Structural Engineers*, Prentice-Hall, Englewood Cliffs, NJ, 1991.
7. Timoshenko, S. and Young, D.H., *Elements of Strength of Materials*, 5th ed., Van Nostrand, Princeton, NJ, 1968.
8. Reissner, E., On final bending of pressurized tubes, *Journal of Applied Mechanics, ASME*, 386–392, 1959.
9. American Society for Testing Standards, *Classification of Soils for Engineering Purposes*, ASTM Standard no. D2487, American Society for Testing Standards, West Conshohocken, PA.
10. Water Pollution Control Federation, *Gravity Sanitary Sewer Design and Construction*, WPCF Manual of Practice no. FD-5, Water Pollution Control Federation, Washington, D.C., 1982.
11. Mohitpour, M., Golshan, H., and Murray, A., *Pipeline Design and Construction: A Practical Approach*, American Society of Mechanical Engineers, New York, 2000.

14 Pipeline Operations, Monitoring, Maintenance, and Rehabilitation

14.1 GENERAL OPERATION OF PIPELINES

There is a huge difference between operating a pipeline and operating a pipeline company, with the former being only a part of the latter. To operate a pipeline company, one must deal with not only the pipeline system, but also the business aspects of the company — including finance, accounting, marketing, human resources, legal, and public relations. It is beyond the scope of this book to deal with operating a pipeline company. This chapter will be confined to operating pipelines.

Operating a pipeline requires an understanding of the purpose of the pipeline, how the pipeline was designed and constructed, the codes and standards that govern the operation of the pipeline, the operational history of the pipeline, and the pipeline's current status. It also requires a good knowledge of pipeline engineering and many other related fields, such as corrosion control, automatic control, fluid mechanics, structural engineering, machine maintenance, etc. Therefore, it takes engineers and technicians of different disciplines and training working together as a team in order to keep a modern pipeline system running and maintained in good condition. Unqualified operators and/or inadequate training often result in improper operation of the pipeline and damage to the system, or frequent unscheduled shutdown. The use of modern computers and automatic control systems has greatly decreased the number of technical personnel needed to run a pipeline system; however, the knowledge and training required by the technical personnel who run a pipeline have been increasing steadily.

How to operate a given pipeline system depends on the purpose of the pipeline. All the operational strategies and details must be designed to achieve this purpose. For instance, simply stated, the Trans-Alaska Pipeline (now called Alyeska Pipeline) was constructed to bring the crude oil that exists in the North Shore of Alaska to the southern port of Valdez for further transport by ship to the contiguous states of the U.S. Once the purpose of a pipeline is defined, the strategy and operational details must be laid out in an operation/maintenance (O&M) manual. The strategy is normally planned around a set of operational parameters such as the discharge through the pipe, mean velocity, temperature range of the fluid transported, maximum and minimum pressures at various locations along the pipe, pump speed and head, valve closing speed, etc. These operational parameters are determined not only by the original design but also by the changing demands of the population it serves. For instance, as described in Chapter 2, the Big Inch Pipeline was designed during World War II to transport crude oil from East Texas to the Northeast states, but it was converted later to transport natural gas. Due to this change, the purpose of the pipeline changed, and the operational strategy and parameters also changed. This shows that the operational strategy and parameters that govern a pipeline may change with time, and the current situation governs the ongoing operation of the pipeline.

Based on operational strategy and parameters, and the existing equipment as well as other considerations, a detailed operational procedure can be and must be laid out and stated clearly in an **operations manual**. The manual should not only describe the routine procedure, but also what needs to be done in various emergency situations, such as when a leak in the main is detected. The manual will be needed not only for the operational crew of the pipeline, but also for design of the flow charts and computer programs used for automatic control of the pipeline system under both routine and emergency conditions. Separately, the maintenance needs of the pipeline must be carefully assessed, and the schedule and procedure for maintenance must be clearly spelled out in a maintenance manual.

14.2 AUTOMATIC CONTROL SYSTEMS

Most modern pipeline systems are automated and controlled by computers. The degree of sophistication of the automation/computer system depends on the size and complexity of the pipeline and its operation. Large, long, and complex pipeline systems require sophisticated automatic control systems. They often include three parts: (1) a **SCADA (supervisory control and data acquisition)** system that serves as the brain, to collect, receive, and process data, including on-line data processing, as well as sending command signals out to control various equipment, including those in remote locations; (2) the communication media that links the SCADA to the remote stations (e.g., booster pump stations), by use of satellites, microwave towers, fiber optic lines, or dedicated telephone lines; (3) the local control units, often referred to as the **RTUs (remote terminal units)**, which send data on local stations to the SCADA and receive instruction from SCADA for remote control of equipment.

The SCADA can be further divided into three parts: (1) a user-friendly **console** to interface with the operator (human), (2) a **server,** which contains the database needed for operating the pipeline, and (3) the **MTU (master terminal unit)** that communicates with the RTUs. Each RTU at a given station, such as a pump station, may be a personal computer. The RTU receives data from each piece of the equipment located in the station, transmits the data to the SCADA, and executes orders from the SCADA to control each piece of equipment, often through what is called a **PLC (programmable logic controller)**. Programming the PLCs and programming the computers to perform the complex tasks that are required of a pipeline system is a job for specialists. It is usually done by outside consultants rather than in-house personnel.

Automatic control of a modern pipeline system relies on accurate online data collected by a multitude of instruments including flowmeters and various transducers and sensors (such as pressure sensors, temperature sensors, and vibration sensors), most of which were discussed in Chapter 10. The instrument or transducer output signal (voltage) is usually an analog signal that must be first converted to a digital signal before it can be analyzed by any digital computer or PLC. Because the analog signal is usually small, it must first be conditioned (amplified) by a signal conditioner before feeding into the a-to-d (analog-to-digital) converter of the computer or PLC. To minimize the noise that may affect the signal, the signal conditioner must be placed as close as practical to the source of the signal (i.e., the instrument or transducer). Furthermore, the cable that transmits the signal must be shielded, grounded, and isolated from any possible source of electromagnetic noise such as a transformer, an AC power line, a motor, a generator, or a pump. Insufficient attention to such details often results in poor data acquisition by the computer, which in turn affects the automatic control of the pipeline system.

14.3 INTEGRITY MONITORING AND LEAK DETECTION

14.3.1 INTEGRITY MONITORING

Pipeline integrity monitoring includes a variety of measures taken to monitor the condition of the pipeline including its immediate environment, in order to determine or head off damage to the pipe and its associated equipment, maximize the efficiency and safety of the pipeline, minimize potential accidents and service interruptions due to pipeline neglect, and safeguard company and public interests. The following is a partial list of measures that should be included in pipeline integrity monitoring.

- Leak detection by using a variety of measures to be discussed in the next subsection.
- Inspection pigs used to examine pipes for dents, corrosion (loss of metals), and possible cracks. Chapter 10 contains a discussion of such pigs. Since this is an area of intensive current research and development, further improvement in pig-based sensors and devices to detect pipe damage, including micro cracks, is expected in the near future.
- Visual inspection of pipe exterior for any exposed pipe or exposed portion of a pipeline.
- Underwater inspection of pipe exterior for submarine pipes by using divers or special submarines carrying photographic equipment.
- Remote sensing by satellites for early detection of encroachment by heavy vehicles traveling on or across a pipeline right-of-way, or detection of other conditions that may threaten the pipeline's integrity, such as a flood, a landslide, or ground subsidence.

- Line patrols flying over pipeline right-of-way to detect problems or potential problems.
- Daily checking of pumps and other rotating machines used in running the pipeline.
- Checking of pressure regulators and pressure-relief valves.
- Checking of control valves and check valves.
- Checking the calibration of flowmeters, pressure transducers, and other sensors.

14.3.2 LEAK DETECTION

The U.S. has an enormous network of aging pipelines for transporting water, sewage, petroleum, and natural gas. The average age of these pipelines is over 30 years; some are over 100 years. A large number of them are either seriously corroded, or in disrepair in other ways. Many of them are leaking. To replace all of this aging underground infrastructure with new pipeline at once would be enormously costly for the public, posing unacceptable financial burden. The only sensible solution is to renovate, repair, and rejuvenate most of the pipe over many years on a rotational basis, in the same way other infrastructure such as highways are repaired and maintained. Eventually, all pipelines must be abandoned and replaced with new ones when they can no longer be repaired, or cost more to repair than to rebuild.

Pipeline leaks and ruptures have many causes. The most common cause is third-party damage, which results from accidents, for example, caused by excavation conducted by a third party, such as a contractor working on a land-development project, who is unaware that there is a buried pipeline in the work area. The second largest cause is corrosion, which in time causes leaks, and sometimes rupture. This is especially serious for high-pressure metal pipes, especially steel pipes used for transporting oil, natural gas, and liquid fertilizers such as ammonia, as compared to low-pressure plastic pipes. High-pressure concrete pipes also can corrode over time because of the steel embedded in the concrete — the steel reinforcement (bars and pretensioned wires) and steel cylinders. Corrosion-caused leaks and rupture are most common in aging pipes that have corroded for many years, some even with cathodic protection, which is never 100% effective. Other causes of pipeline leak or rupture include materials defect and outside forces, such as caused by soil movement due to earthquakes, washouts due to flood, landslides, frost, lightning, ice, snow, high winds, and operator error.

The danger or seriousness of pipeline leaks and rupture depends mainly on the type of fluid transported through the pipe. For instance, water and air are both mostly benign fluids. Therefore, any pipeline that transports water or air, or uses water or air to carry solids, such as capsule pipelines, are safe and environmentally friendly. When a water pipeline leaks, the consequence is solely or mainly economic — wasting potable water, and wasting energy used to pump water through pipe; it does not affect public safety, nor does it damage the environment. It is a cause for concern and requires action, but it is seldom an emergency. When a water pipe ruptures, even though it may not threaten life or the environment, it can cause serious disruption of lives, and hence requires immediate repair. The leakage or rupture of a crude-oil or petroleum-product pipeline is more serious because it poses a threat to both the environment and public health due to the pollution of surface water, groundwater, and/or soil. Product pipelines such as those transporting gasoline are more dangerous than crude-oil pipelines, due to the high volatility of the product and the resultant danger of fire and explosion. The same can be said about sanitary sewers, and sewers that carry industrial wastes. Even though the government does not classify sewers as hazardous pipelines, their hazardous nature should be recognized by the engineers dealing with sewers, and should be treated with the same degree of caution as crude petroleum pipelines. Sanitary sewers may even explode due to accumulation of methane, the same basic ingredient of natural gas. Explosion can occur in parts of a sewer with gravity flow when a free surface (open-channel flow) is present, and when the sewer is improperly vented. Storm sewers and drainage pipes, on the other hand, are designed to allow leakage. Such leakage poses no economic, public health, or environmental problem. However, storm sewers must be designed and operated with great concern for public safety. Improper design or improper maintenance of storm sewers can threaten lives, such as drowning children or even adults swept into or fallen into a storm sewer without a grate, or drowning caused by a localized flood due to inadequate flood-handling capacity of the storm sewer. For natural gas pipelines, the greatest threat is property and life lost from explosion and fire resulting from pipeline leaks or rupture due to the highly flammable nature of natural gas. Therefore, the utmost concern must be given to natural gas pipeline safety, and any leak or rupture of a natural gas pipeline is a life-threatening emergency. Finally, highly dangerous gases or liquids, such as cyanide, and highly radioactive wastes, such as those existing in nuclear weapons plants, are often transported by pipelines for relatively short distances. Such pipelines pose a high risk to the workers of those plants, and to the neighbors of such plants. They must be designed and operated with extreme care and leaks or ruptures of such pipes must be prevented at all costs. One may wonder why pipelines are used for transporting such highly dangerous chemicals. The answer is that a pipeline is the safest mode of transport, and it minimizes human contact with the hazardous materials. Using other modes for transporting these materials would pose even greater risks to workers and the public.

From the foregoing discussion, it can be seen that the need for preventing and detecting leaks and rupture of pipelines varies greatly with the type of fluid transported by the pipeline. However, whatever the need is and whatever type of pipeline one is dealing with, some common or similar methods exist to detect pipeline leaks. They are described briefly in the following sections.

14.3.2.1 Mass-Balance Method

The mass-balance method, also referred to as the **materials balance method**, is straightforward. It uses the continuity equation of one-dimensional flow between an upstream point and a downstream point to calculate the amount of flow due to leakage or rupture. For instance, if in a natural gas pipeline the mass flow rate upstream is $\dot{m}_1$ and the mass flow rate downstream is $\dot{m}_2$, and if there are no branches to divert the flow between the two points, then the leakage flow rate is simply $\dot{m}_L = \dot{m}_1 - \dot{m}_2$. Use of this method requires accurate measurements of the mass flow rates $\dot{m}_1$ and

$\dot{m}_2$, for the leakage flow is determined from the difference between the two quantities. Since even the newly calibrated flowmeters can have more than 0.5% errors, it is difficult to detect leakage flow from this method when the leakage rate $\dot{m}_L$ is much less than 1% of the flow in the pipe. This limits the usefulness of this method to large leakage flow — 1% or more of the flow rate in pipe. Another problem is that since flowmeters are spaced at great distances apart along a pipeline, sometimes more than 100 km apart, the method does not indicate where the leak is located within such a long distance. Other techniques are also needed to confirm and locate the leak. Within these limitations, the mass balance method is very useful and reliable, and it is widely used in current practice. Two things can improve the usefulness of this technique: (1) further improvement of the accuracy and reliability of flowmeters and their calibration methods, and (2) use of more flowmeters in a pipeline so that spacing between them can be reduced. The spacing between such flowmeters for hazardous liquid pipelines and natural gas pipelines should be much closer to each other in urban or residential neighborhoods, and in environmentally sensitive areas such as river crossings, so that leakage can be quickly pinpointed.

14.3.2.2 Pressure-Drop Method

Leakage in a pipeline can also be detected by sudden pressure drop along the pipe measured by pressure transducers. Again, the usefulness of this method is limited by the accuracy of pressure transducers and by the spacing between the transducers. Because the pressure drop along a pipe having turbulent flow is proportional to the square of the discharge Q, the relative pressure drop caused by leakage is double that of the relative drop of the discharge due to the same leakage. For example, if a leak causes a 2% decrease in discharge Q or mass flow rate $\dot{m}$ in the pipe, the corresponding pressure drop will be 4% and so forth. Due to this amplification factor, and due to the fact that pressure transducers cost less than flowmeters for large pipes, and can be tapped relatively easily along the pipe at close intervals, the pressure-drop method is useful especially when the spacing between transducers is small. In spite of this, the method has its own shortcomings compared to the flowmeter (mass-balance) method in that it needs more frequent calibration, it is more susceptible to local disturbance such as caused by imperfect tapping, and it requires more frequent maintenance.

14.3.2.3 Computational Modeling of Pipeline Systems

Both the mass-balance method and the pressure-drop method discussed above have advantages and disadvantages, and have about the same accuracy and reliability. Because of this, and because of the fact that most pipelines already have both flowmeters and pressure transducers, it makes sense to use both methods to enhance the reliability of leak detection — to confirm real leaks and to reduce false alarms. Both methods can and should be incorporated into a common SCADA program that runs the pipeline. This is usually done by setting up a system of equations on the computer, based on fluid mechanics and input data pertaining to the pipeline system, to predict the velocity V, discharge Q, pressure p, temperature T, and density ρ (for gas flow) at many locations along the pipeline. The predicted values are compared with the measured values to determine if some abnormality exists in the pipeline system. Whenever a leak, rupture, or overheating condition exists, the measured values at certain locations will differ noticeably from that of the computed normal values. From the difference, a leak or rupture of the pipe can be detected, and its approximate location can be determined. Emergency measures, such as shutting a valve and stopping a pump, can then be taken by the SCADA automatically, or through manual override. This explains in a nutshell the principles involved in such computational monitoring of pipeline integrity and safety.

In addition to the foregoing methods, other methods for leak detection based on fluid mechanics have also been proposed, such as by recording the high-frequency noise generated by leaks, or sending ultrasonic waves into the pipe and measuring the change of wave speed caused by leaks. However, since these methods are still in the early stage of development, more research and development is needed before they can be used reliably and easily.

14.3.2.4 Visual and Photographic Observations

Often, leaks have been discovered through visual observation of pipes or their immediate surroundings. This can be done not only for exposed pipes but also for underground and submarine pipelines. For instance, one can patrol the right-of-way of a buried petroleum or natural gas pipeline; withered vegetation in an area above the pipe may indicate leakage in that area. In contrast, grass above a certain area of the water pipeline growing more vigorously than elsewhere along the pipe may indicate a water leak from the pipe.

Likewise, by sending a diver or special submarine to patrol an underwater natural gas pipeline when bubbles are seen to rise from a certain spot on the pipe, a leak can be pinpointed. Such visual observation or discovery by divers or a submarine should be followed by taking photos of the scene for further analysis.

14.3.2.5 Ground-Penetrating Radar

A ground-penetrating radar is capable of detecting the spilled natural gas and petroleum that exists in the soil above the leak point of a buried pipeline. By moving this radar along the right-of-way in an all-terrain vehicle, the pipeline can be surveyed for leaks, and the location of any leak can be pinpointed.

14.3.2.6 Pigs

Pigging is a good way to find leaks because the pig moves inside the pipe in close contact with the pipeline. Smart pigs that carry special sensors (acoustic or electromagnetic sensors) can detect leaks. Unfortunately, to date insufficient research and development have been done in this area to develop the full potential of pigs for leak detection. More future research is needed in this area. The current generation of smart pigs detect wall thinning or loss of metal due to corrosion; however, they are insensitive to the existence of small leakage flow through cracks.

14.3.2.7 Dogs

The same Labrador retrievers used by law enforcement agencies to sniff out illegal drugs and explosives can be trained to detect natural gas leaks, provided that a special odorant is mixed with the gas. The dogs have proven to be very reliable, with a success rate of detecting leaks in excess of 90%. They can detect the scent at concentrations as low as 10^{-18} molar (1 part per billion billion). The dogs are reliable, and demand no reward other than room and board. They have proven to be successful not only under ordinary conditions, but also in extremely cold weather when the pipeline was 12 ft underground with an additional 3 ft of snow above. The dogs are more reliable and cost less than many sophisticated high-tech methods, but because they need training and care, their use in North America has been curtailed or discontinued in recent years.

14.4 INTEGRITY MANAGEMENT PROGRAM

In recent years, the U.S. government and many states have promulgated numerous laws and regulations on pipeline safety. Some of the provisions of these laws and regulations are not only costly for the operation of pipelines but also ineffective in improving safety. This has caused great concern to the pipeline industry. In response, the industry has reached a consensus and proposed to the government that each pipeline operator (i.e., pipeline company) should prepare its own pipeline integrity management program (PIMP) for review and approval by the U.S. Department of Transportation (DOT), which has jurisdiction over pipeline safety. Each PIMP can then identify the most serious risks unique to the company's pipelines so that priorities in risk-reduction measures can be set and manpower and financial resources can be used most effectively to minimize risks and to maximize safety.

The pipeline industry's concerns and suggestions were well taken by DOT and the U.S. Congress. For instance, the Pipeline Safety Act of 2000 (House of Representative Bill 5361), requires that each operator of hazardous liquid and/or natural gas pipelines develop a PIMP and submit it to DOT for review. DOT either accepts the PIMP or requests changes. The statute also requires the following from pipeline operators: (1) periodic inspections of pipelines at least once every 5 years in areas of high population or environmental sensitivity; (2) enhanced training and establishment of a certification program for employees who perform safety-sensitive pipeline functions; (3) expanded programs to inform government agencies and to educate the public on pipeline safety, including the operator's PIMP; (4) increased fines for the spill of oil and other hazardous liquids — $1,000 per barrel, with a maximum of $100,000 per violation and $1 million for a series of violations; (5) expanded states' roles in pipeline safety; (6) funding of nine Regional Advisory Councils to enhance citizen involvement in pipeline safety; and (7) enhanced funding for DOT's pipeline safety program so that the agency can perform its job in pipeline safety more effectively.

In November 2000, DOT issued pipeline safety rules, including the following: (1) any operator who owns at least 500 mi of oil or hazardous liquid pipelines must follow the rule (this includes about 87% of liquid pipelines in the U.S.); (2) each applicable operator must identify the pipeline segments that could affect a high consequence area, where **high consequence area** is defined to include unusually sensitive areas (e.g., protected wilderness areas), cities in excess of 50,000 people, other areas with populations under 50,000, or a commercially navigable waterway (this must be done within 9 months from effective date); (3) each operator must develop and submit a written integrity management program that addresses the risks to each pipeline segment that could affect a high consequence area (this must be done within 12 months of effective date); (4) each operator must complete and submit a baseline assessment plan, to assess the effectiveness of the measures taken to enhance safety (50% of this element must be completed within $3^1/_2$ years, 100% within 7 years, and ongoing testing on a 5-year schedule); (5) complete and submit the ongoing integrity management program (this must be done within 12 months of effective date).

Each pipeline integrity management program shall contain the following components: (1) a process for identifying the pipeline segments that could affect a high consequence area; (2) a baseline assessment plan; (3) an analysis that integrates all available information about the integrity of the entire pipeline and the consequences of a failure; (4) criteria for repair actions to address integrity issues raised by the assessment plan and information analysis; (5) a continual process of assessment and evaluation to maintain pipeline integrity; (6) identification of preventive and mitigation measures for protecting the high consequence area; (7) methods to measure the program's effectiveness; and (8) a process for the review of integrity assessment results and for information analysis.

The foregoing DOT rules are for oil and other hazardous liquid pipelines. A similar set of rules are being promulgated by DOT for the safety of natural gas and other hazardous gas pipelines — transmission lines. The only difference from the rules on hazardous liquid pipelines is in the definition of **high consequence area**. It is defined as housing areas, as well as facilities where people have restricted mobility or facilities that would be difficult to evacuate, such as hospitals, retirement homes, and prisons. Moreover, near larger pipelines operating under higher pressure, the high consequence areas would be expanded.

14.5 RISK-BASED MANAGEMENT

The kind of pipeline integrity management program preferred by industry is risk-based, which means giving highest attention to those items that present the highest risks. Risk is not simply the probability of occurrence of a damaging event. It is rather such probability multiplied by the severity or consequence of such events. If a risk is to be assessed scientifically, the probability of occurrence of the damaging event must be determined or at least estimated, and the severity or consequence of occurrence must be quantified as a cost item in dollars. Then, the risk can be calculated from $R = PC$, where R is risk in dollars P is probability, which is a dimensionless number less than or equal to 1.0, and C is cost in dollars. For instance, if the probability of explosion of a 5-mi segment of a natural gas pipeline passing through a city is 1% during the next 50 years, and if it explodes the anticipated damage includes 20 lives lost (at $1 million per person) and $80 million of property damage (total damage would be $100 million), the risk of such an explosion in the next 50 years will be $R = 0.01 \times \$100$ million = $1 million. Once this is known, to avert such a costly incident or tragedy in the next 50 years, the pipeline operator may want to consider two alternatives. The first alternative is to improve the condition of the pipe so that the probability of occurrence can be lowered to 0.001. Suppose that this will cost $5 million. Due to this improvement, the risk is now reduced to $R = 0.001 \times \$100$ million = $100,000. So, by spending $5 million, the pipeline company saved only $900,000, which may not appear to be a good investment from the standpoint of risk management alone. However, by spending the $5 million for infrastructure improvement, the economic life of the structure has been extended for 10 years. If the revenue of each year is $5 million, this means a total additional income of $50 million will be generated in the future due to extended pipeline life. This benefit far exceeds that derived from safety improvement. This shows that when considering investment on pipeline improvement, the benefit derived from extended pipeline life may far exceed that derived from safety improvement or risk reduction. What appears to be a bad investment from the safety standpoint can actually be an excellent investment for the company in the long run, based on overall economic considerations. This calculation should make pipeline operators more willing to invest in pipeline improvements.

Now consider a second alternative, which is to invest only $100,000 to educate the residents of the city through which the pipeline passes, and $100,000 for installing some warning equipment such as gas detectors and sirens, which can be used to warn the residents of evacuation in the event of a gas leak. If by investing the $200,000 ten lives can be saved in the event of an explosion then the expected benefit is $10 \times 1,000,000 \times 0.01$ or $100,000. This shows that by adopting the second approach, spending of $200,000 will result in a $100,000 reduction of risk. The second approach is a pure safety investment, which does not improve the life of the pipeline. It is not a good investment for this case. The foregoing calculations show that one cannot make good management decisions by separating risk investment from investment to improve the pipeline and extend its life. Both safety benefit and other benefits should be considered and combined when making sound management decisions. Because risk-based management is new to the pipeline industry, it will be of interest to see how pipeline companies will use it to make safety-related decisions, and to justify them in the integrity management program report that they must submit to DOT. Good discussions of risk-based management for pipelines can be found in many publications such as Reference 1, and also many Web sites.

14.6 PIPELINE MAINTENANCE

There are two types of pipeline maintenance — routine maintenance and renovation. Routine maintenance is needed to keep the pipe functioning smoothly, but it does not necessarily cause a significant increase in the useful life of the pipeline. Routine maintenance of pipelines includes not only maintaining the pipe itself but also all the equipment needed for the pipeline to function properly. This includes the pumps, the motors, the valves, the flowmeters, the transducers, and many other items. They all need to be serviced and adjusted according their individual maintenance schedules. Any equipment that malfunctioned must be repaired and put back to service as soon as possible. Damaged or worn-out parts must be replaced. The pipe itself must be maintained according to an established schedule. This includes periodic checking of the cathodic protection system (for steel pipes), pressure testing of all pipes for possible leaks for aging, corroded, or otherwise damaged pipes, and inline maintenance (pigging) periodically to clear out debris (using cleaning pigs) and to detect possible damage (using inspection pigs). All these measures, plus many other measures not mentioned here, are part of routine maintenance.

Renovation (rehabilitation), on the other hand, includes more costly measures that are taken when the pipe's exterior is already badly corroded by the soil that surrounds the pipe, or when the pipe interior is seriously damaged by internal corrosion or encrustation. In the case of a seriously corroded steel pipe or cast iron pipe for conveying water, or seriously damaged sewer pipes made of concrete, the solution could be installing an *in situ* lining, or a smaller new pipe inside the old pipe. Many such new technologies for renovating pipes have been developed in recent years. They are usually referred to as trenchless technologies, for they do not require digging a long trench for laying a pipe, and hence are more environmentally friendly. Through such a renovation, the pipe is given a second life, and should function well for another 30 years or longer. So far, the trenchless technologies for renovating pipes has been used mainly for water and sewer pipes. However, it is gradually finding its way into the renovation of corroded steel pipes used for transporting natural gas and petroleum. As such technology improves through research and development, no doubt it will find increasing use for renovating high-pressure steel pipes, so that the thousands of miles of badly corroded steel pipes in the U.S. and in other nations can be renovated and given a second life. Different trenchless rehabilitation technologies or techniques are discussed in the next section.

14.7 TRENCHLESS REHABILITATION METHODS

Most nations in the world have an aging pipeline infrastructure that requires rehabilitation, repair, and selected replacement. In the U.S., when any cross-country oil or natural gas pipeline is badly corroded and leaks, the most common solution is to decommission the pipeline and build a replacement line, using the same pipeline right-of-way. For pipelines in urban and other densely populated areas, such as sewer lines and water lines in cities, it is difficult and costly to build new lines. Pipeline renewal using **trenchless technologies** (i.e., technologies that require minimum digging or open cut) is preferred to decommissioning and building replacement new pipelines. Major types of trenchless technologies to rehabilitate aging pipelines are now described.

14.7.1 CURED-IN-PLACE LINING

This method involves using a flexible fabric tube that can be folded (deflated) and inserted into a leaking old pipe to form a lining. By pulling the folded flexible tube into the pipe with a winch, or inverting it under fluid pressure, the tube can advance deeply into the buried pipe. Fluid pressure is used to cause the tube to inflate and attach firmly to the pipe wall. The tube, containing a thermosetting resin and a catalyst, can then be inflated by a heated fluid, such as hot water or steam. The heat causes the resin to set, forming a strong lining to protect the old leaky pipe. As it is the case with most other trenchless rehabilitation methods such as slip-lining, the pipe interior must be thoroughly cleaned and dried before inserting the fabric liner into the pipe. Types of thermosetting resin used in such linings include polyester, vinylester, and epoxy. Polyester is the most commonly used, and it has higher resistance to acids than epoxy resins do. Epoxy resins are adhesive to pipe and resistant to fluids having high pH values. Vinylester has superior corrosion resistance at high temperature. With such a lining, the pipe diameter is slightly reduced.

14.7.2 SLIP-LINING

Slip-lining involves the insertion (sliding) of a flexible new pipe into an old pipe to be rehabilitated. The new pipe must be significantly smaller in diameter than the old pipe, or else difficulties will be encountered in sliding the new pipe into the old pipe, especially if bends exist in the pipe. Thus, sliplining always involves a significant reduction of pipe size and the corresponding reduction of the hydraulic conveyance — the rate of flow that can be conveyed by the pipe for a given pressure gradient. Grouting in the annular space between the two pipes is needed to securely fasten the two pipes together.

14.7.3 PIPE-BURSTING

Old underground pipe can be replaced with new pipe of the same diameter or somewhat larger diameter without having to dig a long trench and dig out the old pipe. This is done by *in situ* bursting of the old pipe, and subsequent installation of the new pipe using one of two types of machine. The first type involves inserting into the old pipe a pneumatic mole that contains a pair of breaker (cracker) arms. These breaker arms are hydraulically operated; they can burst the old pipe and push the fragments outward. Following that, a new pipe of the same diameter as, or greater than the diameter of the old pipe is pulled into the space created by the pipe bursting. The second method involves using a microtunneling machine (see Chapter 12) with the ability to burst pipes. As the microtunnel is formed, the old pipe is removed, and the new pipe is installed.

14.7.4 PIPE-SHRINKING

Pipe-shrinking involves taking a flexible plastic pipe such as HDPE or PVC, and compressing and deforming it into a U-shaped cross section, which results in a much smaller cross-sectional area than that of the original round pipe. This deformed and shrunk pipe is easily pulled into a pipe of a larger diameter. Then the lining pipe is inflated into its original round shape and size by using heat and internal pressure. For HDPE pipes, shrinking can be done without change of shape or deformation. By applying a radial inward pressure on the pipe, the pipe shrinks. The shrunken pipe can easily be slip-lined into the cleaned old pipe. After the liner is in place and the inward pressure is removed, the pipe gradually relaxes and expands, restoring to its original size and shape. This forms a tight fit between the liner and the old pipe. Applying internal pressure to the liner speeds up the recovery process.

14.7.5 PATCHING AND SEALING

For the repair of local damage to a pipe caused by accidents or impacts during construction, the hole, cut, or puncture can be repaired by patching (bandaging) the hole from outside the pipe. Special patches are available commercially for such repairs. They differ for different types of pipes. Holes can also be repaired from inside the pipe by chemical grouting, which involves injecting resins through the holes. First, a forming bladder is placed inside the pipe at the location of the hole that needs to be sealed. Then, the resin is injected between the inflated bladder and the pipe until the outside surface of the pipe and the surrounding soil are saturated with the resin. After the resin is set, the hole is sealed and the bladder can be removed.

More detailed discussion of pipeline operation and maintenance can be found in handbooks such as Reference 2, and in codes and standards of professional organizations, such as Reference 3. More about trenchless technologies for pipeline rehabilitation can be found in Reference 4.

REFERENCES

1. Willke, T.L., Shires, T.M., Cowgill, R.M., and Selig, B.J., U.S. risk management can reduce regulation, enhance safety, *Oil and Gas Journal*, June 16, 37–40, 1997.
2. Nayyar, M.L., *Piping Handbook*, 6th ed., McGraw-Hill, New York, 1967.
3. ASME B31.8 Code, *Gas Transmission and Distribution Piping*, American Society of Mechanical Engineers, New York, 1995.
4. Iseley, D.T. and Najafi, M., *Trenchless Pipeline Rehabilitation*, National Utility Contractors Association, Arlington, VA, 1995.

Appendix A
Notation

ENGLISH

a pipe radius

a aspect ratio of capsule (capsule length divided by capsule diameter)

a_1, a_2, a_3 arbitrary constants

A cross-sectional area of pipe

A_c cross-sectional area of capsule

A_j cross-sectional area of jet

A_o cross-sectional area of orifice

A_o cross-sectional area of electromagnetic pump

A_p cross-sectional area of piston

A_s surface area of reservoir

b blockage ratio

b pump impeller thickness

b depth of electrode penetration into ground in soil-resistivity measurement

b_1, b_2, b_3 arbitrary constants

B Boltzman's constant

B a function of Mach number defined by Equation 3.46

B magnetic flux intensity

B_d effective width used for calculating earth load on conduits

B_1, B_2, B_3 arbitrary constants

c specific heat capacity

c_p specific heat capacity at constant pressure

c_v specific heat capacity at constant volume

C celerity of water hammer waves in pipe

C cohesion coefficient of soil used to determine earth load on conduits

C_1, C_2, C_3, C_4 arbitrary constants

C_1', C_2', C_3' arbitrary constants

C_A volume concentration of solids in slurry at pipe axis (centerline)

C_c contraction coefficient

C_d discharge coefficient

C_d load coefficient in Marston's equation for buried conduits

C_D drag coefficient

C_H Hazen-Williams coefficient

397

C_p compressibility coefficient
C_t celerity of water hammer waves in surge tank
C_T volume concentration of solids in slurry near pipe top
C_v velocity coefficient for flow meters
C_v mean volume concentration of solids in a pipe
C_w mean weight concentration of solids in a pipe
d_s size or diameter of solid particles in pipe
ds infinitesimal surface area
$d\vec{s}$ vector of infinitesimal surface area ds
D inner diameter of pipe
D_b inner diameter (bore) of magnetic flowmeter
D_c capsule diameter
D_d capsule end disk diameter
D_m mean diameter of pipe
D_n nominal diameter of pipe
D_o outer diameter of pipe
D_p pump impeller diameter
e absolute roughness
E bulk modulus of fluid
E_o voltage (used when V is needed to denote velocity)
E_p modulus of elasticity of pipe material
EGL energy grade line
EI energy intensiveness
f Darcy-Weisbach resistance (friction) factor
f line frequency
Δf frequency shift due to Doppler effect
f' Fanning's resistance (friction) factor
f_{em} electromagnetic force per unit volume
f_m friction factor of solid–fluid mixture in pipe
F force
$\vec{F}$ force vector
F_D drag force
F_e external force on piston
F_f contact friction force
F_L densimetric Froude number for calculating limit-deposit velocity
F_L lift force on a capsule in pipe
F_p piston force
F_x x-component of force
F_y y-component of force
g gravitational acceleration
G gas gravity
G seismic acceleration (i.e., number of gravitational acceleration g)
h total head
h head or height of liquid
h_a atmospheric pressure head
h_{em} head of electromagnetic pump

h_j piezometric head at pipe junction
h_L head loss
h_p pump head
h_s static head at pump location
h_t turbine head
h_v vapor pressure head
H water height (elevation) in a reservoir above pipe exit
H pump head
H Hedstrom number
H_m magnetic field intensity
HGL hydraulic grade line
ΔH pressure head rise in pipe caused by valve closure
ΔH_s pressure head rise in pipe caused by slow closure of a valve
i specific internal energy
i pressure gradient of fluid in pipe ($\Delta p/L$)
i_m pressure gradient of solid–fluid mixture in pipe ($\Delta p_m/L$)
I electrical current
$I.D.$ inside diameter of pipe
J current density
k adiabatic exponent (c_p/c_v)
k capsule body diameter ratio (i.e., capsule diameter divided by pipe I.D.)
k_d capsule disk diameter ratio (i.e., capsule disk diameter divided by pipe I.D.)
K local head loss coefficient
K consistency index of the power-law fluid
K bedding constant
ℓ variable distance along a pipe
ℓn natural logarithm
log common logarithm
L pipe length
L spacing between electrodes used in soil resistivity measurement
L spacing between pipe supports
L' length of pipe flow entrance region
L_c capsule length
L_e equivalent pipe length for head loss calculation
L_{em} length of electromagnetic pump
L_p piston stroke length
m attenuation ratio of surge tank
m number of moles per unit weight
m mass
$\dot{m}$ mass flow rate (dm/dt)
M Mach number
M_i molecular weight of constituent i
M_o limiting Mach number
n Manning's coefficient
n number of moles per unit weight ($n = 1/m$)
n power-law exponent for non-Newtonian fluid

n_p number of poles in an electric motor
N angular speed (rpm, rad/s, etc.)
N_c number of capsules in pipe
N_{lr} loading ratio (weight of solids in pipe divided by weight of fluid in pipe)
N_s specific speed of pump
N_{sp} specific pressure ratio (pressure drop of mixture divided by pressure drop of fluid)
N_t number of capsules in a capsule train
$NPSH$ net positive suction head
p pressure of fluid at a given point
p pump pressure (discharge pressure minus suction pressure)
p_b pipe buckling pressure (external pressure that causes pipe to buckle)
p_c critical pressure
$\bar{p}_c$ pseudocritical pressure
p_e external pressure
p_i internal pressure
p_o limiting pressure
p_p piston pressure
p_s static pressure in pipe
Δp pressure rise due to water hammer in pipe
Δp pressure drop along pipe over distance L
Δp_c pressure drop across a capsule
Δp_s water hammer pressure due to slow closure of valve
Δp_t pressure rise due to water hammer in a pipe protected by a surge tank
P power (brake horsepower)
P_e power delivered to piston by an external force
P_L power loss
P_i power input of pump
P_o power output of pump
P_r reduced pressure (p/p_c)
Q rate of heat loss through unit length of pipe
Q volumetric discharge of fluid through pipe
Q_c rate of heat loss through pipe of length L
Q_m volumetric discharge of mixture through pipe
Q_s volumetric discharge of solids through pipe
ΔQ discharge correction factor used in the Hardy Cross method
r radial distance from pipe centerline
r compression ratio (pressure after compression divided by pressure before compression)
r_i impeller radius
r_o plug flow radius for Bingham plastic fluid in pipe
r_o rotor radius
$\Re$ Reynolds number
$\Re_c$ critical Reynolds number
R engineering gas constant
R electrical resistance

R_b bend radius
R_H hydraulic radius
R_x x-component of force by pipe on fluid
R_y y-component of force by pipe on fluid
s surface area
S surge height in a surge tank
S density ratio (solid density divided by fluid density)
S slip — an electrical quantity defined as $(V_s - V_m)/V_s$
S_e energy slope $(S_e = h_L/L)$
S_o maximum surge height in a surge tank
S_p linear speed of piston
t time (variable)
t_o time to drain a reservoir
t_p time when pump is on
Δt cycle time of pump
T temperature
T torque
T tensile force
T_c valve closure time
T_c critical temperature
$\bar{T}_c$ pseudocritical temperature
T_m torque of motor
T_p torque of pump
T_r reduced temperature (T/T_c)
ΔT temperature change $(T_2 - T_1)$
u local or point velocity at time t and at a distance y from wall
u shorthand for $\bar{u}$ starting Section 2.3.1
u' turbulent (fluctuating) component of u
$\bar{u}$ temporal mean of local velocity u
u_* shear velocity $(u_* = \sqrt{\tau_o/\rho})$
u^+ dimensionless local mean velocity $(u^+ = \bar{u}/u_*)$
U tangential velocity at pump impeller tip
V mean fluid velocity across pipe $(V = Q/A)$
V voltage
V_a mean fluid velocity in capsule-pipe annulus
V_c pipe centerline velocity
V_c capsule velocity in pipe
V_d differential velocity between fluid and capsule (or pig)
V_i incipient velocity
V_L limit-deposit velocity
V_L lift-off velocity of capsule in pipe
V_m motor linear speed
V_m meridian (radial) velocity component of centrifugal pump blades
V_o steady-state mean flow velocity in pipe
V_o critical velocity of capsule flow
V_p mean velocity of particles moving through pipe

V_p average piston velocity
V_p velocity of pig in pipe
V_r velocity relative to blade tip of a centrifugal pump
V_s speed of water surface decrease in a reservoir or tank ($V_s = -dH/dt$)
V_s settling velocity of solids in fluid
V_s synchronous speed of electric motor
V_t tangential velocity component of blade tip of a centrifugal pump
w molecular weight
w work per unit mass (specific work)
$\dot{w}$ weight flow rate (weight per unit time)
W weight
W_c capsule weight
W_c earth load (force) per unit length on a buried conduit or pipe
x longitudinal distance along pipe in flow direction
x_i mole fraction of component i of a gas mixture
x_o distance along a pipe to produce limiting condition
X distance from valve subjected to maximum water hammer pressure
ΔX horizontal deflection of pipe cross section under vertical load
y distance from pipe wall perpendicular to wall ($y = 0$ at wall)
y mole fraction
y_i mole fraction of component i in a gas mixture
y^+ dimensionless distance from wall ($y^+ = \rho\, u_* y/\mu$)
Y yield number of Bingham plastic fluid through pipe
ΔY vertical deflection of pipe cross section under vertical load
z elevation
z supercompressibility factor (also called compressibility factor)

GREEK

α energy correction factor
α angle of pipe incline (relative to a horizontal plane)
α thermal expansion coefficient of solid material
β momentum correction factor
β angle between V_r and U
β clearance ratio, $(A - A_c)/A$
γ specific weight of fluid (weight per unit volume)
γ_s specific weight of solid particle (weight per unit volume)
δ pipe thickness
Δ differential (e.g., $\Delta T = T_2 - T_1$, or $\Delta p = p_1 - p_2$)
ε void ratio of solids in pipe
ε volume reduction ratio in mixing two liquids
ε dimensionless factor in Equation 2.80
ε_m magnetic permeability of fluid
η contact friction coefficient
η efficiency of pump and other machines
η_m motor efficiency

η_p pump efficiency
θ angle of pipe bend
θ angle between electric current I and voltage V (i.e., phase angle)
κ von Karman constant ($\kappa = 0.40$ for pipe flow of Newtonian fluids)
λ linefill rate of capsules
μ dynamic viscosity of fluid
μ_1 viscosity of gas at one atmospheric pressure
μ_p Poisson's ratio of pipe material
ν kinematic viscosity of fluid ($\nu = \mu/\rho$)
π 3.1416
ρ density of fluid in pipe
ρ electric resistivity of material
ρ_a density of gas at standard atmospheric condition
ρ_m density of solid–fluid mixture
ρ_s density of solids
σ stress
σ_t tensile stress in pipe
σ_T thermal stress in pipe
Σ summation sign
τ shear in flow at radius r
τ_o shear in flow at pipe wall
τ_y yield stress of non-Newtonian fluids with yield
ω angular velocity (rad/s)
ω_m angular velocity of motor (rad/s)
ω_p angular velocity of pump (rad/s)

OTHERS

$\rightarrow$ arrow sign placed above any vector quantity
$\propto$ proportionality sign
$\forall$ volume sign
$\dfrac{d}{dt}$ derivative with respect to time

Appendix B
Conversion between SI and English (ft-lb-s) Units

LENGTH

1 km = 1000 m, 1 m = 100 cm, 1 cm = 10 mm, 1 m = 100 mm = 10^6 μm
1 mile = 5280 ft, 1 ft = 12 inches, 1 inch = 2.540 cm = 25.40 mm
1 m = 3.28 ft, 1 ft = 30.48 cm, 1 mi = 5280 ft = 1.609 km

VOLUME

1 liter = 1000 cc, 1 cc = 1 cm^3, 1 m^3 = 35.32 ft^3
1 ft^3 = 7.481 American gal = 6.229 British gal
1 American gallon = 3.785 liters

VELOCITY

1 knot = 1.15 mph, 1 mph = 1.467 fps, 1 m/s = 3.281 fps
1 mph = 1.609 km/h = 0.447 m/s

MASS

1 kg = 1000 g = 2.205 lbm, 1 slug = 32.2 lbm = 14.60 kg, 1 lbm = 453.5 g

DENSITY

1 kg/m^3 = 0.001941 $slug/ft^3$, 1 $slug/ft^3$ = 32.2 lbm/ft^3 = 0.5154 g/cc

FORCE

1 kg = 1000 g, 1 g = 981 dyne, 1 kg = 2.205 lbf,
  1 lbf = 454 g = 4.448×10^5 dyne
1 N (Newton) = 10^5 dyne, 1 lbf = 4.45 N

PRESSURE AND SHEAR

1 kg/m^2 = 0.1 g/cm^2 = 98.1 dyne/cm^2, 1 psi = 144 psf,
  1 psf = 478.8 dyne/cm^2 = 47.9 Pa
1 Pa (Pascal) = 1 N/m^2 = 10 dyne/cm^2, 1 bar = 10^5 Pa = 14.51 psi = 0.987
standard atmospheric pressure

WORK, ENERGY, AND HEAT

1 dyne-cm = 1 erg = 10^{-7} joule, 1 kg-m = 9.81 joules, 1 N-m = 1 joule,
1 ft-lb = 1.357×10^7 erg, 1 kWh = 2.66×10^6 ft-lb, 1 Btu = 778 ft-lb = 252
small cal, 1 cal (small cal) = 4.18 joules, 1 large cal = 1000 small cal

POWER

1 kW = 1000 W, 1 Hp = 550 ft-lb/s = 746 W, 1 kW = 1.34 Hp
1 ft-lb/s = 1.356 W, 1 W = 1 N-m/s

DYNAMIC VISCOSITY

1 poise = 1 dyne-s/cm^2, 1 centipoise = 10^{-2} poise
1 lb-s/ft^2 = 47.88 N-s/m^2 = 478.8 poise

KINEMATIC VISCOSITY

1 stoke = 1 cm^2/s, 1 ft^2/s = 929 stoke = 0.0929 m^2/s, 1 m^2/s = 10.76 ft^2/s

ELECTRIC UNITS

1 watt = 1 amp-V, 1 ohm = 1 V/amp

TEMPERATURE

$$K = °C + 273°, \quad °R = °F + 460°, \quad °C = \frac{5}{9}(°F - 32°), \quad °F = \frac{9}{5}°C + 32°$$

Appendix C
Physical Properties of Certain Fluids and Solids

TABLE C.1
Physical Properties of Certain Liquids at Atmospheric Pressure (ft-lb Units)

Liquid	Temperature T (°F)	Density, ρ (slug/ft³)	Dynamic Viscosity $\mu \times 10^5$ (lb-s/ft²)	Vapor Pressure p_v (psia)	Bulk Modulus $E_v \times 10^{-5}$ (psi)
Ethyl alcohol (ethanol)	68	1.53	2.49	0.85	1.54
Gasoline	60	1.32	0.65	8.0	1.9
Glycerin	68	2.44	3130	2×10^{-6}	6.56
Mercury	68	26.3	3.28	2.3×10^{-5}	41.4
SAE 30 oil	60	1.77	800		2.2
Seawater	60	1.99	2.51	0.256	3.39
Water	60	1.94	2.34	0.256	3.12

TABLE C.2
Physical Properties of Certain Liquids at Atmospheric Pressure (SI Units)

Liquid	Temperature T (°C)	Density, ρ (kg/m³)	Dynamic Viscosity $\mu \times 10^3$ (N-s/m²)	Vapor Pressure $p_v \times 10^{-3}$ (N/m² abs.)	Bulk Modulus $E_v \times 10^{-9}$ (N/m²)
Ethyl alcohol	20	789	1.19	5.9	1.06
Gasoline	16	680	0.31	55	1.3
Glycerin	20	1260	1500	14	4.52
Mercury	20	13,600	1.57	1.6×10^{-4}	28.5
SAE 30 oil	16	912	380	—	1.5
Seawater	16	1030	1.20	1.77	2.34
Water	16	999	1.12	1.77	2.15

TABLE C.3
Physical Properties of Water as a Function of Temperature at Atmospheric Pressure (ft-lb Units)

Temperature, T (°F)	Density, ρ (slug/ft³)	Dynamic Viscosity, $\mu \times 10^5$ (lb-s/ft²)	Vapor Pressure, P_v (psia)	Bulk Modulus of Elasticity, $E_v \times 10^{-5}$ (psi)	Speed of Sound, C (ft/s)
32	1.940	3.732	0.0885	2.87	4603
40	1.940	3.228	0.122	2.96	4672
50	1.940	2.730	0.178	3.05	4748
60	1.938	2.344	0.256	3.13	4814
70	1.936	2.037	0.363	3.19	4871
80	1.934	1.791	0.507	3.24	4819
90	1.931	1.500	0.698	3.28	4960
100	1.927	1.423	0.949	3.31	4995
120	1.918	1.164	1.692	3.32	5049
140	1.908	0.974	2.888	3.30	5091
160	1.896	0.832	4.736	3.26	5101
180	1.883	0.721	7.507	3.18	5195
200	1.869	0.634	11.52	3.08	5089
212	1.860	0.589	14.69	3.00	5062

TABLE C.4
Physical Properties of Water as a Function of Temperature at Atmospheric Pressure (SI Units)

Temperature, T (°C)	Density, ρ (kg/m³)	Dynamic Viscosity, $\mu \times 10^4$ (N-s/m²)	Vapor Pressure, $p_v \times 10^{-4}$ (N/m²)	Bulk Modulus of Elasticity, $E_v \times 10^{-9}$ (N/m²)	Speed of Sound, C (m/s)
0	999.9	17.87	0.0611	1.98	1403
5	1000	15.19	0.0872	2.05	1427
10	999.7	13.07	0.123	2.10	1447
20	998.2	10.02	0.234	2.17	1481
30	995.7	7.98	0.424	2.25	1507
40	992.2	6.53	0.738	2.28	1526
50	988.1	5.47	1.233	2.29	1541
60	983.2	4.67	1.992	2.28	1552
70	977.8	4.04	3.116	2.25	1555
80	971.8	3.55	4.734	2.20	1555
90	965.3	3.15	7.010	2.14	1550
100	958.4	2.82	10.13	2.07	1543

TABLE C.5
Physical Properties of Certain Gases at Atmospheric Pressure (ft-lb Units)

Gas	Temperature, T (°F)	Density, $\rho \times 10^3$ (slug/ft³)	Dynamic Viscosity, $\mu \times 10^7$ (lb-s/ft²)	Gas Constant, $R \times 10^{-3}$ (ft-lb/slug/°R)	Specific Heat Ratio, k	Specific Heat Capacity, c_v (Btu/lbm/°R)
Air	59	2.38	3.74	1.716	1.4	0.171
Carbon dioxide	68	3.55	3.07	1.130	1.29	0.155
Helium	68	0.326	4.09	12.42	1.66	0.753
Hydrogen	68	0.163	1.85	24.66	1.41	2.44
Methane	68	1.29	2.29	3.099	1.31	0.403
Nitrogen	68	2.26	3.68	1.775	1.40	0.177
Oxygen	68	2.58	4.25	1.554	1.40	0.157

TABLE C.6
Physical Properties of Certain Gases at Atmospheric Pressure (SI Units)

Gas	Temperature, T (°C)	Density, ρ (kg/m³)	Dynamic Viscosity, $\mu \times 10^5$ (N-s/m²)	Gas Constant, $R \times 10^{-2}$ (J/kg/K)	Specific Heat Ratio, k	Specific Heat Capacity, c_v (J/kg/K)
Air	15	1.23	1.79	2.869	1.4	717
Carbon dioxide	20	1.83	1.47	1.889	1.29	647
Helium	20	0.166	1.94	20.77	1.66	3125
Hydrogen	20	0.0838	0.884	41.24	1.41	10087
Methane	20	0.667	1.10	5.183	1.31	1685
Nitrogen	20	1.16	1.76	2.968	1.40	744
Oxygen	20	1.33	2.04	2.598	1.40	654

TABLE C.7
Physical Properties of Air at Standard Atmospheric Pressure

Ft-lb-s Units				SI Units			
Temperature, T (°F)	Density, $\rho \times 10^3$ (slug/ft³)	Dynamic Viscosity, $\mu \times 10^7$ (lb-s/ft²)	Speed of Sound, C (ft/s)	Temperature, T (°C)	Density, ρ (kg/m³)	Dynamic Viscosity, $\mu \times 10^5$ (N-s/m²)	Speed of Sound, C (m/s)
−40	2.94	3.29	1004	−20	1.40	1.63	319
−20	2.81	3.34	1028	0	1.29	1.71	331
0	2.68	3.38	1051	5	1.27	1.73	334
10	2.63	3.44	1062	10	1.25	1.76	337
20	2.57	3.50	1074	15	1.23	1.80	340
30	2.52	3.58	1085	20	1.20	1.82	343
40	2.47	3.60	1096	25	1.18	1.85	346
50	2.42	3.68	1106	30	1.17	1.86	349
60	2.37	3.75	1117	40	1.13	1.87	355
70	2.33	3.82	1128	50	1.11	1.95	360
80	2.29	3.86	1138	60	1.06	1.97	366
90	2.24	3.90	1149	70	1.03	2.03	371
100	2.20	3.94	1159	80	1.00	2.07	377
120	2.13	4.02	1180	90	0.972	2.14	382
140	2.06	4.13	1200	100	0.946	2.17	387

TABLE C.8
Physical Properties of Certain Solids (Pipe Materials) at 70°F

Pipe Material	Young's Modulus, E_p (psi)	Poisson's Ratio, μ_p	Specific Gravity, S	Thermal Expansion Coefficient, $\alpha \times 10^6$ (1/°F)
Steel	3.0×10^7	0.3	7.14	6.5
Ductile iron	2.4×10^7	0.28	7.08	6.2
Cast iron	1.34×10^7		7.14	5.8
Copper	1.6×10^7	0.3	8.94	9.8
Aluminum	1.05×10^7	0.33	2.71	1.3
Asbestos cement	3.4×10^6	0.3	–	4.5
Concrete	$5.7 \times 10^4 \times f_c^{1/2}$	0.3	2.1–2.7	7.0
PVC	4×10^5	0.38	1.4	30

Note: f_c is the 28-day compressive strength of the concrete.

Index

Milton Keynes UK
Ingram Content Group UK Ltd.
UKHW031139141024
449569UK00024B/1218